THE
HANNAH RESEARCH INSTITUTE
1928–1978

THE HANNAH RESEARCH INSTITUTE

1928–1978

Essays on the scientific work of the Institute

Edited by

J. H. MOORE *and* J. A. F. ROOK

THE HANNAH RESEARCH INSTITUTE
AYR

Published by The Hannah Research Institute
Ayr KA6 5HL, Scotland

ISBN 0 905151 27 5

Printed by T. & A. Constable Limited, Edinburgh

CONTENTS

PREFACE

The editors hope that this volume will serve to mark the contribution that has been made to agricultural science by the Hannah Research Institute during the first 50 years of its existence. It was considered that this contribution could best be illustrated by a collection of essays on research activities for which the Institute has gained a considerable reputation. These essays have been assembled into two groups; the first of these contains articles by former members of staff on topics that constituted a major part of the earlier research programme of the Institute, whereas the essays in the second group are by present members of staff on subjects that form part of the current research interests. With some justification, it may be argued that this collection of essays on selected topics does little to acknowledge the effort and enthusiasm that have been devoted over the years to other areas of investigation. However, some impression of the breadth and depth of the Institute's contribution to research may be gained from the list of publications included at the end of this book. This list is preceded by the names of the members of the senior staff and of the Joint Committee of Management and the Council of the Institute from 1928 to 1978. In this book the historical features relating to the establishment of the Institute in 1928 and its subsequent development and organization are mentioned only briefly; a full account of these aspects is given in a companion volume written by Dr J. A. B. Smith, the Director of the Institute from 1950 to 1970.

Dr Smith was intimately associated with the conception and the initial planning of this book, and the valuable help and advice he has freely given at all stages is gratefully acknowledged. The editors are indebted also to Mr B. A. Nettlefold, the Librarian of the Institute, and Mr N. A. D. McCance, the Secretary of the Institute, for the valuable assistance they have given in the preparation of this volume. Financial support towards the publication of the volume was gratefully received from The Aberdeen and District Milk Marketing Board, The Company of Scottish Cheese Makers Ltd, The Express Dairy Company Ltd, The Milk Marketing Board of England and Wales, Northern Foods Ltd, The North of Scotland Milk Marketing Board, Scottish Agricultural Industries Ltd, The Scottish Milk Marketing Board, The Scottish Milk Trade Federation and Unigate Ltd.

J. H. M.
J. A. F. R.

FOREWORD

The Hannah Dairy Research Institute was founded 50 years ago in a manner which neatly illustrates the importance of opportunism even to such sober bodies as Parliamentary Commissions. Given the need for another centre comparable with the National Institute for Research in Dairying at Reading, the south-west of Scotland was an obvious preferred area for location. The region was, as it continues to be, one of the most important and efficient dairy farming districts in the country and was already the site of one of the three Agricultural Colleges serving Scotland. Moreover, it lies within the catchment area of the University of Glasgow, one of the few British universities of those days with the range of expertise and facilities to advise and, if need be, to assist a newly established Institute on any specialized line of research it wished to follow. These criteria delineated the region in which the Institute would be established. The actual year of foundation and its precise location, however, were determined by Mr J. M. Hannah when he gave the nation the manor and estate at Auchincruive for use as a centre for agricultural research and education. The Development Commission promptly declared the estate to be the site of the Institute which was designated the 'Hannah' in honour of the magnificent benefaction.

Since 1928, the Hannah Research Institute, as it is now called, has become internationally famous for its scientific investigation into the production and utilization of milk. Its growth has inevitably led to changes in the internal organization of the Institute. Yet its style of government by a Council consisting of nominees of the University of Glasgow, the West of Scotland Agricultural College, and the Secretary of State for Scotland, remains unchanged. The University is privileged to have these tangible links with an Institute which has achieved so much distinction in agricultural research. It has tried to show its high regard for the researches undertaken at the Institute by creating the Hannah Chair for the Director of the Institute and Honorary Lectureships for the Heads of Departments. The esteem accorded the work of the Institute by Council members representing the West of Scotland Agricultural College and the Secretary of State for Scotland is, I know, as great as that of nominees from the University of Glasgow. I am, therefore, writing on behalf of the entire Council when I congratulate the Institute on the occasion of its fiftieth birthday. We are all confident that the Institute will celebrate its centenary with an equally illustrious period of achievement to review and at least as distinguished a programme of continuing research to describe as are recorded in this Jubilee volume.

ALWYN WILLIAMS

January 1978

COUNCIL OF THE HANNAH RESEARCH INSTITUTE, 1978

ALWYN WILLIAMS, PhD, DSc, FGS, FRS (*Chairman*)
Professor J. S. GILLESPIE, MB, ChB, PhD
Professor R. G. WHITE, MA, DM, FRSE
(*Nominated by the University of Glasgow*)

J. CALDWELL, Esq
J. J. M. HANNAH, Esq, CBE, NDA, CDA (*Vice-Chairman*)
G. L. WILSON, Esq
(*Nominated by the West of Scotland Agricultural College*)

Professor F. D. GUNSTONE, BSc, PhD, DSc, FRIC, FRSE
W. W. WATSON PEAT, Esq, CBE
(*Nominated by the Secretary of State for Scotland*)

Secretary
N. A. D. MCCANCE, Esq, AMBIM

STAFF OF THE INSTITUTE, 1978 (January)

Director:

Professor J. A. F. Rook, PhD(Glasgow), DSc(Wales), FRIC, FIBiol, FRSE

Deputy Director:

J. H. Moore, MSc(Wales), PhD, DSc(Reading), FRIC

APPLIED STUDIES

Head of Department:

W. Banks, BSc, PhD(Edinburgh)

Scientific Staff:

J. Abbot, NDDT
Jean M. Banks, BSc(Strathclyde)
S. Basmaeil*, BSc(Riyadh), MSc(Aberdeen)
M. E. Castle, BSc(Agr)(Leeds), PhD(Reading), NDA, NDD, MIBiol
J. S. Chalmers†, BSc(Agr)(Aberdeen)
D. G. Chamberlain, BSc(Reading)
J. L. Clapperton, BSc, PhD(Dunelm)
M. W. Griffiths, BSc(North East London Polytechnic), PhD(Leicester)
N. C. Kelly, BSc, BAgr(Belfast)
Elizabeth Macdaid
Sarah H. A. McLauchlan
D. D. Muir, BSc, PhD(Edinburgh)
I. H. L. Ormrod‡, BSc(Sheffield)
J. D. Phillips, BSc(St Andrews)
D. Reid, BSc(Agr), PhD(Glasgow), MS(Wisconsin)
W. C. Retter§, BSc(AgrSci)(Leeds)
W. Steele, BAgr(Belfast), PhD(Reading)
N. H. Strachan, HNC
A. W. M. Sweetsur, PhD(Strathclyde), ARIC
P. C. Thomas, BSc, PhD(Wales)
M. K. Wait, NCA

* Postgraduate scholar, Government of Saudi Arabia.
† Postgraduate scholar, Department of Agriculture and Fisheries for Scotland.
‡ Postgraduate scholar, SRC/CASE jointly with the University of Bradford.
§ Postgraduate scholar, Imperial Chemical Industries Ltd, Agriculture Division.

J. C. D. White, BSc, PhD(Glasgow), FRIC
Agnes G. Wilson
Fiona J. Anderson (Mrs), BSc(Glasgow), Hazel M. Beattie, ONC, Una A. D. Connell (Mrs), D. S. W. Cooney, Agnes A. Cumming, Ann H. Docherty, Carolyn Downie, Jean Grant, Margaret T. McGhee (Mrs), Nancy McGregor, Catherine McD. Milligan, Eleanor A. Noble, ONC, Lynn G. Russell, ONC, Sheila J. Wyllie (Mrs), NDD

Farm and Estate Manager:
J. N. Watson, NDDH, CDD

Animal Attendants:
D. H. Ford, J. R. Munro, S. Roberston, BSc(Newcastle), J. Storrie

Farm Staff:
T. Brown, A. Finlayson, A. G. Garven, J. McCulloch, W. Martin, P. McInally, BEM, E. Nixon, W. Robertson, J. Steell, D. A. Wilson

Departmental Secretary:
Barbara A. M. Watt, NDA

BIOCHEMISTRY DEPARTMENT

Head of Department:
J. H. Moore, MSc(Wales), PhD, DSc(Reading), FRIC

Scientific Staff:
Grace H. McG. Breckenridge, ONC
R. E. Brice, BSc(East Anglia)
W. W. Christie, BSc, PhD(St Andrews)
R. A. Clegg, MA(Cantab), PhD(Bristol)
J. W. Czerkawski, BSc(London), PhD(Leeds)
J. T. Drummond, BSc(Strathclyde)
Catherine G. Faulds (Mrs)
Margaret L. Hunter, MSc(Strathclyde), LRIC
A. K. Martin, BSc, PhD(Dunelm), ARIC
I. M. Morrison, BSc, PhD(Edinburgh), ARIC
R. C. Noble, BSc, PhD(Dunelm)
D. R. Paterson
Pamela J. Poole (Mrs), BSc(Nottingham)
J. H. Shand, BSc(Aberdeen)
G. R. Tanner*, BSc(Liverpool)
Eleanor Taylor (Mrs)
I. G. West, ONC

* Postgraduate scholar, Agricultural Research Council.

A. G. Williams, BSc, PhD(Wales)
Susan E. Withers (Mrs), BSc(Edinburgh)
Susan C. Eyley (Mrs), Anne Halbert, Joyce M. McBlane (Mrs), Ruth J. McCartney, Marie Morrison (Mrs), HNC, Jan Quig (Mrs), HNC, Jeannette V. Wilson, MSR

Animal Attendants:
A. C. McBlane, J. R. McDill, SDA, SDD, C. E. Park

Departmental Secretary:
Elizabeth L. Atkinson, SHNC

CHEMISTRY DEPARTMENT

Head of Department:
D. G. Dalgleish, BSc, PhD(Edinburgh)

Scientific Staff:
W. D. Annan, MA(Cantab), PhD(Edinburgh)
T. Carolan, BSc(Strathclyde)
D. T. Davies, BSc(Wales), PhD(Glasgow)
C. Holt, MSc(Bristol), PhD(Leeds)
D. S. Horne, BSc, PhD(Edinburgh)
A. J. R. Law, BSc(Glasgow), MIBiol
W. Manson, BSc, PhD(Glasgow)
J. A. Nimmo, HNC
T. G. Parker, BSc(Glasgow), PhD(Strathclyde)
Elspeth Paterson*
D. W. West, BSc, PhD(London), ARIC
Linda Baird (Mrs), HNC, J. A. H. Bryson (Mrs), Christy Halbert (Mrs), BSc(Glasgow), Celia Mitchell, Mary C. Wallace (Mrs), Patricia B. Wallace (Mrs), ONC

Departmental Secretary:
Marion M. Stewart

PHYSIOLOGY DEPARTMENT

Head of Department:
M. Peaker†, BSc(Sheffield), PhD(Hong Kong)

Scientific Staff:
Susan M. Butler (Mrs)‡, BSc(Leeds)
D. T. Calvert, BSc(St Andrews), PhD(Glasgow)
P. L. Clarke, BSc, PhD(Aberdeen)
A. J. Downie, ONC

* Postgraduate scholar, Agricultural Research Council.
† Takes up appointment on 1st July 1978.
‡ Postgraduate scholar, Medical Research Council.

Anne Faulkner, BSc, PhD(Liverpool)
D. J. Flint, BSc(Liverpool)
A. J. Henderson, BSc(St Andrews)
D. McE. Jenkinson, BSc, PhD(Glasgow)
Colette D. R. Jones, BSc(London)
D. H. Lloyd*, BVetMed, MRCVS(London)
R. M. Mabon, AIST
J. A. McLean, BSc(Birmingham), PhD(Glasgow)
Shona M. Park, BSc(Glasgow)
P. A. Sinnett-Smith†, BSc(Leeds)
G. E. Thompson, BSc(Hull), PhD(Western Ontario)
E. M. Thomson‡, BSc(Glasgow)
R. G. Vernon, BSc, PhD(Birmingham)
Elizabeth Y. Brechany, HNC, Lesley J. Brown, Pamela N. Carson, ONC, J. Davidson, Lynn K. Dunabie (Mrs), E. Finley, HNC, S. J. Gibson, Margaret C. Holland, HNC, Helen T. Kennedy, Irene Sommerville, ONC

Animal Attendants:
T. F. Hutchison, Jacqueline C. Bean (Mrs), Sandra M. Paton (Mrs)

Departmental Secretary:
Joyce Reid

ADMINISTRATIVE STAFF

Secretary of the Institute:
N. A. D. McCance, AMBIM

Administrative Assistant:
Janet Howie, SNC

Office Assistants:
Janet H. Graham (Mrs), Ethel D. Jenkins, Margaret W. Scott (Mrs)

Librarian/Information Officer:
B. A. Nettlefold, BA(Strathclyde), ALA, AIInfSc

Director's Secretary:
Marilyn A. Beatson, MA(Glasgow), DipSecStuds(Strathclyde)

Assistant to Director's Secretary and Assistant Librarian:
Elizabeth J. McM. Whisker

ENGINEERING AND MAINTENANCE SERVICES

Head:
J. Ross, CEng, MIMarE

* Animal Health Trust Wooldridge Farm Livestock Research Fellow.
† Postgraduate scholar, SRC/CASE, jointly with University of Nottingham.
§ Postgraduate scholar, Agricultural Research Council.

Officer-in-charge Instruments:
W. Nisbet

Services Engineer (Mechanical):
F. K. Maxwell, BOTEng

Services Engineer (Electrical):
A. G. Scott, ONC

Skilled Craftsmen:
E. P. Barrs, J. B. Dick, J. M. Henderson, P. Holden, G. T. Murray

Assistant:
A. MacD. Purvis

Caretaker:
C. D. Duncan

Cleaners:
Mrs A. S. Finlayson, Mrs M. Johnstone, Mrs M. H. H. Logan, Mrs J. Lucas, Mrs C. McLauchlan, Mrs E. J. Stewart

CANTEEN

Mrs J. H. Murdoch

INSTITUTE OFFICERS

Animals Officer:
G. E. Thompson, BSc(Hull), PhD(Western Ontario)

Isotopes Officer:
R. C. Noble, BSc, PhD(Dunelm)

Safety Officer:
W. Manson, BSc, PhD(Glasgow)

Training Officer:
J. Abbot, NDDT

HONORARY FELLOWS OF THE INSTITUTE

Professor R. C. GARRY, OBE, MB, ChB, DSc, FRCP(Glasgow), FRSE (1972)
E. C. OWEN, BA, BSc, DSc, PhD (1972)
J. A. B. SMITH, CBE, PhD, DSc, LLD, FRIC, FRSE (1972)
Sir GEORGE WILSON, KBE, MA, LLD (1972)
R. WAITE, MScTech, PhD (1973)
Miss MARGARET M. HUGHES, MBE (1975)
Sir Charles H. Wilson, MA, LLD (1977)

CONSULTATIVE PANEL FOR THE INSTITUTE

Dr T. R. Ashton, Technical Director, Express Dairy Group Services Ltd

Mr M. Boyle, General Manager, Aberdeen and District Milk Marketing Board

Mr W. Martin Cox, Consultant Food Technologist

Dr R. M. J. Crawford, Head of Department of Dairy Technology, West of Scotland Agricultural College

Dr D. B. Gammack, Chief Scientist, Unigate Technical Centre

Dr J. F. Gordon, Express Dairy Group Services Ltd

Dr E. Green, Director, Technical Division, Milk Marketing Board of England and Wales

Mr K. Green, Carnation Foods Co Ltd

Professor J. Hawthorn, Department of Food Science and Nutrition, University of Strathclyde

Mr J. A. Lockhart, Dairy Manager, Scottish Farmers Dairy Co Ltd

Mr I. A. McAlpine, Dairy Manager, Scottish Milk Marketing Board

Mr C. McCall, Development Manager, Milk Group Scotland, Co-operative Wholesale Society

Mr L. Pattison, Scottish Milk Trade Federation

Mr J. Russell, Manager, The Company of Scottish Cheese Makers Ltd

Mr R. A. Russell, OBE, Manager and Secretary, North of Scotland Milk Marketing Board

Mr J. H. Smith, Technical Services Manager—Cheese, Unigate Foods Ltd

Dr H. Fore, Food Science Adviser, Agricultural Research Council, and a representative from the National Institute for Research in Dairying, Reading, attend meetings of the Consultative Panel by invitation.

BIOGRAPHICAL NOTES ON PAST MEMBERS OF STAFF WHO HAVE CONTRIBUTED TO THIS VOLUME

Blaxter, Sir Kenneth L., BSc(Agric), PhD, DSc, NDA(Hons), FIBiol, FRSE, FRS
Head of Nutrition Department, Hannah Research Institute, 1948 to 1965. Director, Rowett Research Institute from 1965.

Bligh, John, BSc, PhD
Member of staff of Physiology Department, Hannah Research Institute, 1952 to 1956. Member of staff of Physiology Department, Institute of Animal Physiology, Babraham, Cambridge 1956 to 1977. Director of the Institute of Arctic Biology, Fairbanks, Alaska from 1977.

Edwards, S. John, DSc, FRCVS
Veterinary Pathologist, Hannah Research Institute, 1934 to 1938. Head of Bacteriology Department, Agricultural Research Council's field station (later Institute for Research on Animal Diseases), Compton, Berkshire, 1938 to 1967.

Holmes, William, BSc(Agric), PhD, DSc, NDA(Hons), NDD, FIBiol
Appointed to staff of Hannah Research Institute, 1944, and Head of Grass and Dairy Husbandry Department, 1947 to 1955. Professor of Agriculture, Wye College, University of London, from 1955.

Owen, Edwin C., BA, BSc, PhD, DSc
Head of Nutrition Department, Hannah Research Institute, 1939 to 1947 and of Biochemistry Department, 1947 to 1970. Retired 1970. Senior Visiting Research Fellow, University of Ibadan, Nigeria, 1971 to 1972.

Smith, James A. B., CBE, PhD, DSc, LLD, FRIC, FRSE
Head of Biochemistry Department, Hannah Research Institute, 1936 to 1946. Acting Director, 1944 to 1946. Lecturer in Applied Biochemistry, Department of Biochemistry, University of Glasgow, 1946 to 1947. Deputy Director, 1948 to 1950, and Director, 1950 to 1970, Hannah Research Institute. Retired 1970.

Waite, Robert, MScTech, PhD
Appointed to staff of Hannah Research Institute, 1939. Released for military service, 1941 to 1946. Head of Technical Chemistry Department, 1946 to 1971, and of Applied Studies Department, 1971 to 1973. Retired 1973.

INTRODUCTION

THE HANNAH RESEARCH INSTITUTE, 1928 to 1978

J. A. F. Rook

The movement of people away from the land and the depression of rural industry led, in 1909, to the passing of an Act of Parliament to promote the scientific development of agriculture, forestry and fisheries under the guidance of a Development Commission. One of the major successes of the Commissioners was the formation of a sound system of investigation in agricultural science based on state-aided but independent research institutes, which were established usually in, or in association with, an appropriate university. These institutes formed the nucleus of the present-day agricultural research service, now under the scientific control of the Agricultural Research Council.

The Commissioners recognized that agricultural science was too broad to be studied as a single subject and individual institutes were given responsibility for research in a particular subject area. Thus, in 1912, they established at the University College (later University) of Reading the first British dairy research institute, the National Institute for Research in Dairying. The choice of site was influenced by the fact that the British Dairy Institute, which was set up at Aylesbury in 1888 by the British Dairy Farmers Association to provide practical and scientific education in dairying, had been moved to Reading where better teaching facilities were available.

Increasingly it became apparent that it was beyond the resources of a single institute to undertake research on all the topics that required attention, and in their 18th Report the Development Commissioners proposed that a second institute concerned with dairying should be established. There appeared to be strong reasons for placing it in south-west Scotland, an area where, at that time, there was no research institute, where the dairy industry was of great importance and where the farmers were likely to support research and put its results to practical use. Moreover, there were specific Scottish problems which required attention, and the Reading institute was too fully occupied to investigate them, and seen as too remote to influence Scottish farming practice.

In 1927, therefore, the Secretary of State for Scotland set up a Joint Committee of Management with a view to giving effect to the proposals for the establishment of the new institute. The Committee consisted of eight members, three nominated by the University of Glasgow, three by the West of Scotland Agricultural College and two by the Secretary of State for Scotland, and was under the chairmanship of Sir

Donald MacAlister of Tarbert, the then Principal of the University of Glasgow.

At their first meeting on 15th March 1927, the Committee considered the post of Director and in September of that year approved the appointment of H. H. Green, Sub-Director in the Division of Veterinary Education and Research of the Union of South Africa, with effect from 1st August 1927. The first duty accorded to him was to represent the Committee at the British Empire Conference on Agricultural Research which was held in October, and with the Conference discussions as a background, Dr Green drew up outline proposals for research and presented the financial implications for a quinquennial period. He then returned to South Africa, to await the resolution of decisions concerning a site and the funding of the Institute.

Unfortunately, on his return to South Africa, Dr Green fell ill and eventually had to resign his appointment. E. P. Cathcart, Regius Professor of Physiology at the University of Glasgow and a member of the Management Committee, agreed to act as Interim Director until October 1928 and then continued as Honorary Director, with N. C. Wright as Assistant Director, until Dr Wright became the first permanent Director in June 1930.

PHYSICAL RESOURCES

Approval by the Treasury for the establishment of the Institute was given in a letter from the Board of Agriculture dated 2nd September 1927, and at the same time an initial maintenance grant of £1500 was sanctioned. Later, approval was given to an annual maintenance grant not in excess of £2000 or two-thirds of the actual expenditure, whichever was less, on the condition that one-third of the costs was met from local sources. Fortunately, these limits were not imposed rigorously and the maintenance grant provided by the Department of Agriculture and Fisheries for Scotland, which in 1977/78 amounts to £956 480, currently accounts for more than 99% of the costs.

Crucial to progress was the acquisition of a suitable site. It was felt that it should be in close proximity to the University of Glasgow and, if possible, shared with the Dairy School of the West of Scotland Agricultural College which was then in Kilmarnock. Possible locations at Pollok and Houston were identified and the site at Houston was inspected and found suitable, but a decision was deferred. In January 1928, however, the Joint Committee was informed that Mr J. M. Hannah of Girvan Mains had gifted to the nation, for the purposes of agricultural education and research, the estate of Auchincruive, together with the mansion-house, near to the town of Ayr, and that it had been proposed by the Development Commissioners that both the Dairy Research Institute and the Dairy School for Scotland should be situated at Auchincruive.

Of the 350 ha of the estate, about 80 ha were allocated to the Institute and a capital grant of £15 000 for new buildings was provided through the Board of Agriculture. The decision was taken, however, to purchase for the Institute additional land adjacent to Auchincruive, consisting of the 17 ha of Kirkhill farm and the 21 ha of the lands of Slatehall, and to add to this a single field of 12 ha from the Auchincruive estate, the purchases to be funded from the sale of other land. Arrangements were made for the transfer of the land with effect from November 1929, and plans and estimates of cost were drawn up for a laboratory block at Kirkhill. In recognition of the generosity of Mr Hannah, it was agreed to name the Institute the Hannah Dairy Research Institute.

To allow the early appointment of staff, temporary accommodation was provided in the Physiological Institute at Glasgow University, and later an office and laboratory were acquired in the Sandgate, Ayr, and one member of staff was, for a while, given accommodation at the Royal Technical College, Glasgow. The laboratory block was completed and occupied early in 1931 and the Institute was formally opened, by the Rt Hon. William Adamson, LLD, MP, then Secretary of State for Scotland, in April of that year. Shortly afterwards the Joint Committee was incorporated, a procedure which ensured autonomy and continuity in the government of the Institute, and became the Council of the Institute.

In 1951, the original laboratory block was extended, more than doubling the floor area, and in 1959 a new block, providing canteen and recreational facilities and a lecture hall, was completed. The major facilities provided, over the years, for experimental work with animals have included a metabolism house for cows, erected in 1932 and extended in 1951, a calorimetry block in 1958 and a climatology laboratory in 1959 (with funds kindly provided by the Colonial Office), both of which were extended in 1965, one to provide a post-mortem room and isotope facilities, the other accommodation for a gradient-layer calorimeter, and a small-animal laboratory in 1976. Currently, a new laboratory block is under construction.

On the transfer of Kirkhill farm to the Institute, modifications were made to the farm buildings, a new byre to accommodate a milking herd of twenty cows was erected and a 'fine herd of thirty pedigree Ayrshire cows, together with two well-bred bulls and a full complement of young stock' was established. The animals were all tuberculin-tested, and the farm was licensed to produce Certified and Grade A (TT) milk. Later improvements to the farm are dealt with in the article by J. N. Watson in this volume (p. 11).

THE STAFF AND THEIR RESEARCH

1928 to 1950

The first permanent member of staff to be appointed, on the recommendation of the Interim Director, was Norman C. Wright, MA(Oxon), PhD(Cantab) to the post of Physiologist to the Institute in September 1928. He had been an Agricultural Research Scholar from 1922 to 1924, a Research Assistant at the National Institute for Research in Dairying from 1924 to 1926 and a Commonwealth Fund Fellow at Cornell University, New York and the United States Department of Agriculture, Washington, DC from 1926 to 1928, and was ideally suited to the post. As previously mentioned, he was to become Assistant Director and in 1930 the first permanent Director. Once in post, he quickly assembled others around him.

Three projects had been proposed for immediate consideration: an investigation of the incidence of tuberculosis in dairy herds and the percentage of reacting cows giving tuberculous milk, a joint investigation with the Royal (Dick) Veterinary College, Edinburgh, on milk fever and allied diseases, and studies on the physiology of milk secretion. In the absence of adequate laboratory facilities, investigations had to be largely in the nature of surveys of existing conditions in various branches of the dairy industry. Nevertheless, a valuable early contribution was made and several bulletins and reviews quickly emerged. The first publication was of a report on *Milk surplus and milk residues* by A. MacNeilage Jr, which the Institute undertook to edit and print at the request of the Scottish Milk and Health Association.

With additional full-time appointments, which in the period to 1930 included A. B. Fowler, BSc(Agric)(Aberdeen), as an Animal Husbandry worker, L. Jordan, MRCVS, as Veterinary Pathologist, L. A. Allen, BSc, PhD(London), MSc(Reading), as Bacteriologist and Miss M. E. Stephen, NDD, as Dairy Assistant, and with improved facilities, research investigations quickly got under way, as reflected in the output of research communications. Staff continued to expand and in 1939 at the outbreak of war there were nine permanent scientific staff, with work proceeding in physiology under N. C. Wright, in nutrition under E. C. Owen, in biochemistry under J. A. B. Smith, in veterinary pathology under P. S. Watts, in animal husbandry under A. B. Fowler, in bacteriology under Constance Higginbottom and in technical chemistry under R. Waite. Early in the war, Waite left for service with the armed forces. The rest of the staff responded to the national emergency by giving a greater emphasis to the shorter-term, practical aspects of their work; of particular relevance are the investigations into farm self-sufficiency, protein substitutes for dairy cows, bovine mastitis and the preparation and storage of dried milk. In 1942, Dr Fowler was temporarily seconded as Chief Executive Officer to the Agricultural Executive Committee for South Ayr and Dr Wright was, from 1944 to 1946, seconded on three occasions for special service overseas.

In 1947, Dr Wright was seconded again, to the post of Chief Scientific Adviser to the Ministry of Food, and in 1950 that position was made permanent and he resigned as Director of the Institute. He had guided the Institute for 20 years with notable success. He was to become in 1958 Deputy Director-General of the Food and Agriculture Organization of the United Nations and was knighted in 1963, having received the CB in 1955. He was succeeded as Director by J. A. B. Smith, who had been head of the Biochemistry department from 1936 to 1946 and had served, in Dr Wright's absence, as Acting Director from 1944 to 1946. After a short period as Lecturer in Applied Biochemistry in the University of Glasgow Dr Smith had returned as Deputy Director on Dr Wright's secondment.

1950 to 1970

In the late 1940s and early 1950s, under Dr Smith's guidance, there was a further substantial expansion of staff in association with the extension of the laboratories and the provision of new facilities. The earlier administrative grouping of staff according to their training and research interests was retained but the departments were expanded. R. Waite had returned to the Institute but A. B. Fowler and P. S. Watts had resigned and, because of the resignation of Dr Wright and the promotion to Director of Dr Smith, four new senior appointments had to be made. Two were internal promotions, W. Holmes as head of Crop and Stock-Husbandry and Farm Management and J. D. Findlay as head of Physiology, and two were from outside. In 1948, K. L. Blaxter, initially in receipt of a Senior Award of the Agricultural Research Council, was appointed as head of the Nutrition Department, as E. C. Owen had previously transferred to the Biochemistry Department, and in 1951 P. S. Blackburn was appointed as head of the Veterinary Pathology Department. Dr Holmes resigned in 1955 on his appointment to the Chair of Agriculture in the University of London, Wye College, Kent, and was replaced by M. E. Castle. In 1965, Dr Blaxter became Director of the Rowett Research Institute, Bucksburn, Aberdeen, and he was followed, as head of the Nutrition Department, by J. H. Moore.

This period was a most rewarding one. The Institute's considerable reputation in many of its established areas of work was strengthened and international recognition was obtained for work in new areas, especially that of Blaxter in energy metabolism and of Findlay and his colleagues in environmental physiology. The work itself is described in detail in later chapters and will not be enlarged on here.

In 1959, Dr Smith was awarded the CBE. In 1967 Dr Blaxter was elected a Fellow of the Royal Society and in 1977 received a knighthood in the Queen's Jubilee honours list.

1971 onwards

When Dr Smith retired in 1970, the scientific staff numbered thirty-three and the total staff had increased to about 100. Dr Owen had recently retired and both Dr Blackburn and Dr Higginbottom had died in post a few years earlier. The opportunity was taken, therefore, to reorganize the departmental structure. Against the background of increasing specialism in agricultural research and the national need to direct research to the solution of industrial problems it was decided to narrow the research interests of the Institute and, organizationally, to achieve a cohesion of effort without prejudicing opportunities for original work. Four departments were established, three specialist and one applied, and in an attempt to reflect more fully its functions, the Institute was renamed 'THE HANNAH RESEARCH INSTITUTE for studies relating to the production and utilization of milk'.

The three specialist departments, Biochemistry, Chemistry and Physiology, were given responsibility for basic research into scientific problems of importance either to milk production or to milk utilization. The Applied Studies Department was given a wider role: its primary function is to carry out research with well defined practical objectives but it also acts as a liaison group, on the one hand maintaining close links with the specialist departments and assisting with the practical exploitation of their research findings, on the other keeping abreast of problems and opportunities within industry. A fuller account of the present structure of the departments and of their work is given in the final chapter of this volume (p. 254).

LINKS WITH OUTSIDE BODIES

Through the representation of the University of Glasgow on the Council of the Institute, firm links between the two institutions were assured but these were strengthened when Dr Smith, following his return to the Institute from his post in the University of Glasgow, was appointed an Honorary Lecturer in Biochemistry. In 1970, this distinction was accorded also to Dr Moore, who became an Honorary Lecturer in Biochemistry and to Dr Findlay, who became an Honorary Lecturer in Physiology. In discussions preceding the retirement of Dr Smith, a fuller teaching link was established when the University Court created the Hannah Chair which is held along with the Directorship of the Institute and agreed that the status of Honorary Lecturer should be conferred on all Heads of Department. It was also agreed in principle that the Institute should contribute to a proposed new degree in animal science and Institute staff now have prime responsibility for the University degree course in Animal Nutrition, the teaching for the degree being shared with several University departments.

The location of the West of Scotland Agricultural College, which has now transferred all its teaching departments to the Auchincruive campus, and the representation of the College Governors on the Council of the Institute similarly have facilitated working links between the College and the Institute. These were strengthened in 1971 when joint Working Groups on milk production and milk utilization were formed to co-ordinate research and development work. Many of the research projects concerned with milk utilization are now done collaboratively with the Dairy School of the College and milk processing facilities are shared. An opportunity for collaborative work with College staff on problems concerning milk production was made possible in 1975, with the establishment of the Crichton Royal Development Farm at Dumfries. The Farm is administered by the College but is controlled by an Advisory Committee of the College Governors, on which there are staff members of both the College and the Institute, and the research and development work is planned by a joint College and Institute staff committee.

From the establishment of the Institute, personal contacts between its staff and members of the farming community and the dairy trade have been good but, to provide a more formal channel, in 1972 a Consultative Panel was formed to advise on priorities in the work on milk utilization. The Panel includes representatives of industrial, marketing and academic groups with an interest in dairy products and food, and staff members of the National Institute for Research in Dairying, Reading, and the Agricultural Research Council scientific adviser to the Secretary on food, attend meetings in the role of assessors.

Over the years a fertile contact with other research groups has been achieved through the many visiting workers who have spent periods of a few months to several years at the Institute. These people have contributed greatly, not only to the research work but also more generally to the life of the Institute and many life-long friendships and associations have been established.

The articles in this volume have been written by past and present members of senior staff and it is to them, and to the many other members of staff that they represent, that the major credit must go for the success that the Institute has achieved. Due acknowledgement must be given, however, to the many distinguished people who, over the years, have given freely of their time and effort to serve on the Council of the Institute, to the

Agricultural Research Council and its headquarters' staff who since 1931 have offered scientific advice, and to the Department of Agriculture and Fisheries for Scotland (formerly the Board of Agriculture) who have had responsibility for the funding and administration of the Institute from its inception. The efforts of administrative staff go largely unrecognized and unpublicized but without exception the Department staff have dealt sympathetically with the special needs of a scientific research institute and they have been ably supported by the administrative staff of the Institute itself. The Institute and the country at large are greatly indebted to them.

THE INSTITUTE FARM

J. N. Watson

Kirkhill farm, which now extends to 66·8 ha, is an important Institute facility for supplying animals and feedingstuffs for an ever-increasing range of experiments. Some details of the farm, including a description of the grassland management and milk production since 1951, have been published (Castle, 1960; Castle & Watson, 1964; Castle *et al.*, 1972), but a brief review of the changes which have occurred since the inception of the Institute may be of interest.

The original farm of 48·6 ha and the farm buildings were taken over in November 1929. At that time there was a poor byre, holding only eighteen cows, which was renovated and ventilated to bring the cow accommodation up to the standards of the day. A new byre, to hold twenty cows, was built on the end of the original building with a bull box and feed store between the byres. Thus, by 1932, the Institute had thirty pedigree Ayrshire cows, two bulls and a full complement of young stock. The farm produced and bottled certified and Grade A (TT) milk, which was sold locally. A valuation in 1933 placed a value of £5400 on the land and buildings; the price of a small tractor at the present time!

In 1939 the land at Mount Hamilton, 8·1 ha, was leased to the Institute and eventually purchased in 1948 with the house. A further 1·6 ha of Glebe land were rented from 1952 onwards which brought the total area to 58·3 ha. Finally, in 1975 and 1977, 12·1 ha were added to the east side of the farm which is about 2·5 km from the main Institute buildings. Over the years 3·6 ha of land have been lost to buildings and roads, leaving 66·8 ha now available to the farm. To ease the movement of stock on the farm, a new road was constructed from the farm buildings to near Mount Hamilton in 1976.

The farm lies about 30 m above sea level on a raised beach and is mainly a well-drained medium loam. There are clay cappings on some of the small hill tops and a rock fault in three fields which comes within 20 cm of the surface. Coal was mined from under the farm, and in three or four areas the land has subsided. These are all relatively small problems, and in general the soil type is uniform with the fields varying in size from 1·2 to 3·6 ha. Because of the climate, maximum use is made of grass, and cropping is now mainly grassland with only two or three fields sown with barley.

By 1955 the original byre had been modified so that twenty eight cows could be used for experimental work, but it was difficult to measure individual intakes of feed accurately. The byre could hold fifty cows but none of the stalls were single stalls and were thus not ideally suited for accurate experimental work. To overcome this problem, a new byre for experimental feeding work was built in 1962 on the site of the old one. This new building held sixty-four cows, and half of them were housed in carefully designed single stalls which allowed a precise measurement of food and water intake. For 10 years this byre was adequate for all the experimental work being done with dairy cows. Half the herd calved in the spring and were used on grazing experiments, and the other half calved in the autumn and were used for winter feeding experiments.

In 1971, owing to an increased demand for experimental animals, plans were made for extra farm buildings to allow the herd size to increase to 120 cows. This expansion included calf and young stock accommodation, a machinery shed, feed store and silos, and most important of all, a new dairy complex. This consisted of another byre to hold fifty-six cows in individual stalls, a milking parlour, a covered yard for weighing silage, a cattle weigh-bridge, a further feed store and loose boxes. Cows were milked in the new parlour in August 1975, and both the byres were full of cattle on experiments in 1976. Thus, between 1966 and 1976, the total number of cattle on the farm doubled from approximately 100 to 200 animals. The breed of cow has not changed since the Institute was founded, and descendants of the original Ayrshires described by Wright (1931) are still in the herd.

The Institute now has ideal dairy buildings for housing about 200 cattle for experimental purposes. The area of farmland, 66·8 ha, is small for such a large number of animals, but, by intensive management it provides grazing, silage and some hay and barley for the numerous experiments. Kirkhill farm has seen many dramatic changes in the last 50 years, and will no doubt continue to be at the centre of future developments in dairy farming for the next half-century.

REFERENCES

Castle, M. E. (1960) *Farm. Rev.* **14**, 18–20
Castle, M. E. & Watson, J. N. (1964) *J. Brit. Grassl. Soc.* **19**, 343–348
Castle, M. E., Macdaid, E. & Watson, J. N. (1972) *J. Brit. Grassl. Soc.* **27**, 87–92
Wright, N. C. (1931) *Ayrsh. Cattle Soc. J.* **3**, 1–4

FORMER RESEARCH TOPICS

THE PRODUCTION AND UTILIZATION OF GRASS AT THE HANNAH INSTITUTE, 1935 to 1955

W. HOLMES

IN his preface to 'The production of artificially dried grass' (McNair & Fowler, 1942) the Director, N. C. Wright, wrote, 'Publications from this Institute have repeatedly stressed the outstanding importance of grass and grassland products as sources of food for British livestock. Thus it has been shown (Wright, 1940) that grazing provides roughly half the country's total requirements of feedingstuffs, while hay – latterly supplemented by grass silage – raises the figure to nearly two-thirds.'

This early statement emphasizes that the Hannah Institute has been concerned with the production and utilization of grass from the outset. It is seldom realized that J. McNair and A. B. Fowler conducted some of the pioneer studies on the intensive fertilizer treatment of grassland in the United Kingdom. From 1932 to 1936 inclusive, duplicated plots cut and recorded five times each year were treated to a range of fertilizer applications ranging from a control treatment receiving no fertilizer, to the annual application of 360 kg N, 68 kg P_2O_5 and 75 kg K_2O per hectare. The mean yields recorded ranged from 4800 kg dry matter (DM) per ha for the control treatment to 11 550 kg per ha for the highest level of fertilizer indicating a response of 18·8 kg DM/kg N, a figure which has been just about the average response in nearly all subsequent work at the Hannah Institute and elsewhere in Britain.

During the period 1940 to 1950 the Institute farm was run on the basis of complete self-sufficiency in feedingstuffs, and grassland contributed a major portion of the diet of the dairy herd (Holmes, 1952, 1953).

Detailed experimental work on grass production was resumed in 1945. It was shown that grass herbage was likely to be more productive than green cereals or vetches (Holmes, 1948). The range of responses to fertilizer inputs recorded by McNair & Fowler (1942) was confirmed and the importance of the adequate provision of fertilizers when grass crops were removed for conservation was stressed (Holmes, 1949, 1951). The soils on which the experiments were conducted had low potassium reserves and a high capacity for phosphorus fixation. In consequence when 520 kg N/ha per year were applied for 3 successive years with and without mineral fertilizers, dramatic differences in the yield and botanical composition of the sward were recorded (Holmes, 1951). With no mineral application the yield of DM in the third year was 5200 kg/ha, and when phosphorus and potassium were applied the yield was 10 900 kg DM/ha.

Detailed mineral analyses of the herbage demonstrated the adverse effects of potassium fertilizer on the magnesium content of the herbage, and in general supported the growing evidence on the importance of ionic balance on the mineral content of herbage (Stewart & Holmes, 1953). This particular series of experiments continued and the results were reported again after 6 years (Holmes & MacLusky, 1954) and after 12 years of continuous study (Castle & Holmes, 1960). The corresponding yields in the 12th year were 6700 kg/ha where no minerals were applied and 11 600 kg DM/ha with minerals. This experiment led to a co-ordinated series of experiments on the effects of fertilizer on herbage production, conducted at six sites in Scotland and reported by Reith *et al.* (1961, 1964). These substantially confirmed the earlier results but showed that soil reserves, especially of potassium, varied widely with consequent effects on yield.

The studies referred to so far had been conducted on grass swards dominated by perennial ryegrass (*Lolium perenne*). But in view of the widespread interest in the late 1940s in new 'bred' strains of grasses an experiment was conducted to compare various cultivars of perennial ryegrass, cocksfoot (*Dactylis glomerata*), timothy (*Phleum pratensis*), meadow fescue (*Festuca pratensis*), and mixtures of ryegrass with cocksfoot, and of timothy with meadow fescue. The response to fertilizer nitrogen and the compatibility of the various grasses grown with clover was studied. Cocksfoot gave the highest yields and nitrogen responses, followed by timothy–fescue swards and timothy swards, with perennial ryegrass swards giving the lowest yields (Holmes & MacLusky, 1955).

However, when the grasses were grown with white clover and without fertilizer nitrogen the S26 cultivar of cocksfoot exerted a much more depressing influence on production than timothy or perennial ryegrass or indeed Scotia cocksfoot. The importance of competition for light in the management of clover was emphasized by these results.

These studies on grass had been conducted initially with a view to providing information relevant to the production of dried grass, but they also stressed the potential productivity of grass for general use, and by stressing the input/output relationships of fertilizers they contributed to the recent developments of special-purpose fertilizers for grassland.

The artificial drying of grass was conducted at the Institute during the period under review and field scale trials confirmed that liberal use of balanced fertilizers could sustain high yields of grass of high nutritive value (Castle & Holmes, 1960).

DRIED GRASS AS A FEEDSTUFF

The utilization of dried grass in the diet of dairy cows was studied in several feeding experiments. Dried grass made in spring or in autumn was compared with normal concentrates in diets that included hay and roots. Small but non-significant differences in milk yield were found in favour of spring-cut dried grass compared with that harvested in autumn (Holmes, 1956). Subsequently, dried grass made in spring or in autumn from cocksfoot or from timothy–meadow fescue swards was compared as a major component of the diet of milking cows. No significant differences were found between any of the sources of dried grass and all resulted in satisfactory milk yields and milk composition (Campling & Holmes, 1958).

THE REPLACEMENT OF CONCENTRATES BY CONSERVED FORAGE

Although the Institute had a special interest in dried grass, it was obvious that silage-making was likely to have a much wider impact. In experiments conducted by Reid & Holmes (1956) a diet of high quality forage (including dried grass and silage) was offered *ad libitum* in addition to concentrated feeds containing 22 or 14% crude protein and offered at either 0·2 or 0·4 kg per kg milk. The differences in total intake of nutrients were small. Concentrates on average increased DM intake by 0·32 kg per kg of concentrate offered. Milk yields were increased by the higher level of concentrates but an estimate of costs and margins at the time indicated that the low protein concentrate offered at the lower rate yielded the highest financial margin over feed costs.

In a subsequent experiment (Holmes *et al.*, 1957) concentrates were offered at 0, 0·2, 0·4, and 0·6 kg per kg milk and grass silage was offered *ad libitum*. Milk yields, milk solids-not-fat, milk protein and DM intakes increased with increased level of concentrate feeding. The increase in DM was 0·37 kg per kg concentrate offered. The economically optimal level of concentrate feeding with the available forage was 0·4 kg per kg milk.

APPETITE OF DAIRY COWS

These studies included direct estimates of the appetites of individually fed cows, information of vital importance in the planning of feeding systems. Similar studies to estimate the intake of grazing cows had been conducted in the work described by Waite *et al.* (1950, 1952). A detailed study of the intakes of individually grazing cows was also reported by MacLusky (1955). These estimates of intake all ranged between 100 and 130 g DM/kg $W^{0\cdot75}$ depending on milk yield and dietary quality, values which have been supported by later work.

GRAZING MANAGEMENT

A popular article entitled 'Doubling our grassland yields' (Holmes, 1949) emphasized the potential productivity of grass and the important contribution which it could make to Britain's supplies of feedingstuffs. It referred also to the use of the electric fence – a recent innovation – for rationing of high quality grass to grazing cows.

The evidence from experiments with cut grass together with the development of the readily moveable electric fence led to the development of strip grazing in which milking cows were persuaded to simulate a mowing machine by being alloted a daily ration of herbage by means of the electric fence. These studies led to a fuller appreciation of the potential from grazed grass (Holmes *et al.*, 1952) and led also to some of the earlier studies of the nutrient intake of grazing cows (Waite *et al.*, 1952) and on the behaviour of grazing cattle (Waite *et al.*, 1951). The use of the electric fence to ration the herbage facilitated the attainment of higher stocking rates and much of the benefit of strip grazing was indeed associated with this, but in a later experiment dairy cows were grazed at similar stocking rates under strip grazing within a rotation, or under free grazing, and it was shown that not only was the grazing simpler to manage under a rotational paddock or strip grazing system, but productivity per hectare was raised by about 10% (Campling *et al.*, 1958). The practice of applying fertilizer to blocks within a free grazing system, which has recently been publicized, was developed for these experiments.

The responses to fertilizer application recorded on small plots led also to a field comparison with grazing cows conducted for 3 years during which grass received fertilizer at an average of 212 kg N/ha or 59 kg N/ha. The yields of milk were similar per cow, but per hectare they were 5484 kg or 3853 kg, a response of 10·7 kg milk per kg N, and in DM terms 22 kg per kg N. From this, predictions were made of the fertilizer inputs required to make the attainment of 11 200 kg of milk per hectare (1000 gallons per acre) possible (Holmes, 1954). These levels of production have now been reached on the field scale at the Institute (Castle & Watson, 1975) and elsewhere.

CONCLUSIONS

Early work at the Institute demonstrated clearly the potential productivity of grassland in response to fertilizers. Subsequent work has enabled a detailed estimate of the response curve to fertilizer to be calculated (Reid, 1970).

The nutritive value of dried grass was shown to be high for milking cows and substantially in accord with estimates based on chemical analysis. The contribution of conserved forages to winter milk production was demonstrated.

The application of strip grazing methods to the grazing of pasture facilitated an increase in stocking rates and production per hectare. The response to fertilizer nitrogen applied to grazed pastures was shown to agree with estimates based on cutting experiments.

These studies all contributed to the substantial increase in productivity of British grassland of about 80% which has taken place since the 1930s (Wright, 1940; National Economic Development Office, 1974).

REFERENCES

Campling, R. C. & Holmes, W. (1958) *J. Dairy Res.* **25**, 147–153

Campling, R. C., MacLusky, D. S. & Holmes, W. (1958) *J. Agric. Sci.* **51**, 62–69

Castle, M. E. & Holmes, W. (1960) *J. Agric. Sci.* **55**, 251–260

Castle, M. E. & Watson, J. N. (1975) *J. Br. Grassl. Soc.* **30**, 1–6

Holmes, W. (1948) *J. Agric. Sci.* **38**, 425–436

Holmes, W. (1949) *Scott. Agric.* **28**, 205–212

Holmes, W. (1951) *J. Agric. Sci.* **41**, 70–79

Holmes, W. (1952) *Scott. Agric.* **32**, 83–90

Holmes, W. (1953) *Scott. Agric.* **32**, 211–216

Holmes, W. (1954) *J. Br. Grassl. Soc.* **9**, 17–27

Holmes, W. (1956) *J. Dairy Res.* **23**, 13–20

Holmes, W. & MacLusky, D. S. (1954) *J. Agric. Sci.* **45**, 129–140

Holmes, W. & MacLusky, D. S. (1955) *J. Agric. Sci.* **46**, 267–286

Holmes, W., Waite, R., Fergusson, D. L. & MacLusky, D. S. (1952) *J. Agric. Sci.* **42**, 304–313

Holmes, W., Reid, D., MacLusky, D. S., Waite, R. & Watson, J. N. (1957) *J. Dairy Res.* **24**, 1–10

MacLusky, D. S. (1955) *Proc. Br. Soc. Anim. Prod.* 45–51

McNair, J. & Fowler, A. B. (1942) *Hannah Dairy Res. Inst. Bull.* no. 8.

National Economic Development Office (1974) *UK Farming and the Common Market: Grass and Grass Products*, National Economic Development Office, London

Reid, D. (1970) *J. Agric. Sci.* **74**, 227–240

Reid, D. & Holmes, W. (1956) *J. Dairy Res.* **23**, 159–168

Reith, J. W. S., Inkson, R. H. E., Stewart, A. B., Holmes, W., MacLusky, D. S., Reid, D., Heddle, R. G., Clouston, D. & Copeman, G. J. F. (1961) *J. Agric. Sci.* **56**, 17–29

Reith, J. W. S., Inkson, R. H. E., Holmes, W., MacLusky, D. S., Reid, D., Heddle, R. G. & Copeman, G. J. F. (1964) *J. Agric. Sci.* **63**, 209–219

Stewart, A. B. & Holmes, W. (1953) *J. Sci. Food Agric.* **4**, 401–408

Waite, R., Holmes, W., Campbell, J. I. & Fergusson, D. L. (1950) *J. Agric. Sci.* **40**, 392–402

Waite, R., MacDonald, W. B. & Holmes, W. (1951) *J. Agric. Sci.* **41**, 163–173

Waite, R., Holmes, W. & Boyd, J. (1952) *J. Agric. Sci.* **42**, 314–321

Wright, N. C. (1940) *Emp. J. Exp. Agric.* **8**, 231–248

STUDIES ON MILK POWDER

R. Waite

INTRODUCTION

Commercial production of milk powder in Europe and North America began about the end of the nineteenth century, but until the 1930s the quantities involved were relatively small. The war years saw a great increase in powder manufacture in Canada, the United States of America, New Zealand and Australia which continued after peace was restored, so that by 1960 it had reached about 1·6 million tonnes annually from the chief dairying countries and 3·15 million tonnes in 1968–70 (New Zealand Dairy Board, 1971). About 80% of this was in the form of skim-milk powder, 11% full-cream powder and the rest as dried whey or buttermilk. In the United Kingdom, the annual production for many years was about 70 000 tonnes.

Milk powder is manufactured either by drying milk as a thin film on a pair of rotating, steam-heated rollers or by spraying concentrated milk into a current of hot air. In roller drying, the dried milk is removed continuously as a thin sheet which is then ground and sifted to the required degree of fineness. This produces a solid particle of irregular shape and varied size that has received a moderately severe heat treatment. Spray-dried powder, on the other hand, consists of smaller, hollow spheres which have suffered little heat damage and in which the lactose solidifies as a non-crystalline, hygroscopic glass. The chief differences between the two types are the greater solubility of the spray product in cold water and, in the full-cream powder, its greater susceptibility to fat oxidation. In the roller process some concentration of milk solids is obtained when the milk boils in the trough formed by the twin rollers, but in spray drying a concentration of about 40% solids must be achieved before spraying to give a product of suitable bulk density for packing. The milk is concentrated under a vacuum after an initial preheating to kill pathogenic bacteria and to inactivate unwanted enzymes. As will be seen late, preheating (70 to 88°C for 20 sec) also has the beneficial effect of causing some molecular re-arrangement of the sulphur-containing proteins so that they act as antioxidants.

The milk powders that we shall be considering have typical percentage compositions as shown on page 21.

The main themes of work over the years have been the solubility of milk powders; the prolongation of the storage life of spray-dried whole-milk powder; the deterioration of high moisture content skim-milk powder;

the wettability of powders and the bacteriology of all types of milk powder.

	Full-cream powder	Skim-milk powder
Fat	27	1
Protein	26	36
Lactose	37	51
Minerals	7	8
Water	3	4

THE START

Soon after the Institute was founded N. C. Wright commissioned two comprehensive reviews, one of the existing knowledge on the condensing and drying of milk by Allen (1932), the Institute's first bacteriologist, and the other on the engineering aspects of the same processes by Scott (1932*a*). L. A. Allen's paper dealt with the chemical and bacteriological aspects of the subject and drew together what was then known on the composition and properties of milk in relation to condensing and drying, the bacterial flora and the possible defects in manufacture and storage. Dr Wright was fortunate to find in the Engineering Department of the Royal Technical College, Glasgow, A. W. Scott who was already interested in problems of heat transfer, and collaboration was arranged between the two laboratories. Scott's review dealt with the fundamentals of heat transmission and evaporation and then applied the results to condensing and drying. He ascertained the conditions for maximum thermal efficiency and considered how far these had to be modified for the production of satisfactory products. In collaboration with Wright, Scott also reviewed the design, operation and efficiency of pasteurizing plants (Scott & Wright, 1935) which were then coming into prominence as the untreated milk era drew to an end. All these reviews succeeded in their purpose and were a mine of useful information to both industry and research for many years.

Scott continued his interest in the subject with the construction of a small laboratory spray-drying machine (1931), an examination of the cooling effect of a stream of hot air impinging on a water surface, as in spray drying (1932*b*) and a consideration of the effect of the elevation of the boiling point in concentrated milk on the heat requirements of evaporators (1933). Some of the milk powder made on Scott's machine was used by Allen in a study of the effect of milk solids in bread making (Allen & Bell, 1931) and it is interesting to see that their findings were confirmed, and not appreciably bettered when the subject cropped up again 30 years later in the laboratories of the Ministry of Food!

THE SOLUBILITY OF MILK POWDERS

The art of roller drying is to ensure that the milk film is completely dry just before it is removed from the rollers, which, to allow a slight margin of safety, inevitably results in some overheating of the already dry solids. Wright measured the solubility of the proteins in a number of samples of roller-dried powders, found that they differed widely in this property and decided that here was a problem to be investigated. In his first paper (Wright, 1932), a masterly combination of practical and scientific thinking, he showed that the sensitivity of milk proteins to heat was not a constant factor but increased as water was removed up to a critical point where only 12 to 14% moisture remained. At the temperature of the rollers, 130°C, milk proteins in this moisture range became irreversibly denatured in 0·01 sec (the total drying time on the rollers being about 1·3 sec), whereas at low moisture contents (3 to 5%), holding milk solids at the same temperature produced an insolubility that could be reversed by reconstituting the powder in warm (50°C) water. These results led to the use of two temperatures for the measurement of the solubility of milk powders, 20 and 50°C, protein insoluble at 50°C being considered irreversibly denatured while the difference in value between the solubility at the two temperatures represented the extent of dry overheating (Howat & Wright, 1933). This concept provided a most useful tool with which to diagnose insolubility problems in the milk powder industry.

Measurement of protein insolubility was too time-consuming and a simpler method was required, particularly for industry. In the United States of America, where the spray-drying system of powder manufacture was widely used, an index of solubility had been devised based on the volume of sediment spun down from powder reconstituted in water at 20°C in a standard manner. This worked reasonably well for spray-dried powders but gave anomalous results with roller-dried material, probably because in some powders insolubility was the result mainly of irreversible denaturation while in others the proportion of reversibly denatured protein was higher (Howat *et al.*, 1939). To provide a more sensitive method that could be used for all types of powder, a quantitative measurement of the solids remaining in suspension after centrifuging powder reconstituted at 20 and 50°C was adopted (Howat *et al.*, 1939) and successfully employed in much of the work to be described. In this method, the fat of full-cream powders was included in the 'soluble' fraction but it was noticed that the fat layer formed during centrifuging sometimes appeared to contain protein; this was particularly apparent when some roller-dried powders were reconstituted at 20°C. In such powders of low solubility as much as 17% of the total N could appear in the fat layer although in more soluble powders it would amount only to about 3%.

From these same powders, however, the insoluble sediment could contain 20 to 26% fat, giving a fortuitous cancelling of errors. Spray-dried full-cream powders, with their higher solubility were much less prone to this trouble, not more than 2 to 3% of protein or fat appearing in the wrong fraction (Waite & White, 1949).

THE STORAGE PROPERTIES OF FULL-CREAM SPRAY-DRIED POWDER

The approach of war, obvious by 1938, had led the British Government to consider stock-piling essential foodstuffs among which milk powder ranked high. Because of its high solubility and better flavour when reconstituted, spray-dried whole-milk powder was clearly preferable to the roller dried material and, by then, more spray-drying plants were in operation. The immediate problem was the short time such powder could be stored before fat oxidation rendered the taste so unpleasant as to make the powder unusable. A good quality powder might keep for 8 to 9 months before this happened, but in general a life of 6 months was more realistic. The Institute's study of milk powder was, therefore, concentrated on how to increase appreciably the keeping quality of spray-dried whole-milk powder, and in this they had the constant and fruitful collaboration of C. H. Lea of the Low Temperature Research Station, Cambridge, and, from time to time, of workers at the National Institute for Research in Dairying, Reading, and of E. L. Crossley of the University of Reading. When the time came for large-scale trials, commercial firms agreed readily to the use of their plant.

The proportion of unsaturated fatty acids in milk fat is small but it requires the presence of only minute traces of fat-oxidation products to give milk powder an objectionable tallowy taste, which often develops before any measurable chemical change. The air present inside and outside the spray-dried particle lends itself to fat oxidation and results in the formation of peroxides at the fatty acid double bonds. In the work that followed, assessment of the flavour of the reconstituted powder by a panel of tasters (which included the Secretary of the Institute, Miss M. M. Hughes, as an unbiased member of the 'public'), measurement of the peroxide value and the change of composition of the air in the can of powder were the criteria for the assessment of the course of fat oxidation during storage.

Commercial spray-dried whole-milk powder was at that time packed in rectangular tinplate containers holding 10 kg powder and it was realized that if extra precautions were taken these cans could be rendered airtight. The first approach therefore to the prevention of milk-fat oxidation was to remove the offending air by vacuum and replace it with an inert gas, nitrogen, the process known as gas packing and already used commercially on a small scale for certain special products. When spray-dried powder was gas packed it was found that although the oxygen content inside the container could be reduced to 1% or less by the application of 3 cycles of vacuum–nitrogen filling, it gradually increased to about 5% over a period of 7 to 10 days by desorption of air from within the particles. A study of the oxidation of fresh butterfat showed that for complete prevention of tallowiness the available oxygen had to be reduced to the equivalent of 0·23% oxygen within the can, but that for a storage life of 2 to 3 years, an oxygen content of about 1·5% would be acceptable. This meant that a second gas packing was required.

To determine, in as short a time as possible, the effect of various oxygen concentrations within the can on keeping quality, the powders, packed in both air and nitrogen in small cans, were stored at 37 and 47°C as well as at room temperature. In the range 15 to 47°C, the rate of fat oxidation had a Q_{10} value of 2.2, so that storage for 6 months at 37 or 47°C was equivalent to 3 or 5 years at room temperature (the cans stored at 15°C were, in fact, examined regularly for 3 to 4 years). It was found that during storage of the air-packed powder the development of tallowy taste was accompanied by a fall in the oxygen content of the air in the can and a rise in the carbon dioxide content and that these reactions were accelerated by a rise in temperature. Peroxides formed after an induction period and rose to a maximum from which they declined as oxygen disappeared. The development of tallowy flavour was not observed in gas-packed powder in which the oxygen content was 1 to 2% and was only slight so long as it was below 3 to 4%. Prolonged storage at the higher temperatures introduced a cooked flavour which made assessment of tallowiness more difficult, but the solubility of the gas packed powders would be virtually unchanged after several year's storage so long as the moisture content did not exceed 3 to 4% (Lea *et al.*, 1943), published some time after the results had been applied commercially).

An attempt to decrease the amount of oxygen available for fat oxidation by compressing the powder into blocks was unsuccessful, the high pressure required producing material difficult to reduce to a fine powder and causing the release of readily oxidizable fat on the surface.

Much more successful was a simplified method of gas packing investigated some years after the war (Abbot *et al.*, 1961) in which the cans of spray-dried whole-milk powder were given 1 cycle of vacuum before pressure was restored with a mixture of 95% nitrogen and 5% hydrogen. Inside the can a small pellet containing palladium catalysed the reduction of any remaining oxygen in the can and continued to do so as the air from within the particles came into equilibrium. This process reduced and kept the oxygen content to about 0·1%, eliminated peroxide formation and prevented development of tallowiness, thus considerably simplifying the long-term storage of full-cream powder at much less cost.

With gas packing established, thoughts turned to the protection of milk fat that might be obtained by increasing the production of natural anti-oxidant from the sulphur-containing proteins during the preheating stage of spray-dried powder manufacture. It was thought also that more attention to the cleanliness of the milk supply might be fruitful. This led to a large collaborative experiment with Cambridge and Reading (Mattick *et al.*, 1945), in which milk was clarified then preheated at 88°C instead of the more usual 74°C. Improvement in hygienic quality of the raw milk brought about a slight improvement in keeping quality of powder made from milk preheated at 74°C but none when 88°C was used. The higher preheating temperature, however, increased greatly the resistance of the powder to oxidation and the usual accelerated temperature trials indicated a storage life of 2 years, a marked improvement. Also, it reduced the amount of viable bacteria in the powder, whilst having no deleterious effect on solubility or vitamin content, the only disadvantage being some intensification of the boiled-milk flavour normally found in milk reconstituted from powder.

To determine whether the improved keeping quality could be obtained at a preheating temperature lower than 88°C, Findlay *et al.* (1946) investigated the range 70 to 90°C, but again found that 88°C was the lowest temperature compatible with enhanced storage quality. When an industrial plate preheater was developed operating at temperatures above the atmospheric boiling point of milk, the Institute accepted an invitation to assess the quality of powder produced from such milk, but found little advantage over a temperature of 88°C (White *et al.*, 1952).

The use of added antioxidants to delay the onset of tallowiness in full-cream milk powder had always been attractive but had, in this country, always been illegal. Nevertheless, it was considered worth while to accumulate some knowledge of the efficacy of this class of compounds, if possible of those occurring naturally. Small batches of powder containing the substance under investigation were prepared on a laboratory spray-drying machine constructed at the Institute (Waite, 1940). This used the same spraying system as in Scott's model but employed a larger drying chamber, more flexible heating and a more satisfactory method of powder collection. (No one anywhere in the Institute was ever in doubt as to whether milk powder was being made, the chugging of the air compressor, the banging on the chamber wall to prevent powder lodging and the anguished cries of the operator when the milk spray suddenly became a stream being easily recognized). We knew, of course, that results obtained on such a small scale might not be identical with those in commercially produced powder and control powders without any addition or containing a proven antioxidant were always prepared for comparison. A great variety of antioxidants or metal inactivators was examined, oat flour (for the germ oil), hydroquinone (Waite, 1941;) ascorbic, citric and maleic acids and esters of gallic acid (Findlay *et al.*, 1945); flavones, butylhydroxy-anisole and nordihydroguaiaretic acid (Abbot & Waite, 1962*a*), the tocopherols (Abbot & Waite, 1965), and ascorbyl palmitate (Abbot, 1971). Many of these compounds, at a concentration in the powder of 0·01 %, delayed the production of tallowy flavours but by far the most effective was dodecyl gallate, followed by nordihydroguaiaretic acid and one of the tetrahydroxyflavones. Addition of ethyl gallate, when the process was transferred to the commercial scale, led to a three- to four-fold improvement in keeping quality and if this was combined with milk preheated at the higher temperature (88°C), the improvement was enhanced considerably (Findlay *et al.*, 1945; White *et al.*, 1947).

BACTERIOLOGY OF MILK POWDERS

Because of the heat treatment milk receives en route to becoming powder and the low moisture content of the product, it might be expected that problems of bacterial spoilage would be rare, and this has proved to be so. Nevertheless, a poor quality milk supply or the continuous running of a drying plant for many hours can result in the build-up of viable thermoduric organisms in the powder. This was particularly true before the higher preheating temperature was adopted and in a survey of spray-dried powder from eight factories Nichols (1939) found great variation between samples, the mean plate counts ranging from 10^3 to 10^8 viable bacteria per gram of powder. Acid-producing bacteria and heat-resistant streptococci formed the bulk of those organisms, which tended to die out during storage. Powder from the same factories 5 years later when the higher preheating temperature was in use, had a lower count, 10^3 to 10^5 organisms per gram (Higginbottom, 1948*a*), The decrease in bacterial content of powder during storage was further investigated by Higginbottom (1944*b*, 1953) who noted considerable variation in this property between powders and, finding that the humidity at which powders were stored could markedly influence the bacterial content, linked some of the variability with observed differences of moisture content in the powder. She suggested that, in equity, all powders should be examined at some fixed time soon after manufacture, e.g. after 14 days. As would be expected from their methods of production, roller-dried powders contained relatively few viable bacteria, usually less than 10^3 per gram (Higginbottom, 1944*a*), a factor which no doubt contributed to its choice as a Government sponsored 'National' baby food in this country.

The rapid dye reduction tests used for the assessment of the hygienic quality of liquid milk were found to be unsuitable for milk reconstituted from powder, the results being poorly correlated with those given by the standard plate count or with the keeping quality of the reconstituted milks (Higginbottom, 1948*b*).

SKIM MILK POWDER

Moisture content and nutritive value

Although the problems of long-term storage described so far have concerned the fat in full-cream powder it was recognized from early work that the protein fraction could contribute off-flavours described as 'cardboardy' or, when well developed, 'gluey', which were associated with higher than usual moisture contents (Findlay *et al.*, 1945). About this time, workers at the National Institute for Research in Dairying reported a 20% loss in biological value of the protein in a supply of spray-dried skim-milk powder which they had kept for 3 years during which the moisture content had risen from 3 to 7%. This gave rise to a thorough investigation of the changes, physical, chemical and nutritive, which occurred when skim-milk powder was stored at low, medium and high moisture contents and involved collaboration once more with Reading and Cambridge. Powders containing 2·8, 4·7 and 7·3% moisture were stored in sealed cans in either air or nitrogen at 20, 28·5 and 37°C and analysed periodically in considerable detail for nearly 2 years (Henry *et al.*, 1948). The two factors chiefly concerned with the spoilage that occurred were a high moisture content and a high storage temperature, which together produced an increase in off-flavour and a darkening of the colour of the powder, a fall in pH and in the solubility not only of the powder as a whole, but of the lactose fraction as well, an accelerated absorption of oxygen and generation of carbon dioxide, a decrease in free amino-N, and an increase in the reducing power of the powder and in the amount of non-dialysable material. Gas packing in nitrogen did not prevent these changes in the high moisture powder. All these reactions were accompanied by a fall in the biological value of the protein and an apparent loss of lysine; supplementation of the original 7% moisture powder with lysine completely restored its biological value. In the powders containing 2·8 and 4·7% moisture, many of the reactions listed occurred only very slowly or not at all even when the powders were stored at 37°C. It was concluded that the deterioration apparent in the high moisture powder was the result of a reaction involving the free amino groups of the milk protein, largely the ε-amino groups of the lysine residues, with the potential aldehyde group of lactose to form a sugar–protein compound. The reaction appeared to proceed in two stages, primary combination without discoloration or loss of solubility followed by a second stage which led to the observable deterioration. These interesting results threw considerable light on the general problem of non-enzymic browning in foodstuffs and reinforced the long established commercial practice of producing milk powder with a low moisture content and, by adequate packaging, ensuring that it did not absorb atmospheric moisture in transit or storage.

'Instant' milk powder

'Instant' skim-milk powder is made by bringing about a surface aggregation of the fine particles of spray-dried powder to form a granule. This is achieved by raising the moisture content of dry powder to about 12% or by running the dryer so that the powder emerges with this amount of moisture; the damp, aggregated powder is then dried back to the normal 3 to 4% moisture. From the work already described, it was apparent that there would be some danger of spoilage if the heat received by the moist powder was excessive or, perhaps more likely, if the final moisture content remained too high. These points were investigated soon after firms in this country became interested in the process and in one commercially produced 'instant' powder of 6·6% moisture stored at 37°C there was discoloration, off-flavour development, loss of solubility and a rapid rise in protein-reducing value after 4 weeks, equivalent to about 20 weeks at room temperature. A second 'instant' powder made from the same bulk powder but dried down to 3·7% moisture had storage properties as good as the original powder. As before, packing in nitrogen did not prevent this type of deterioration (Abbot & Waite, 1962*b*). With the universal spread of 'instant' powders, there was need to define more carefully the terms in general use, wettability, dispersibility, sinkability, solubility and others and to agree on methods of measuring such qualities, a task in which international co-operation has been effective (Abbot & Waite, 1966).

Full-cream milk powder can be aggregated by the same process used for skim milk but results in an increase of surface 'free' fat which, on re-drying, oxidizes readily with production of tallowy flavour (Waite, 1968).

COMMITTEES

No account of the work of the staff of the Institute on milk powder would be complete without some mention of that largely postwar activity known as attending committees. So far as they concerned questions of milk powder, these were convened by the British Standards Institution, by the Milk and Milk Products Technical Advisory Committee (a Government inter-departmental committee) and by the International Dairy Federation, each usually spawning sub-committees and 'working parties'. It can be recorded that the Institute has taken its fair share in these deliberations.

CONCLUSION

In trying to assess the contribution that one or two workers in a small research institute can make to the problems of the milk powder industry or to the initiation of new processes, it would appear that they are probably best employed when investigating the more basic physical and chemical changes that take place in milk, after its production at the farm, through intermediate processing to the manufacture of the product. Most, if not all, of these reactions will have been initiated by the application or removal of heat. Problems arising in the milk industry in general and in milk powder production in particular have been largely those of engineering, the design of machines and their control mechanisms for dealing with a heat-sensitive material, as Wright no doubt recognized when he obtained the collaboration of the Glasgow engineers.

REFERENCES

Abbot, J. (1971) *J. Soc. Dairy Technol.* **24**, 182–183
Abbot, J. & Waite, R. (1962*a*) *J. Dairy Res.* **29**, 55–61
Abbot, J. & Waite, R. (1962*b*) *Proc. 16th Int. Dairy Congr.* **B**, 943–951
Abbot, J. & Waite, R. (1965) *J. Dairy Res.* **32**, 143–146
Abbot, J. & Waite, R. (1966) *Proc. 17th Int. Dairy Congr.* **E/F**, 203–207
Abbot, J., Waite, R. & Hearne, J. F. (1961) *J. Dairy Res.* **28**, 285–292
Allen, L. A. (1932) *Hannah Dairy Res. Inst. Bull.* no. 3
Allen, L. A. & Bell, J. (1931) *J. R. Tech. Coll. Glasg.* **2**, 550–564
Findlay, J. D., Smith, J. A. B. & Lea, C. H. (1945) *J. Dairy Res.* **14**, 165–175
Findlay, J. D., Higginbottom, C., Smith, J. A. B. & Lea, C. H. (1946) *J. Dairy Res.* **14**, 378–399
Henry, K. M., Kon, S. K., Lea, C. H. & White, J. C. D. (1948) *J. Dairy Res.* **15**, 292–363
Higginbottom, C. (1944*a*) *J. Dairy Res.* **13**, 308–323
Higginbottom, C. (1944*b*) *J. Dairy Res.* **13**, 324–328
Higginbottom, C. (1948*a*) *J. Dairy Res.* **15**, 277–279
Higginbottom, C. (1948*b*) *J. Dairy Res.* **15**, 280–284
Higginbottom, C. (1953) *J. Dairy Res.* **20**, 65–75
Howat, G. R. & Wright, N. C. (1933) *J. Dairy Res.* **4**, 265–272
Howat, G. R., Smith, J. A. B., Waite, R. & Wright, N. C. (1939) *J. Dairy Res.* **10**, 498–514
Lea, C. H., Moran, T. & Smith, J. A. B. (1943) *J. Dairy Res.* **13**, 162–215
Mattick, A. T. R., Hiscox, E. R., Crossley, E. L., Lea, C. H., Findlay, J. D., Smith, J. A. B., Thompson, S. Y. & Kon, S. K. (1945) *J. Dairy Res.* **14**, 116–159
New Zealand Dairy Board (1971) *10th Annu. Rep.*
Nichols, A. A. (1939) *J. Dairy Res.* **10**, 202–230
Scott, A. W. (1931) *J. R. Tech. Coll. Glasg.* **2**, 456–461
Scott, A. W. (1932*a*) *Hannah Dairy Res. Inst. Bull.* no. 4
Scott, A. W. (1932*b*) *J. R. Tech. Coll. Glasg.* **2**, 620–630
Scott, A. W. (1933) *J. R. Tech. Coll. Glasg.* **3**, 116–120
Scott, A. W. & Wright, N. C. (1935) *Hannah Dairy Res. Inst. Bull.* no. 6
Waite, R. (1940) *Chem. Ind. (Lond.)* **59**, 659–661
Waite, R. (1941) *J. Dairy Res.* **12**, 178–183

Waite, R. (1968) *Hannah Dairy Res. Inst. Rep.*, 52–58
Waite, R. & White, J. C. D. (1949) *J. Dairy Res.* **16**, 379–389
White, J. C. D., Smith, J. A. B. & Lea, C. H. (1947) *J. Dairy Res.* **15**, 127–139
White, J. C. D., Waite, R., Hawley, H. B., Clark, J. G. & Henry, K. M. (1952) *J. Dairy Res.* **19**, 339–347
Wright, N. C. (1932) *J. Dairy Res.* **4**, 122–141

DISEASES OF DAIRY CATTLE, 1929 to 1939

S. J. EDWARDS

IN writing a short account of disease investigation at the Hannah Research Institute, one is forced to ask why the subject was included in its original programme of research, bearing in mind that already there were other centres in England and Scotland concerned exclusively with this subject. Moreover, there was no precedent for including a veterinarian on its staff at the older established centre, the National Institute for Research in Dairying, Reading.

The decision to do so arose largely from the fact that the Medical Research Council (MRC) had made available a grant to support an investigation into the feasibility of eradicating bovine tuberculosis from dairy herds. Today, it would seem surprising that a scheme for the control of cattle disease should have been sponsored by a body controlling medical research as it would now fall more appropriately within the broad scope of agricultural research. It should be remembered, however, that the Agricultural Research Council (ARC) was at that time in its early formative days and had not as yet promoted any major disease investigation. Further, of course, the serious hazard to human health of bovine tuberculosis must be considered, a hazard which at that time was responsible for about 3000 fresh cases of tuberculosis in children annually in the British Isles arising from the consumption of infected milk. Wright (1929 showed that up to 8·5% of milk samples taken from bulk milk as supplied to Scottish cities could be infected with tubercule bacilli but that not one sample from tuberculin-tested herds was found to be infected. Apart from the public health aspect of bovine tuberculosis, there was also the serious depreciation in the health of the cattle population with consequent losses in milk yield, altogether amounting to about £10 million annually (Jordan, 1930; Beveridge, 1960).

THE 'TB' ERADICATION SCHEME

After inspecting various areas in the country where the experimental scheme could be carried out, the MRC selected an area in Ayrshire, an intensely dairying county with a high population of cattle. Moreover, there was already a good nucleus of tuberculin-tested herds in the county resulting mainly from the encouragement of the veterinary staff of the local authority together with the keen desire of the Ayrshire farmer to rid his herd of tuberculosis.

The first veterinarian appointed to take charge of the scheme in 1929 was L. Jordan, MRCVS. The scheme included thirty typical dairy farms enclosed in an area near Mauchline clearly bounded by public roads. In these herds, all the cattle were tuberculin-tested at regular intervals, reactors either being disposed of or segregated on the farm. Continuous supervision was exercised also on the renovation of buildings, methods of disinfection and the proper erection of boundary fences. The main principles in the elimination of infection were:

1. to repair or renovate existing premises together with thorough disinfection,
2. to isolate reactors both at pasture and in the byre,
3. to rear young stock away from adult stock and eventually to replace the latter with home-bred heifers.

At the start of the scheme the infection rate in individual herds varied from 2 to 69% but after 3 years of operation, the proportion of infected cows in the thirty herds had been reduced from 25 to 9% (Jordan, 1933) and twenty of the herds were free from tuberculosis. Two years after the termination of the experimental scheme, the herds were tested again to determine if the farmers had kept up the initiative of the scheme. It was found that twenty-five of the herds were then free of reactors, while in the remaining five no substantial progress had been achieved because of the initial high number of reactors and the consequent reluctance of the owner to attempt eradication.

This experiment showed that, for the first time in this country, a large number of herds in an area could be freed from tuberculosis and, further, that the example set by the farmers in such an area provided a stimulus for eradication in a large number of herds contiguous with the original area. As quoted from an article written at the time (Edwards, 1936), 'the results noted above show that the area has acted as a leaven from which eradication has spread on a remarkable scale'. This scheme together with the enthusiastic work of the veterinary staff of the local authority, led to the outstanding achievement of the county of Ayr being declared the first tuberculosis-free area in the country.

Thus, at the time of the transfer of responsibility for the supervision of animal health and milk production from the local authority to the Ministry of Agriculture on the passing of the Agriculture Act of 1937, an excellent blueprint was placed before the Ministry for the eradication of tuberculosis. But the Attested Herd Scheme developed at that time contained none of the basic principles by which this progress had been achieved, and since the scheme contained only the attraction for the farmer of an extra bonus on milk from tested herds, the result achieved was mainly to convert the designation of herds already free (i.e. Certified or Tuberculin Tested herds) to that of 'Attested Herds'.

The further progress made by the Ministry of Agriculture in the eradication of tuberculosis from the country was, on the whole, not too successful between 1937 and 1950 despite payments of vast sums of money in compensation for slaughtered animals and payments for the bonus on milk produced by attested herds. It was not until the introduction in 1950 of the Area Eradication Plan, which included the principles of free testing and advice to the farmer, as had been adopted by the Hannah Institute, that real progress was made, so that by 1960 the whole country was freed from bovine tuberculosis. Thus, the name of the Hannah Research Institute should always be associated with a pioneer effort to control a major disease of cattle having serious hazards to human health – an effort which was sponsored by the MRC. It should be placed on record that one of the greatest enthusiasts for the scheme was the Council's Secretary, the late Sir Walter Fletcher.

PASTEURIZED MILK IN CALF REARING

Before tuberculosis in cattle was eradicated, the only effective means of ensuring the safety of the public milk supply from infection with the tubercule bacillus was by pasteurization. But in spite of the evidence that this process did not seriously affect the chemical composition of milk (Blackwood *et al.*, 1936) there still remained considerable prejudice against pasteurized milk from the nutritional standpoint.

Perhaps the most convincing evidence of the value of pasteurization was provided by an experiment carried out jointly by members of the staff on the relative value of raw and pasteurized milk in the feeding of calves (Wilkie *et al.*, 1937). Thirty-six calves were fed for 12 weeks on raw milk bulked from ten farms, and thirty-seven the same milk pasteurized by the Holder process. At the end of the experiment, there were no significant differences in the weight gains of the two groups of calves, indicating that for the calf, the natural consumer of cows' milk, pasteurization had had no ill effects on growth and development. Of greater importance, however, was the demonstration, at the end of the experiment, that pasteurization had destroyed the tubercule bacilli which were regularly present in the raw milk. In twenty-three of the thirty-six calves receiving raw milk, tuberculosis of the thoracic cavity, mainly of the bronchical and mediastinal glands was found, whereas not one calf fed on the same milk pasteurized was affected.

An interesting observation on the pathogenesis of tuberculosis was that the glands of the intestinal tract were not affected; this was an unexpected finding since the path of introduction was by ingestion (Edwards, 1937).

CONTROL OF BRUCELLOSIS

A special report prepared by the Institute in 1931 showed that disease accounted for nearly 50% of the annual disposals of the milking stock from the dairy farm, the principal causes being reproductive disease, including abortion due to *Brucella abortus* infection, and diseases of the udder, commonly known as mastitis. The significance of these data was later confirmed in 1934 by the Committee on Cattle Diseases of the Economic Advisory Council which estimated that the productive life of the dairy cow at that time was no more than two lactations. It is not surprising therefore that the consequent losses fell heavily on the Ayrshire farmer whose stock was, in the main, dairy cattle.

In 1934, the Institute, therefore, collaborated with other centres in a scheme to control *Br. abortus* infection in a number of dairy herds under the general supervision of a committee of the ARC. Twenty-five herds were enlisted, being in the main pedigree herds whose owners were anxious to eliminate the disease with a view to exporting blood-tested stock to Canada and the United States of America. Blood samples were collected from all cows and heifers at regular intervals and submitted to the agglutination test. So far as facilities permitted on the farm, reactors were isolated from those passing the test, both at pasture and while housed, especially at calving. After 4 years, it was found that contagious abortion could be controlled in those herds with a moderate number of reactors and it seemed that the presence of a degree of infection conferred a state of natural resistance in the herd to the disease. On the other hand where, as occurred in a few herds, no initial infection was found, such natural protection was absent with the result that the introduction of infection from outside sources brought about abortion storms with almost calamitous losses through premature calvings and reduction of milk yield. In spite of these drawbacks to a scheme based on segregation alone, some fundamental information on the epizootiology of *Br. abortus* infection was provided from a total of about 100 herds. This information should have provided a basis for all subsequent attempts to control the disease in this country. Unfortunately, these valuable results were never published by the ARC due partly to printing difficulties caused by the onset of the Second World War, and partly to the diversion of attention to the use of vaccines in controlling the disease. At that time, their use was not allowed in licensed herds because of the danger of excreting living *Br. abortus* in the milk of vaccinated cattle with the consequent risk to human health of undulant fever. Fortunately, however, research had been going on, both in this country and in the United States of America, to discover strains of low virulence for vaccination, which were free from the objection that they might persist in the tissues of the bovine and be excreted in the milk. One of these vaccines was the American Strain 19 which had been found to produce a good immunity without becoming localized in the udder of vaccinated cattle; the other was the British Strain 45(20), discovered by McEwen (1940), and described by him as a rough strain which did not elicit agglutinins in vaccinated cattle. To confirm the value of these two strains, the ARC purchased a farm on the Berkshire Downs, in order that a trial should be conducted in strictly controlled conditions, using animals of known health history such as were not available on the ordinary farm. Although the experiment was conducted under the uneasy conditions of wartime Britain – the premises being bombed on one occasion – the results showed that the strains were valuable in protecting heifers against experimental infection without the disadvantage of setting up permanent infection. This work, which was a direct continuation of that conducted by the writer at the Hannah Institute and by workers at other centres, provided the information required by the Ministry of Health for withdrawing its embargo on the use of live vaccines in 1942, so that they could be used during and after the war. The calfhood vaccination scheme was soon established, and the losses to the dairy industry caused by brucellosis at a time of postwar economic stress were greatly minimized. A further policy by the Ministry of Agriculture to eradicate brucellosis has since been developed by the introduction of the Accredited Herd Plan but the progress made has been slow. This is largely due to the failure of the Animal Health Division to recognize three important principles: firstly, the simultaneous control of all the herds within an area; secondly, the application of a blood-testing programme to determine the degree of infection; and thirdly, the use of vaccination according to the degree of infection present in the herd. The conventional test and slaughter programme now practised by the Ministry is not rational since it takes no account of these principles.

BOVINE MASTITIS

The disease which has probably accounted for the largest single loss among dairy herds is mastitis and at one time it was calculated that up to 10% of cows were affected. It was of serious concern to the Ayrshire farmer whose main source of income was from the production of milk.

Early in 1934, a number of farmers approached the Institute for assistance in dealing with the mastitis problem with the result that investigations were made in a few herds. In order to provide laboratory facilities, the ARC was approached for financial assistance and a sum of £100 was granted annually, on the proviso that work was supervised by a committee composed of members of the Scottish Board of Veterinary Research.

Previous to this, mastitis had been studied extensively at the Research

Institute, Royal Veterinary College, London, by Minett, Stableforth and Edwards who had been concerned with its causes, diagnosis and methods of control. Their recommendations were put into practice in the Ayrshire herds and satisfactory results were obtained in the prevention of the spread of infection – success depending on the thoroughness with which hygienic measures at milking were applied. Attempts were also made to cure a few affected cows, using the therapeutic substances then available such as acriflavine, entozon and other dye solutions. But it could not be claimed that treatment was successful in those days since most of the drugs available were irritant when injected into the udder.

It should be recounted here that the early work at the Hannah Institute provided a stimulus to the ARC to take a wider interest in research into bovine mastitis. In consequence many other centres became involved and the work was co-ordinated by a special committee of the Council. Certain aspects of the disease were studied by a well organized conjoint effort and included research on the distribution of infection and modes of spread of *Streptococcus agalactiae* (ARC, 1944). At the same time, the value of the new therapeutic agents which had been found of great service in the wartime treatment of wounds was investigated by the writer at the Field Station, Compton, Berkshire. No doubt the development of the sulphonamides and of penicillin at that time (Edwards & Brownlee, 1946), proved of great value in reducing the losses caused by clinical mastitis.

The basic principles of correct hygiene supplemented by appropriate therapy were soon established by means of co-ordinated trials carried out by workers on the mastitis committee and these principles are still in use today. Unfortunately, it would appear that basic research as undertaken by that committee has for some reason fallen into abeyance and more attention to the disease is now given by commercial firms whose main interest is in demonstrating the advantages of their particular formulations.

REFERENCES

Agricultural Research Council (1944) *Imp. Bur. Anim. Health Rev. Ser.* no. 2

Beveridge, W. I. B. (1960) *Vet. Rec.* **72**, 810–815

Blackwood, J. H., Morris, S. & Wright, N. C. (1936) *J. Dairy Res.* **7**, 228–237

Economic Advisory Council, Committee on Cattle Diseases (1934) *Report*, His Majesty's Stationery Office, London

Edwards, S. J. (1936) *Vet. Rec.* **16**, 561–566

Edwards, S. J. (1937) *J. Comp. Pathol. Ther.* **50**, 377–382

Edwards, S. J. & Brownlee, A. (1946) *Vet. Rec.* **58**, 335–343

Hannah Dairy Research Institute (1931) in *Annu. Rep. Scott. Milk Rec. Assoc.*

Jordan, L. (1930) *Farm. Stock Breed.* **44**, 768

Jordan, L. (1933) *Med. Res. Counc. (G.B.) Spec. Rep. Ser.* no. 184

McEwen, A. D. (1940) *Vet. Rec.* **52**, 815–819

Wilkie, J., Edwards, S. J., Fowler, A. B. & Wright, N. C, (1937) *J. Dairy Res.* **8**, 311–323

Wright, N. C. (1929) *Br. Med. J.* **2**, 452–454

NITROGEN METABOLISM IN THE RUMEN

J. A. B. Smith

One of the first subjects to be studied at the Hannah Research Institute was the protein requirements of dairy cows. It was realized that if much more were known about the nature of the proteins in different products and their value for milk production, optimum yields could be achieved with greatly increased efficiency in the use of feedingstuffs, and wastage, particularly of the more expensive protein-rich feeds, would be reduced. Work on this subject continued at the Institute from 1930 to 1938 and several publications resulted (e.g. Morris & Wright, 1933*a*, *b*; Morris *et al.*, 1936).

By 1938, however, political events in Europe indicated all too clearly that Britain might once again be involved in a major war, and if that were to happen, feedingstuffs, of which a large proportion came from abroad, would become scarce and expensive, and the country's supplies of milk and meat would suffer. The Director of the Institute, N. C. Wright, proposed, therefore, that the possibility of using simple non-protein nitrogenous substances (NPN), such as urea, as substitutes for part of the protein needed by cows, should be investigated thoroughly.

The possibility that NPN might be utilized by ruminants had been suggested as early as 1891 (see review by McNaught & Smith, 1947), but although the subject had been studied in several countries, much of the work had been badly planned and inadequately controlled, with the result that the conclusions were often conflicting. In fact many of the results lent themselves to entirely opposite conclusions, depending on the bias of those interpreting them. This can be seen from a review by Krebs (1937) in which he cited more than 120 papers, most of which dealt with the growth of cattle and sheep. Krebs was not convinced that substances, such as urea, were converted to protein in amounts that would be important to the ruminant, though he admitted that urea appeared to have some kind of protein-sparing effect. He suggested that this might be explained by assuming that dietary protein neutralizes the organic acids produced in the digestive tract, and that any substance that formed ammonia in the rumen might replace the dietary protein needed for that purpose, and so allow digestion and absorption, which would otherwise be depressed by the acid conditions, to be maintained at normal levels. It is, of course, important to recall that the well known work of A. I. Virtanen in Finland, who maintained lactating cows for long periods on diets which contained NPN as the *sole* source of nitrogen, was not done until some 25 years after the publication of Krebs' review.

In 1938 it was clear that if future research was to make a worthwhile contribution to knowledge, it would have to be planned in such a way as to avoid the pitfalls and deficiencies of earlier work. With this in mind it was decided (i) that an experiment should be done to determine whether, under the carefully controlled conditions that were possible at the Institute, urea could act effectively as a substitute for part of the protein needed by the lactating cow, and (ii) that a detailed study should be made of rumen contents *in vivo* to establish whether significant amounts of NPN are in fact converted to protein by the rumen micro-organisms.

PERSONNEL

In a historical essay, such as this, it is appropriate to give some information about the members of staff who did the work. In the years 1930 to 1938 the person chiefly rseponsible for research on the value of dietary proteins for milk production was Samuel Morris. When Morris resigned in 1938, E. C. Owen of the Rowett Research Institute was appointed Head of the Nutrition Department and took up his duties in 1939. Owen translated the review by Krebs (1937), and later he reviewed the subject himself on several occasions. He also directed and conducted the lactating-cow experiment described below.

J. A. B. Smith, who joined the staff in 1936 to study the biochemistry of milk fat secretion, was asked by N. C. Wright to be responsible for the studies *in vivo* of rumen contents. It happened that soon after his work began Smith read a paper on the subject at a scientific meeting at which M. P. Appleby was present. Appleby had been one of Wright's tutors at the University of Oxford and was now on the Board of Imperial Chemical Industries Ltd, at Billingham. He was impressed with the possible protein-sparing effects of urea, and suggested that ICI might help in three ways, by supplying the urea needed for the work, by giving a grant towards the cost of it all, and by seconding to the Institute an assistant from the laboratory staff at Billingham. These offers were accepted, and a young analyst, R. M. Pearson, was seconded to assist Smith in the work, a most successful arrangement which continued for some 3 years. When Pearson returned to Billingham a young graduate from the Biochemistry Department of the University of Glasgow, Mary L. McNaught (now Mrs J. H. Moore), was appointed as an assistant to Smith.

As the work with rumen contents developed it became clear that its value would be enhanced greatly if a microbiologist with a sound working knowledge of rumen micro-organisms could be found to collaborate in it. Until then, the microbiology of the rumen had not been studied much in Britain, and it was not easy to find anyone with the requisite knowledge and experience. After extensive enquiries, however, it was found that

F. Baker of Guildford had devoted much time to the subject, and he agreed to collaborate.

For the work on rumen contents a steer with a rumen fistula was used. The fistula was made by P. S. Watts, then Veterinary Pathologist at the Institute. Of the others who gave much valuable assistance Miss Agnes G. Wilson, Miss Sarah H. A. McLauchlan and Mr W. Robertson are still members of the Institute staff. Miss Jean Littlejohn, who later became Mrs J. D. Findlay, assisted in the urea-feeding experiment.

THE LACTATING-COW EXPERIMENT

It was believed that results gained by using only a few cows but studying them in detail under well controlled conditions would lead to much more reliable conclusions than had been obtained elsewhere in the past by larger scale, but less well controlled, feeding trials. And so this experiment, under E. C. Owen's leadership, was done with seven lactating cows kept in metabolism stalls where their excreta could be collected. During control periods of 4 to 6 weeks some 25% of the total N of the diet was supplied by blood meal, a product that had been shown earlier at the Institute to be an excellent source of protein for milk production (Morris & Wright, 1933*b*). The control periods were followed by experimental periods of 4 to 6 weeks, in which the blood meal of the diet was replaced by its N equivalent of urea and its energy equivalent of starch. With two of the cows the urea was then withdrawn from the diet and the blood meal restored, but with the remaining five cows the urea period was followed by a 2-week period in which both blood meal and urea were withheld.

The urea diet was eaten readily, no harmful effects were observed and the weights of the cows were well maintained. Two of the cows showed a slight decrease in yield during the urea period, but the decrease was only a very small percentage of the yield that would have been expected if blood meal had been given throughout. With the remaining five cows milk yields were well maintained when urea replaced blood meal, but when urea and blood meal were both withheld, the yields of four of them declined.

The composition of the milk was not affected adversely by urea-feeding. The level of urea in the milk followed closely that of the blood, but it did not exceed 28 mg N/100 ml, and was usually much lower.

The values for N intake and excretion suggested that although much of the urea-N given to each cow was retained, its retention was not as great as that of blood meal N. The utilization of urea-N was calculated to be as high as 88% for one cow and as low as 53% for another, with an average of 75%. This suggested that on average some 25% of the urea-N of the diet passed through the animal without being used. It was realized that some loss of urea-N was inevitable, since some of it was bound to pass rapidly into the blood stream or further along the alimentary tract before it could be used by the rumen micro-organisms, and although some of that N would return to the rumen in the saliva, a portion of it would be excreted in the urine as the blood passed through the kidneys. It is important to note that the amount that appeared to be lost in this way varied so much from one cow to another.

Only those who have conducted well controlled experiments of this type with lactating cows can appreciate fully the problems that are encountered from day to day in the work itself and later in the interpretation of the results, but from these experiments it was clear that urea could act as a partial protein substitute and maintain milk yields at reasonable levels with no ill effects on milk composition or on the health and well-being of the cows (Owen *et al.*, 1943).

WORK INVOLVING RUMEN CONTENTS

Experiments in vivo

In 1939 two of the most important questions about urea-feeding were these: (i) Can it be shown beyond doubt that protein is synthesized in significant amounts in the rumen from NPN, and (ii) if so, are the biological value and the digestibility of the resulting protein such as to make it of value to the host animal?

With the first of these questions in mind it was decided to carry out a series of experiments with a steer that had a rumen fistula. The N distribution in the rumen contents would be determined when the steer was on a diet that contained no added NPN. Urea would then be included in the diet, and its conversion to protein, if such conversion occurred, would be observed by sampling the rumen contents at hourly intervals and determining the N distribution. In theory this sounded a straightforward approach, but after a long series of preliminary experiments it was realized that it was quite impracticable. The contents of the bovine rumen may weigh as much as 75 kg, and while the upper layers of this material have a consistency somewhat similar to that of wet partially chewed grass or hay, lower down in the rumen the solid material is much more finely divided and is bathed in fluid. The composition of samples taken simultaneously from different parts of the rumen showed variations that were much too great for any synthesis of protein that might have occurred to be detected. However, although these experiments on sampling problems showed that the proposed work *in vivo* was not feasible, much valuable information was obtained about the physical, chemical and microbiological nature of the rumen ingesta and how best the more liquid contents could be sampled and analysed (Pearson & Smith, 1943*a*).

Experiments in vitro

Preliminary considerations. Since the approach *in vivo* was not feasible, it was decided to study the ability of rumen contents to synthesize protein from NPN when these contents were incubated in the laboratory under conditions that resembled as closely as possible those existing in the rumen. For most of this work 'rumen liquid', which could be sampled readily for analysis, was used. It was obtained by taking samples from fairly deep in the rumen and straining them through muslin. The resulting liquid was rich in bacteria and protozoa.

It was important that the bacteriology of the samples should be studied as well as the chemistry, but it was found that the main types of rumen bacteria (Baker, 1943) would not grow on any kind of media usually used for plate counts in those days. When C. Higginbottom, bacteriologist at the Institute, tried to apply conventional plate-count procedures the only bacterium that survived and grew on the plates was *Streptococcus bovis.* It was decided, therefore, that the direct counting methods evolved by Baker should be used (Smith & Baker, 1944). When samples of rumen liquid were incubated under conditions similar to those in the rumen, samples taken from the incubation flasks at specified intervals were fixed with formalin and sent to Baker for microbiological examination. He showed that when rumen liquid was incubated at 39°C for long periods, it was only during the first few hours that the microbiological 'picture' in the incubated samples remained similar to that of the fresh rumen liquid. After some 3 to 4 hours, deterioration set in, the general appearance and smell of the rumen liquid began to alter, and considerable changes occurred in the microbiology of the samples. Therefore, in all the ensuing work only those changes in NPN and protein contents that occurred during the first 2 to 4 hours of incubation were regarded as being typical of changes that could reasonably be assumed to take place in the rumen itself.

The urease activity of rumen contents. At an early stage in the work it had been observed that even when as much as 40 g urea were given to a steer in a day, no urea could be detected in the rumen contents, though the levels of ammonia varied considerably. It was decided, therefore, to study the conversion of urea to ammonia in rumen liquid.

In some typical experiments as much as 100 mg urea-N/100 g rumen liquid were converted to NH_3-N in 3 hours, and in discussing this work Pearson & Smith (1943*b*) pointed out that if the weight of rumen contents be taken as 75 kg, it would appear that all the urea ever likely to be given, even to a high yielding cow in full lactation, would readily be converted to ammonia within an hour. The conversion took place over a wide range of conditions, but the optimum temperature was found to be between 40 and 50°C and the optimum pH between 7 and 9. Neither the concentration of urea over a wide range, nor the time that elapsed after a meal before the rumen contents were sampled, had any significant effect on the ability of the rumen liquid to convert urea to ammonia. This urease activity was destroyed by small amounts of quinine, sodium fluoride and boracic acid, all known to be enzyme inhibitors, but it was unaffected by toluene, which inhibits bacterial growth but not enzyme action under the conditions in which it was used.

The fact that over 40 g urea could be converted to ammonia in 1 hour suggested that some ammonium salts could probably be used as a source of NPN in the rumen as effectively as the more expensive urea. It is important to note, however, that the results of this section of the work emphasized how dangerous it could be to overfeed ruminants with urea. If too much of it were given in error at one time, large and possibly lethal amounts of ammonia would be formed immediately. Such an occurrence, which resulted in the death of several cattle, was reported from abroad at about the time when this work was published.

Another consideration that arose from this work was that the utilization of urea would be likely to be much more efficient if in the rumen the urea could be liberated slowly in small amounts throughout the day instead of in relatively large amounts twice a day at feeding times, and it was suggested that coating urea in some way to retard its solution in rumen contents might be helpful. Several samples of urea coated with various proportions of starch, stearic acid and cellulose acetate were supplied by ICI Ltd, but when these coated products were tested in incubations with rumen liquid, all the urea was extracted from them in less than 20 min. The same was true of urea contained in cubed feedingstuffs such as those studied by Snow *et al.* (1945). In fact, no effective method was found by which urea could be liberated slowly into the rumen contents under practical conditions (McNaught, 1948).

The synthesis and breakdown of protein in rumen contents. The experiments discussed under this heading formed the most important section of the work involving rumen contents. They indicated that synthesis and breakdown of protein occur simultaneously in rumen contents, and that whether the amount of protein formed exceeds the amount broken down depends on several factors.

Rumen liquid was incubated *in vitro* at 39°C under various conditions with and without the addition of urea and carbohydrate, and the distribution of N in the rumen liquid was determined initially and at intervals during incubation. From the results of many experiments of that type (Pearson & Smith, 1943*c*), the following observations were made. 1. The content of total N remained unchanged. 2. Urea was rapidly converted to ammonia. 3. Relatively little NPN was present except as urea or ammonia. 4. During the first 3 hours, when the microbiological conditions in the medium resembled closely those of the rumen, synthesis of protein occurred. 5. The decrease that occurred in NPN was mainly in NH_3-N, suggesting that the synthesized protein was produced from ammonia and not to any significant extent from the NPN present in other forms.

For many similar incubations the average amount of protein synthesized in 2 hours appeared to be equivalent to about 8 mg N/100 g rumen liquid. Assuming the total weight of contents in the rumen to be 75 kg, this would be equivalent to a synthesis of some 450 g protein per day, or more than a third of the protein requirements of a cow yielding 15 litres of milk per day.

For many of the incubation experiments samples were 'fixed' in formalin and sent to Baker for microbiological examination. He made direct counts of some of the larger bacteria, many of which stained blue with iodine. He could not make direct counts of the smaller bacteria, which were present in vast numbers, but changes in the total mass of bacteria were measured by centrifuging known volumes of the formalized samples at 3600 rev./min for 20 min, and recording the weights and relative volumes of the sediments so obtained (Smith & Baker, 1944). It appeared reasonable to conclude that in the first 2 hours of incubation the protein synthesis that occurred as measured by chemical analysis, was accompanied by increases in the numbers of the larger bacteria and in the mass of bacteria as a whole. The total weight of sediment often increased by over 300 mg/100 g rumen liquid while the amount of protein increased by only 100 mg. Further work proved that the difference of about 200 mg/100 g was due mainly to the fact that a large proportion of the rumen bacteria synthesized and stored within themselves a starch-like polysaccharide, which was readily determined in the sediments obtained before and after incubation by a gravimetric cuprous oxide method following hydrolysis with HCl.

The effect of adding various substances to rumen liquid on the amount of protein synthesized on incubation was determined. With no addition of any kind the NPN content of the rumen liquid often *increased*, indicating hydrolysis of protein, whereas if 0·3 % starch was added before incubation there was a significant *decrease* in NPN. These and many similar observations suggested that protein synthesis and hydrolysis occurred simultaneously in the rumen, and that for synthesis to predominate it was essential that readily available carbohydrate should be present.

Certain other carbohydrates, such as maltose, galactose and glucose, and also some amino acids, had an effect similar to that of starch in promoting synthesis, whereas when certain other substances, such as glycerol and lactic acid, were added, hydrolysis of protein predominated over synthesis, and the hydrolysis was accompanied by a decrease in the counts of the larger bacteria and in the total weight of sediment obtained on centrifugation (Smith & Baker, 1944). In subsequent work McNaught (1951) showed that the polysaccharides starch and inulin, the trisaccharide raffinose, several disaccharides and some hexoses and pentoses were readily utilized by the rumen bacteria and that in their presence protein synthesis predominated. On the other hand the addition to rumen liquid of oxidation products of glucose, such as gluconic and glucuronic acids, and of a reduction product, such as mannitol, produced no measurable protein synthesis. It appeared, therefore, that the presence of both a potential aldehyde group and a primary alcohol group was essential for synthesis to predominate over hydrolysis. Molecular configuration was also important, since fructose was readily utilized whereas sorbose was not, and L-(+) arabinose gave a marked amount of protein synthesis whereas D-(−) arabinose gave much less.

Since in some countries small amounts of seaweed are included in cattle foods, the incubation technique developed here was used to determine whether some of the carbohydrates of the brown algae and of dried milled samples of *Laminaria cloustoni* and *Ascophyllum nodosum* would enable protein synthesis to predominate over hydrolysis. Of the various products tested, only the polysaccharide laminarin and the frond of *L. cloustoni*, which contains laminarin, promoted synthesis in amounts similar to those obtained with maltose (McNaught *et al.*, 1954*b*).

When an added carbohydrate was utilized during the incubation of rumen liquid and protein synthesis predominated over hydrolysis, the carbohydrate gave rise to various end-products. Experiments were done to determine how much of the carbon of the utilized carbohydrate could be accounted for by these end-products, and whether the end-products with a disaccharide such as maltose, which consists of hexose units, differed from the end-products formed with pentoses. The results for individual experiments varied considerably, but the average results for three experiments with maltose suggested that some 12% of the maltose carbon had become protein carbon, 28% had formed bacterial polysaccharide, 35% had formed volatile fatty acids, 8% had formed lactic acid, 8% carbon dioxide and 3% had formed methane, the total accounted for being 94%. With pentose the results were broadly similar to those obtained with maltose, except that little or no lactic acid was formed, and only about 17% of the pentose carbon appeared to be accounted for by the formation of bacterial polysaccharide. The total pentose carbon accounted for was only some 83%, so it was probable that when pentoses were utilized some compound not studied in these experiments, was formed instead of lactic acid (McNaught, 1951). These results confirmed the view that when rumen liquid was incubated under the conditions used in this work, the type of fermentation that occurred during the first few hours was similar to that which occurs in the intact rumen since the nature of the end-products was so similar.

During the course of the work with rumen contents various experiments were done to ensure that what was interpreted as protein synthesis was indeed the result of bacterial activity, and that it was not due to some strange artefact. Confirmation that it was the result of bacterial activity was obtained in the following ways (Pearson & Smith, 1943*c*; Smith, 1945):

1. When rumen liquid was filtered through a Chamberland filter to remove all micro-organisms, the filtrate was completely inactive and showed neither synthesis nor breakdown of protein on incubation.

2. When several samples from one batch of rumen liquid were incubated at different temperatures in the presence of added urea and starch, synthesis of protein predominated between 30 and 40°C but at 50 to 60°C hydrolysis of protein predominated, and above 70°C no detectable change in N distribution occurred.

3. When toluene, which inhibits bacterial action but not enzyme action, was present in the incubated liquid, conversion of urea to ammonia occurred as normally, but there was marked protein hydrolysis, whereas in control incubations without toluene there was marked synthesis.

4. In incubations with various concentrations of sodium fluoride ranging from 0 to 1 %, protein synthesis occurred in the absence of NaF, but with as little as 0·01 %, synthesis was greatly reduced, and with 0·05 %, hydrolysis of protein predominated. With 1 % NaF all bacterial and enzyme activity was inhibited and no change occurred in N distribution. On the other hand 0·5 % NaCl had no inhibiting effect on the amount of synthesis.

It is appropriate at this point to refer to a study of the effects of adding certain metals to rumen liquid on the growth of rumen bacteria as indicated by the amounts of protein synthesis obtained on incubation (McNaught *et al.*, 1950*a*). The results showed that between 1 and 2 p.p.m. iron permitted good growth and that the bacteria could tolerate 100 p.p.m., but that with 1000 p.p.m. growth was inhibited. It was found also that the bacteria could tolerate 10 p.p.m. copper, somewhat less than 10 p.p.m. cobalt and between 100 and 1000 p.p.m. molybdenum, but that growth was inhibited when these amounts were increased to 25, 1000 and 2000 respectively.

Considerations 1 to 4 enumerated above confirmed that the protein synthesis resulted from bacterial activity. They indicated also that protein hydrolysis, resulting from the death and decomposition of micro-organisms in the rumen, occurred at the same time as synthesis, and that whether synthesis predominated could depend on several factors. This led Pearson & Smith (1943*c*) to suggest that in practical feeding conditions the nature of the dietary protein might well be important in determining whether a net synthesis of protein would occur in the rumen. If, for example a very insoluble source of protein, such as blood meal, were an important constituent of the diet, it might well pass through the rumen virtually unchanged and without losing any of its nutritive value. If, however, a more soluble protein, such as casein, or a very soluble protein, such as gelatine, were included in the diet, these soluble proteins might well tend to be partially hydrolysed in the rumen, and more NPN would be formed than would be utilized. In other words more protein would be broken down than would be synthesized. Evidence for the likelihood of these suggestions being justified was obtained by various incubation experiments. In the presence of blood meal the amount of protein synthesis was the same as in the control, but when casein was added to samples of the same rumen liquid, hydrolysis of protein predominated over synthesis. For example, in one of the casein tests synthesis equivalent to 10·4 mg N/100 g rumen liquid in the control became hydrolysis equivalent to 31·3 mg N/100g in the presence of casein, and similar results were obtained with gelatine.

Pearson and Smith (1943*c*) concluded that if NPN added to a diet were to be effective in practice as a partial protein substitute the following conditions would have to be observed. 1. Since there must be a limit to the amount of NPN that the rumen micro-organisms can use, the diet should contain as little naturally occurring NPN as possible. 2. The main proteins of the diet should be as insoluble as possible in rumen liquid. 3. Readily available carbohydrate must be present. If these conditions were not observed and the diet contained relatively large amounts of naturally occurring NPN and proteins of a relatively soluble nature, it seemed unlikely that there could be a net utilization of NPN. It was believed that these considerations might explain many of the conflicting results reported in the literature as reviewed by Krebs (1937). Workers who used blood meal, for example, as one of the main constituents of the control diet, as did Owen *et al.* (1943) in their experiment with lactating cows, would probably find urea to be of value, whereas in experiments in which the diet contained more soluble proteins, or relatively large amounts of naturally occurring NPN as in roots, the addition of urea to the diet might well be valueless. Until this work had been done at the Hannah Institute, these aspects of the problem had not been clearly appreciated.

Later, Edwards & Darroch (1956), working under the direction of E. C. Owen, calculated that in the rumen of the lactating goat as much as 40 g protein could be synthesized in a day.

The nutritive value of the protein synthesized in the rumen

The work discussed above showed that under favourable conditions there could be a net synthesis of protein in the rumen, but what is the nutritive value of that protein to the host animal? To answer this question two experiments were carried out: the first dealing with the bacteria only and the second with both bacteria and protozoa.

Experiment 1. When freshly sampled rumen liquid was strained through muslin and then centrifuged for 5 min at 2000 rev./min (800 *g*), the supernatant liquid was rich in bacteria but virtually free from protozoa and feedingstuffs residues. On incubating 3 litres of this liquid at 39°C for 4 hours with 0·05% urea and 1% maltose, the NPN content decreased indicating that protein synthesis had occurred. After incubation the liquid was immediately passed through a Sharples centrifuge (14 500 *g*), and the sediment so obtained was washed with ethanol and ether. On microscopical examination the resulting grey powder was found to consist mainly of bacteria and bacterial debris; any contamination that might have occurred with feedingstuffs residues from the rumen was extremely slight.

A total of 130 litres of rumen liquid was treated in this way in fifty-seven incubations and produced 461 g of a bacterial preparation, a yield of 3·6 g/litre. For all the incubations the decrease in the NPN content of the liquid during incubation was recorded, as well as the amount of protein synthesized. It was concluded that of the total protein in the preparation about 42% had been synthesized during incubation, the remaining 58% having been present in the liquid as it came from the rumen. On a dry-matter basis the composition of the preparation was protein 44·4, carbohydrate 40·3, lipids 3·1, crude fibre 0.3, ash 7·1, calcium 0·29 and phosphorus 1·35%. The 40·3% carbohydrate included pentose equivalent to 4·4% of the preparation.

The product was sent to S. K. Kon and K. M. Henry at the National Institute for Research in Dairying, Reading, who had agreed to determine the digestibility and biological value of its protein. The product was given for 6 days to twelve female weanling rats in a diet of which it formed 18·1% of the dry weight, and in which it supplied all the protein at a level of 8%. Metabolic faecal nitrogen and endogenous urinary nitrogen were determined during 6-day periods immediately before and after the experimental period.

The mean biological value for the twelve rats was 88, and the mean true digestibility was 73, giving a net utilization of 64 (McNaught *et al.*, 1950*b*). The value of 88 was high and compared well with values recorded in the literature for the proteins of dried milk and yeast determined by the same procedure, though the value of 73 for digestibility seemed a little low.

Experiment 2. A second sample of dried rumen bacteria, weighing 283 g, was prepared by a modification of the method described for Experiment 1. It had a protein content on a dry matter basis of 47·1%. At the same time a sample of dried rumen protozoa was made by straining rumen contents through muslin, diluting the strained liquid with glucose solution, and allowing the mixture to settle at 39°C in tall glass cylinders, the protozoal layer that formed being separated by decantation. A total of 878 litres of rumen liquid was treated in this way, and gave 447 g of a dried protozoal preparation, a yield of 0·5 g/litre. On a dry matter basis this preparation contained protein 28·9, carbohydrate 67·6, lipid 1·5, ash 2·5 and crude fibre 1·9%.

The digestibility and biological values of the proteins of these two preparations were then determined at Reading as in Experiment 1, and a sample of dried brewer's yeast was tested at the same time for comparison (McNaught *et al.*, 1954*a*). The results are summarized in Table 1 where the values obtained in Experiment 1 are included for comparison.

TABLE 1

The nutritive values of the proteins of rumen bacteria, rumen protozoa and brewer's yeast

Protein source	Biological value	True digestibility	Net utilization*
Rumen bacteria			
Experiment 1	88	73	64
Experiment 2	81	74	60
Rumen protozoa	80	91	73
Brewer's yeast	72	84	60

* Biological value × true digestibility/100.

The digestibility of the protozoal protein was higher than that of the bacteria, and so the net utilization of the protozoal protein was also higher. In a paper published some years ago a worker from overseas, who had made valiant but completely abortive attempts to establish the usefulness of the protozoa in the rumen, ended his discussion of the matter in a tone of utter exasperation when he wrote 'At any rate we can be sure of one thing – the protozoa are certainly not there for nothing'! Since some of the protozoa consume and digest significant proportions of the bacteria in the rumen (Baker, 1943), perhaps the results shown in Table 1 suggest that at least one of their functions is to make part of the bacterial protein more readily available to the host animal.

SUMMARY

The work and what it achieved can be summarized briefly as follows.

1. An experiment with lactating cows confirmed that under favourable conditions urea could act as a substitute for some 25% of the total protein of the diet without ill-effect on the animals themselves or on the composition of their milk, but that there could be considerable individual variation between cows in the efficiency with which the urea was used.

2. The urease activity of rumen contents was studied, and the activity was found to be so great that all the urea ever likely to be given to even a high yielding cow would readily be converted to ammonia within 1 hour.

Attempts to retard this process by coating the urea with starch and other substances were unsuccessful.

3. From experiments with rumen liquid *in vitro* it was concluded that protein synthesis resulting from growth of micro-organisms, and breakdown of protein caused by the death and autolysis of micro-organisms, proceed in the rumen simultaneously, and that for net utilization of added NPN to occur, (*a*) the NPN naturally present in the diet should be as low as possible, (*b*) the protein of the diet should be relatively insoluble in rumen liquid so that it does not readily yield ammonia in the rumen, and (*c*) sufficient readily available carbohydrate must be present. It was concluded that under favourable conditions as much as 450 g protein might be synthesized in the bovine rumen in 24 hours.

4. Dried preparations of rumen bacteria and of protozoa were made, and the nutritive values of their proteins determined. The biological values of the proteins of the two bacterial preparations were high at 88 and 81, but the digestibilities were somewhat low at 73 and 74, whereas for the protozoal preparation the digestibility of the protein was 91 and the biological value 80. It seemed possible that one of the functions of the protozoa might be to make at least part of the bacterial protein more available to the host animal.

REFERENCES

Baker, F. (1943) *Ann. Appl. Biol.* **30**, 230–239

Edwards, D. C. & Darroch, R. A. (1956) *Br. J. Nutr.* **10**, 286–292

Krebs, K. (1937) *Tierernährung* **9**, 394–507

McNaught, M. L. (1948) Ph.D. Thesis, University of Glasgow

McNaught, M. L. (1951) *Biochem. J.* **49**, 325–332

McNaught, M. L. & Smith, J. A. B. (1947) *Nutr. Abstr. Rev.* **17**, 18–31

McNaught, M. L., Owen, E. C. & Smith, J. A. B. (1950*a*) *Biochem. J.* **46**, 36–43

McNaught, M. L., Smith, J. A. B., Henry, K. M. & Kon, S. K. (1950*b*) *Biochem. J.* **46**, 32–36

McNaught, M. L., Owen, E. C., Henry, K. M. & Kon, S. K. (1954*a*) *Biochem. J.* **56**, 151–156

McNaught, M. L., Smith, J. A. B. & Black, W. A. P. (1954*b*) *J. Sci. Food Agric.* **5**, 350–352

Morris, S. & Wright, N. C. (1933*a*) *J. Dairy Res.* **4**, 177–196

Morris, S. & Wright, N. C. (1933*b*) *J. Dairy Res.* **5**, 1–14

Morris, S., Wright, N. C. & Fowler, A. B. (1936) *J. Dairy Res.* **7**, 97–121

Owen, E. C., Smith, J. A. B. & Wright, N. C. (1943) *Biochem. J.* **37**, 44–53

Pearson, R. M. & Smith, J. A. B. (1943*a*) *Biochem. J.* **37**, 142–148

Pearson, R. M. & Smith, J. A. B. (1943*b*) *Biochem. J.* **37**, 148–153

Pearson, R. M. & Smith, J. A. B. (1943*c*) *Biochem. J.* **37**, 153–164

Smith, J. A. B. (1945) *Proc. Nutr. Soc.* **3**, 203–216

Smith, J. A. B. & Baker, F. (1944) *Biochem. J.* **38**, 496–505

Snow, D., Smith, J. A. B. & Wright, N. C. (1945) *J. Agric. Sci.* **35**, 65–71

NUTRITIONAL RESEARCH, 1948 to 1965

K. L. BLAXTER

IN 1948 Dr E. C. Owen exchanged the Headship of the Nutrition Department of the Institute for that of the Biochemistry Department and I was appointed by the Agricultural Research Council (ARC) for a period of 3 years to replace him, joining the permanent staff in 1951. In the period under review twenty graduates, some members of staff, some visitors from overseas, worked with me for short or long times. The Department was, however, never very large. Eventually it reached a maximum of five graduate staff ably assisted by ten technical staff in the laboratories and animal houses, but at the beginning there was simply the empty laboratory. This history is an account of what was accomplished in the field of ruminant nutrition during the period from 1948 until 1965. Generally, emphasis on aspects of this wide subject changed during the period. Initially the concern was with the description of the effects of specific nutritional deficiencies, but later this changed towards delineation of the quantitative aspects of animal responses to diet and to environment with most attention being paid to energy exchanges. There was much overlapping in time of these broad concerns, but most work on the former topic was reduced after 1955 and came to an end in 1959, while quantitative studies in energy metabolism really began in earnest in about 1954. Because of this overlapping a chronological accounting has not been adopted and the work is described under a number of headings which subdivide the very general description. Those who collaborated in the work are evident from the authorship of the publications cited, if not from the text.

CALF NUTRITION AND NUTRITIONAL DISEASE

Surveys of calf mortality undertaken in the immediate postwar period showed it to be extremely high and it was commonplace to ascribe a large part of this mortality (and also an unacceptable morbidity) to nutritional inadequacies. The first work undertaken by the Department was designed to augment knowledge about nutritional disease in calves using the classical methods of giving them semipurified diets. Where it seemed that a syndrome might have a practical implication the studies were developed, mostly in collaboration with the Veterinary Investigation Service in field trials. In this way enzootic muscular dystrophy in cattle was discovered, cod-liver oil, or rather polyunsaturated fat toxicity, was discovered and shown to be a practical reality, hypomagnesaemic tetany was found to be a simple dietary deficiency and thus amenable to nutritional prevention, iron deficiency anaemia was demonstrated to occur in suckled calves and to be readily prevented, and the massive changes in water and electrolyte metabolism that occur in diarrhoea in calves were described, to lead to changes in therapeutic measures. The work included the devising of clinical, biochemical and pathological methods for diagnosis and it can probably be said that on completion of the work the ascribing by advisers of many deaths and diseases in calves to 'nutritional causes' was replaced by precise diagnosis. In addition, a major disease, enzootic muscular dystrophy with an incidence in affected areas of about 25 % and a mortality of about 5 %, was completely controlled. The investigations that led to practical advances are described first; those which were of less significance are then summarized.

Muscular dystrophy

The development of artificial diets for calves (Blaxter & Wood, 1951*a*, *c*) was not without problems and in the early stages of the work a severe muscular degeneration occurred in some animals. This seemed analogous to the disease noted by Goettsch & Pappenheimer (1931) in guinea-pigs given diets deficient in vitamin E, and to one noted by Hjärre & Lilleengen (1936) when they fed heated oxygenated milk to calves and which they ascribed to ascorbic acid deficiency. Wood and I together with P. S. Watts, Head of the Institute's Pathology Department, and A. M. MacDonald, Pathologist to the Royal Hospital for Sick Children, Glasgow, investigated the disease, showed that it was a muscular dystrophy with hyaline degeneration of the muscle cell and a massive depletion of creatine content and that it was associated with the presence in the diet of unsaturated fats, notably lard and more particularly cod-liver oil (Blaxter *et al.*, 1952; Blaxter & Wood, 1952*b*; MacDonald *et al.*, 1952). The disease was prevented by the addition of α-tocopheryl acetate to the diet. Later work (Blaxter *et al.*, 1953*d*) firmly incriminated cod-liver oil; its addition to a diet based on dried skim milk resulted in death or severe dystrophy in a high proportion of animals, with evidence of a seasonal severity, more calves being affected in the late winter than in the early summer when cows receive grass which is much higher in vitamin E activity than normal winter rations. It was then shown that some calves at birth show very slight signs of muscular degeneration judged histologically and biochemically (Blaxter *et al.*, 1953*a*). Dr F. Brown had joined the Nutrition Department in 1950 and had developed methods for determining α-, β-, γ- and δ-tocopherols quantitatively using reverse phase paper chromatography (Brown & Blaxter, 1951; Brown, 1952*a*, *b*) and had successfully applied them to the analysis of feedingstuffs (Brown, 1953*a*). We thought it desirable to obtain unequivocal proof that it was indeed the unsaturated acids of cod-liver oil which were the inducing agents.

Cod-liver oil was saponified and the fatty acids fractionated under somewhat primitive conditions by low temperature crystallization and separation of the lithium salts to yield a non-saponifiable residue, a saturated acid fraction and two unsaturated acid fractions with iodine values of 270 and 110 respectively. These fractions and the original oil were fed to calves; dystrophy developed when the unsaturated fatty acids were given as a whole, when the more highly unsaturated ones were given, but not when any of the other fractions were added to the basal diet (Blaxter *et al.*, 1953*b*) showing that the toxicity could not have been due to hypervitaminosis A or D. Later in studies related to calcinosis in calves (and the dystrophic lesion is often calcified) it was unequivocally shown that cholecalciferol toxicity does not result in changes in the muscle cell but rather in calcification of the arterial and venous system and the kidney (Blackburn *et al.*, 1957). The unsaturated fatty acid toxicity occurred despite the appreciable amounts of α-tocopherol in cod-liver and other fish oils (Brown, 1953*b*). At this stage the work was brought to the attention of the veterinary profession and advisers to the animal industry (Blaxter, 1953*a*, *b*) since the then recommendation by the Ministry of Agriculture was that calves could be reared on reconstituted dried milk supplemented with cod-liver oil – the very diet which resulted in the disease. A number of cases of frank cod-liver oil toxicity were found, but investigation of what seemed to be the same disease in north-east Scotland showed no dietary inclusion of cod-liver oil or of unsaturated fat; the disease occurred in suckled calves in beef herds, and the calves received only their mother's milk (Sharman, 1954). Later a similar disease was found in sheep in the same area (Marr *et al.*, 1956). The disease was enzootic muscular dystrophy and Brown and I decided we should shelve work on the chemistry of the tocopherols (which was at a particularly interesting stage) and concentrate on finding the cause of the field condition.

Three years of field work followed with over 500 beef cows on farms in north-east Scotland. It was shown that the feeding of α-tocopherol to the cow before calving gave good protection and so did administration of 150 mg to the calf; together both were completely effective (Blaxter & Sharman, 1953). Experiments to estimate the effective daily amounts necessary to protect showed, however, that 20 mg DL-α-tocopherol acetate was not sufficient to protect completely, reducing incidence from 20 to 6%. Studies of the tocopherol content of blood in calves and their dams and in the dams' colostrum showed without doubt that the tocopherol status of these herds receiving winter diets of turnips and straw was low, but attempts to produce the disease at the Institute in the absence of unsaturated fat by giving diets devoid of vitamin E activity were unsuccessful.

Furthermore, herds in south-west Scotland receiving virtually the same diet as those in north-east Scotland did not develop the disease (Blaxter, 1957).

Attempts were made to find the missing factor. In collaboration with the Rowett Research Institute milkfats from cows with dystrophic calves were examined by fractional distillation to find they were no more unsaturated than were those from normal animals (Garton *et al.*, 1956), and in the Department Rosemary McGill determined polyethenoid acids in blood lipids of animals in affected areas by alkali isomerization, all to no avail. Concomitant deficiencies of cobalt, iron and magnesium were examined, possibilities of toxicity of petroleum additives, notably tri-orthocresyl phosphates, tested (Cowlishaw & Blaxter, 1955) and the possibility that other compounds with antioxidant activity in diet might be responsible for the difference, prompted an investigation of the role of antioxidants on the incidence of the disease (Blaxter *et al.*, 1953*c*), again without giving any clue to the missing factor or factors. Since no progress was being made field and laboratory work were stopped.

In 1957 Schwarz & Foltz and Patterson *et al.* announced that liver necrosis in the rat could be prevented by dietary selenium. This disease is also prevented by α-tocopherol. It seemed that selenium might be the factor. Because of seasonality of calving patterns in the herds concerned, work could not begin until the spring of 1959. Then it was unequivocally shown that selenium (0·25 mg/animal per day) prevented the disease on farms and was as or more effective than α-tocopherol (Sharman *et al.*, 1959). This was the first demonstration of the role of selenium in preventing muscular dystrophy in cattle; Muth *et al.* (1958), however, had priority since they had shown selenium effective in preventing the disease in sheep in the previous year.

Integrative work followed. Plotting the farm incidence of the disease in north-east Scotland and east Scotland showed a clear association with soils derived from the Old Red Sandstone and certain granites. A very large co-operative trial with 5000 sheep was organized and it was found that the small responses of weight gain were related to the presumptive selenium content of the soils based on geochemical evidence (Blaxter, 1963). Dr Sheila McCallum developed methods for selenium and showed that on land at the Institute artificially rendered seleniferous the turnip accumulated less selenium than any other crop or common weed plant examined. Brown (1953*a*) had earlier shown that turnips were virtually devoid of tocopherols. Now selenium medication of cattle in the affected areas (about one-tenth of Scotland's total area) is common farming and veterinary practice.

Magnesium deficiency

Work on magnesium deficiency in calves arose from the studies on muscular dystrophy. Earlier workers had described the nutritional failure of calves given cows' milk supplemented with iron, copper, manganese and vitamins A and D for protracted periods (Moore *et al.*, 1936, 1938) and had shown that it included hyaline degeneration of muscle, calcinosis of the blood vessels and a fall in serum magnesium. Adding magnesium was not effective in prevention (Huffman *et al.*, 1941). Some of these pathological findings were similar to those found in muscular dystrophy induced by cod-liver oil, others to ergocalciferol toxicity. It seemed sensible to define the clinical and pathological syndrome produced by an uncomplicated magnesium deficiency. This was produced (Blaxter *et al.*, 1954*b*) and the metabolic changes described (Blaxter & Rook, 1954). No calcification or hyaline degeneration of muscle was found. Several interesting findings emerged. Firstly, the animals could deplete their skeletons of magnesium yet maintain the magnesium concentration in their soft tissues. This was later shown (Blaxter, 1955) to be a heterionic replacement of magnesium by calcium ions and the ratio of magnesium to calcium in the terminal vertebrae of the tail was developed as a diagnostic method (Blaxter *et al.*, 1957). This was dignified by the term distal coccygeal biopsy! Secondly, it was found that the availability of magnesium even in milk was extremely low compared with that of calcium, a finding shown later by J. A. F. Rook at the National Institute for Research in Dairying, Reading, to be true also of the cow. The work showed unequivocally that hypomagnesaemic tetany in the calf was a nutritional disorder not an endocrine or neural one, despite the fact that there was no depletion of the intracellular magnesium. The same was also true of the cow. K. Presthegge and I carried out experiments to produce experimental magnesium deficiency in cows by feeding diets of acid-leached hay supplemented with a concentrate containing purified starch and maize gluten. The experiments were successful in that hypomagnesaemic tetany was induced (see Blaxter & McGill, 1956), but the cost of the trials was prohibitive. The work was stopped and, having demonstrated a nutritional origin of the disease in cows, studies were made with the cheaper and more easily disposable calves.

Since magnesium participates in most phosphorylating enzyme systems, a deficiency might affect intermediary energy transfer and Rook and I studied energy metabolism in magnesium-deficient animals to find that heat production was increased in the disease but not the efficiency of utilization of energy (Blaxter & Rook, 1955). Much work on the milking cow syndrome associates the occurrence of hypomagnesaemic tetany with heavy manuring with potassic fertilizers. Potassium added to the diet of calves had no effect on serum magnesium or on bone magnesium even though potassium administration was such as to lead to signs of toxicity (Blaxter *et al.*, 1960). This showed that the effect of potassium was unlikely to be direct but rather through its effect on the magnesium content of herbage (Stewart & Holmes, 1953). Converse experiments were made by Rosemary McGill in which potassium deficiency was produced; no effect on magnesium metabolism was elicited.

As with muscular dystrophy, the work was taken to the stage of field investigation in Caithness and Inverness-shire. It was shown that in suckling beef calves there was invariably a depletion of serum magnesium and of the bone magnesium (Blaxter & Sharman, 1955). Experiments with 160 calves showed that daily supplements of 500 mg magnesium (as the carbonate) prevented the major part of the fall in serum magnesium concentration and was an adequate supplement for the first 2 months of life. This work also showed the reasons for the failure to produce other than a temporary clinical cure by the injection of magnesium in affected animals since the amounts injected failed to meet even the daily requirement of the mineral.

Iron deficiency anaemia

Continuing description of deficiency states likely to be found in the field, iron deficiency was induced. The anaemia was characterized and the concomitant clinical signs described. Again the work was taken into the field in experiments on fifteen farms to show that mild anaemia in suckling beef calves is commonplace and responds to iron (Blaxter *et al.*, 1957). Estimates of minimal iron requirement were made and these have been supported by more recent work (MacDougall *et al.*, 1973).

Other nutritional disorders

A number of other nutritional entities were also studied. Experimental thiamin deficiency was induced and more interestingly what was ascribed to tryptophan deficiency but was probably a multiple amino acid deficiency since it was induced by feeding gelatin as the sole protein source (Blaxter & Wood, 1952*a*). This was the first demonstration of amino acid needs in calves or indeed in ruminants.

Diarrhoea in calves, 'scouring', continues to be a major cause of economic loss. Giving a wide variety of artificial diets necessarily invoked alimentary disturbances, and Wood and I summarized our biochemical and physiological findings (Blaxter & Wood, 1953) and went on to induce scouring by dietary means. This work showed the large loss of electrolytes, fatty acids and purines in the faeces of scouring animals, the effect of glucose in inducing proliferation of bacteria in the lower gut, and it was concluded that it was only if 'strains of coliform organism, against which the calf has no immunity, multiply in the lower gut, then tissue invasion follows and death results.' This approach, which did not ascribe all scouring to a specific pathogen, was contrary to belief at the time, and the work certainly promoted interest in acid-base metabolism in the scouring calf and resolution of the pathogenic coli by Sojka (1965). The revolting and daunting 24-hour task of quantitatively collecting the faeces from acutely scouring calves and analysing them has not been repeated; I am not surprised.

FEED UTILIZATION AND ENERGY METABOLISM

The factors that determine the energy requirements of ruminants and the value of the feeds they consume as sources of energy were areas much neglected in the 1940s and 1950s. There had been a small amount of work done in Germany, the United States of America, Denmark and Australia before the war, but most of this had stopped and had not been restarted. The subject was regarded as of little scientific interest, it being accepted generally that the foundations laid by Rubner, Voit and Kellner in Germany were unshakeable and more specifically in the United Kingdom that Kellner's starch equivalent system of feeding ruminants as modified at Cambridge by Wood and Woodman was sufficient guide to practical feeding. There were, however, some misgivings at that very practical level, particularly about the absence of information on new feeds and improved crop varieties. We therefore slowly turned attention to this somewhat outmoded problem.

By the end of the period the work had led to the creation of an entirely new feeding system for ruminants, had provided measurements of requirements for energy in different animals and in a range of environments typical of the United Kingdom, and had created a framework of explanation for the increments of heat that followed ingestion of food and the hypothesis had been subjected to experimental test.

Methods and techniques

Early work on energy metabolism in the Department involved short-term collection and analysis of expired air. It showed that very young animals did not reach stable metabolic rates on fasting and that amino acid deficient diets increased heat production (Blaxter, 1950*a*, Blaxter & Wood, 1952*a*, *b*), but these short-term measurements were clearly limited in their application to farming problems. Twenty-four-hour measurement was essential. With Mr Howells, the Institute's engineer, and Mr Paterson, who cared for the Department's animals throughout the whole 18-year period, the first respiration apparatus was built as an open-circuit instrument with gravimetric determination of carbon dioxide output and volumetric estimation of oxygen consumption. The apparatus is illustrated in the Institute's Annual Report of 1950, but while it was picturesque and was used to determine the net energy of whole milk

(Blaxter, 1952), it was not very reliable. We then built the first closed-circuit instrument for sheep (Blaxter *et al.*, 1954*a*), and two more, one modified to include a treadmill and with temperature control to 0°C, were built to the same design. These worked very well, but with considerable attrition on our clothing and skins. Jackets and neckties dipped in oil seals and the handling of anhydrous calcium chloride in large quantities were extremely unpleasant.

These three instruments were replaced by five better and safer ones which Wainman and I designed (Wainman & Blaxter, 1958*a*) and later a larger 'intermediate' instrument was built with a double absorption system (Wainman & Blaxter, 1969). The most remarkable instrument of all those we built and modified was a closed-circuit instrument for adult cattle (Wainman & Blaxter, 1958*b*). The massive amounts of water vapour and CO_2 were absorbed and weighed, and despite its size the cow chamber was a very accurate instrument, only being abandoned after 8 years of continuous operation, as a result of corrosion in the heat exchange systems.

The principle in all this work was to rely on gravimetric determinations wherever possible and to avoid measurements based on mass flow of air and gas concentration changes. All gas analysis to ensure normality of air composition in the chambers and to measure accumulated methane was done manually for many years and it was quite late in the period that we adopted thermal conductivity methods for CO_2 and for CH_4 determination and paramagnetic methods for O_2. This analyser system allowed us to develop short-term measurements on a large scale. Mr Joyce, a PhD student from New Zealand, and I devised tracheostomy and mask methods for sheep (Blaxter & Joyce, 1963*b*) and these were used in a variety of studies.

Methods of estimating energy retention based on balances of C and N or of energy where heat production is estimated from the 24-hour respiratory exchange are slow. The possibility of estimating bodily retentions indirectly from water retention was examined (Blaxter & Rook, 1956*a*, *b*) to show that with very high precision in analytical work, using classical gravimetric techniques, bodily retentions of Na and K permitted computation of water retention and thus indirectly fat retention. Tests of the method in cattle against indirect calorimetry gave good results, but in sheep retentions were too high. This was traced to a large excretion of K by the skin of the sheep. The approach was abandoned; it did not provide a simplification.

Before coming to the Institute I had attempted to quantify the relationship between pulse frequency and metabolism in adult sheep (Blaxter, 1948) and Wood and I did the same for calves (Blaxter & Wood, 1951*b*). The work in both instances applied to fasting animals, and Webster (1967) extended this approach to a range of nutritional and metabolic states with sheep, having in view the estimation of metabolism from pulse frequency. Good relationships characteristic of individuals were obtained, but the estimation of metabolism from pulse frequency was never exploited; it seemed more sensible to measure metabolism directly.

The measurement of energy retention by balance techniques is one that demands considerable care in analytical work and in the quantitative collection of excreta. The errors which accrue in such work were analysed from time to time and were reviewed at the end of the period (Blaxter, 1967). Certainly, the precise estimation of N retention caused us most concern, and Martin (1966*b*) studied this in some detail to conclude that even when very great care was taken in preservation of voided urine, urine splashing on the the body led to errors in calculated retention equivalent to 1 to 2·6% of the N voided. It is rather doubtful if everyone conducting animal balance trials exercises that same very great care in collection and preservation of material that we did following Martin's work.

Many people from overseas were interested in the techniques of measuring the respiratory exchange and in the climatic facilities in the Physiology Department. There were numerous short-term visitors from several countries who measured the chambers and sketched them. One telegram from the government of an Asian country simply asked us to send the name of the shop where they could be bought!

The utilization of the energy of the products of ruminant digestion

Work at Cambridge by Sir Joseph Barcroft and his colleagues during the war had shown that the major end-products of the ruminal fermentation of carbohydrate were the steam-volatile fatty acids (SVA) and that they were absorbed and were oxidized by isolated tissues. The question arose whether the energy of the SVAs was as efficiently used as was that of glucose, the end-product of carbohydrate digestion in the simple-stomached animal. D. G. Armstrong and I therefore undertook a series of experiments to measure the heat produced when the acids and glucose were given. Sheep with fistulas of their rumens or abomasums or with indwelling catheters in the atria were infused continuously with solutions of the compounds using pumps constructed from laboratory syringes and construction set parts.

In fasted sheep acetic acid infusion was associated with large increases in heat production and massive acidosis. Both were abolished by the simultaneous infusion of small amounts of propionic acid, but not of *n*-butyric acid (Armstrong & Blaxter, 1957*a*; Armstrong *et al.*, 1957). Abolition of the acidosis did not affect the increment of heat (Armstrong *et al.*, 1961). In the fed animal increments of heat were much greater than those noted in the fasting animal, being large for acetic acid and for mixtures containing large proportions of acetic acid and relatively little propionic acid (Armstrong & Blaxter, 1957*b*; Armstrong *et al.*, 1958).

The utilization of glucose was then studied. Sixty-nine per cent of glucose given intravenously was retained in the body as fat, 62% when it was infused into the abomasum, but only 42% when it was infused into the rumen (Armstrong *et al.*, 1960*b*). Faecal losses increased when glucose was infused into the gut and when the results were expressed per 100 J absorbed, 71% of the glucose given by abomasum was retained but only 54% from the rumen. In fasted animals with washed out rumens 93% of the energy was retained when glucose was given by rumen, but 100% when it was given into the empty abomasum (Armstrong & Blaxter, 1961). These results for the metabolism of the SVAs and of glucose showed unequivocally that the low efficiency of utilization of carbohydrate by the ruminant was associated with the ruminal fermentation, that this was not solely due to the heat evolved in the fermentation, but to differences in the values of glucose and SVAs as energy sources.

When A. K. Martin joined the Department we extended the work to protein using much the same approaches (Martin & Blaxter, 1961, 1963; Blaxter & Martin, 1962). This work again showed through estimates of heat production following ruminal and abomasal infusion in fasting and fed sheep the dominant role of the rumen, that energy utilization for lipogenesis differed from that for maintenance and that when amino acids were absorbed they were associated with similar heat increments in ruminants and non-ruminants alike.

At this stage an integrating hypothesis was proposed, namely that nutrients replaced one another for maintenance in proportion to the free energy as ATP they yielded on dissimilation (Blaxter, 1961*a*, *b*) and that for lipogenesis the value of a nutrient as an energy source for fatty acid chain elongation depended on the quantities of ATP and reduced co-enzymes made available on its dissimilation. The increments of heat found for glucose, the SVAs and for protein agreed with this hypothesis and it could be stated that the ruminant required 92 kJ metabolizable energy to form 1 mol ATP from ADP when typical ruminant diets were given (Blaxter, 1962*a*).

It was essential to test this hypothesis and Martin and I attempted to do so. We first measured in the intact animal the heat increment associated with the synthesis of urea from ammonia and CO_2. The stoichiometry of the Krebs-Hensleit cycle showed that 4 mol ATP were required per mol urea formed, so that theoretically the increment of heat should be 4×92 kJ = 368 kJ/mol of urea. The value found was higher at $4{\cdot}8 \pm 0{\cdot}7$ but not significantly so, the high errors being due to considerable technical difficulties (Martin & Blaxter, 1965). Attempts were then made to measure the increment of heat associated with the biological benzoylation of glycine, a reaction with a known ATP requirement. These were unsuccessful in that benzoic acid proved to be toxic to the sheep (Martin, 1966*a*), but the work uncovered a massive excretion of aromatic acids by ruminants which Martin subsequently examined in some detail with rewarding results (Martin, 1973).

The work on utilization of end products of fermentation and digestion up to this time had all been conducted with fattening sheep. Armstrong and I decided to investigate the effects of the SVAs on the heat production of lactating animals and, as a preliminary to work with cows, we commenced studies with goats (Armstrong & Blaxter, 1965). This work showed a lower efficiency of milk secretion when acetic was infused than for propionic acid and generally much higher overall efficiency for lactation than for fattening. More intriguing, and later confirmed by Moe *et al.* (1970) in the United States of America, was the fact that energetic efficiency of fat deposition in the body in lactation was almost as high as that of milkfat secretion and considerably greater than that noted in the same animals when non-lactating. This finding had considerable implications for the optimal feeding of cows; replacement of the fat depots depleted during early lactation by better feeding in late lactation was not as inefficient a process as had been thought.

Further examination of the metabolism of end products followed. J. E. Vercoe, a visitor from Australia in receipt of an Australian Wool Research Fellowship, studied utilization of formic acid to show, contrary to claims by Reid at Cornell (Bensadoun *et al.*, 1962), that formic acid was quantitatively converted to methane (Vercoe & Blaxter, 1965). Clearly CO_2 was being reduced to CH_4 and we wondered whether other compounds could be given to compete with CO_2 for the H_2 evolved in the ruminal fermentation. Vercoe and I first tried infusion of acrylic acid (see Blaxter & Czerkawski, 1966), with the hope that propionic acid might be formed. This was understandably unsuccessful and because we had noted a fall in the concentration of C_{18} unsaturated acids in rumen liquor when we infused formic acid, we therefore infused linseed oil fatty acids expecting that the well established hydrogenation of the ethylenic bonds of unsaturated acids might be accompanied by a reduction in CH_4 production (Garton *et al.*, 1961). This assumes that CO_2 reduction and fatty acid hydrogenation share a common hydrogen donor. To our amazement, the reduction in methane production was far in excess of expectation; expecting a reduction of 0·25 mol in CH_4 production/mol acid with a single double bond we found values in excess of 3·0. J. W. Czerkawski had joined the Department at this time and a large series of experiments was undertaken to elucidate the effects of fatty acids on CH_4 production. With M. J. Ulyatt from New Zealand it was first shown that the hydrogenation of unsaturated acids in the rumen was virtually complete

(Ulyatt *et al.*, 1966). The methane depression by oleic, linoleic and linolenic acids increased with their unsaturation by 0·24 mol/mol according to expectation (Czerkawski *et al.*, 1966*a*). Unsaturation, however, was only one component; the saturated acids from C_6 to C_{18} also depressed methane production, maximal depression occurring with lauric (C_{12}) acid when the depression was 6·4 mol/mol (Blaxter & Czerkawski, 1966). No depression occurred with the normal paraffins or with the fatty alcohols, but sulphated alcohols massively depressed methane production (Czerkawski *et al.*, 1966*b*). Incorporation of the acids in the diet was as effective as their continuous infusion in depressing methane production (Czerkawski *et al.*, 1966*c*).

Depression of cellulose digestibility accompanied the depression of methane production with some but not all compounds and it appeared that there were real possibilities for increasing the overall efficiencies of feed utilization by judicious use of methane depressants. Dr Czerkawski, in the period after 1965, has continued work in this field (see Czerkawski (1970) for a summary). In the context of evaluating the effects of the end products of digestion on the energy exchange the work showed the high efficiency with which the energy of long-chain fatty acids was retained by the animal. Eighty per cent of their energy was retained in body fat.

The utilization of the energy of normal diets

Parallel with the studies of a more physiological and biochemical nature on the energy exchanges of ruminants, a continuous programme of work on the evaluation of natural feeds as energy sources was undertaken. The first major experiment which examined energy retention in relation to the amount of feed given (Blaxter & Graham, 1955) was in many ways the most significant since it showed without doubt that this relationship was curvilinear, implying that it was impossible to ascribe a single unique additive energy value to a food. This depended on the amount ingested or the level of performance prescribed. It was then shown with both cattle and sheep that feeds did not replace one another in direct proportion to their metabolizable energy content, implying that even if a particular level of performance was predicated, feeds could not be given a unique value; their value depended on the mixture in which they were incorporated (Blaxter & Wainman, 1961*a*). Further work demonstrated that sheep and cattle utilized feed in much the same way and that cattle of different ages behaved similarly (Blaxter & Wainman, 1961*b*; Blaxter *et al.*, 1966*b*, *c*). This work, together with much additional evidence from experiments undertaken with other ends in view (see below) clearly implied that the starch equivalent system was hardly tenable as a method of precise rationing. A new system was proposed in 1960, at the Clive Behrens lectures at the

University of Leeds in which it was shown that the findings with natural feeds accorded with the work on the products of digestion (Blaxter, 1962*a*; see also Blaxter, 1961*a*, *b*), and the system was adopted as the framework for expressing ruminant energy needs by the Working Party on Ruminant Nutrient Requirements (ARC, 1965). Ten years later the system was adopted by the agricultural departments of the United Kingdom (Ministry of Agriculture, Fisheries and Food *et al.*, 1975) and the starch equivalent system of Kellner was abandoned altogether.

Much purely descriptive work designed to illuminate the practical problems of ruminant feeding was undertaken in this period in as systematic a way as other demands on our calorimetric facilities permitted. Graham and I examined the factors which might affect the energy value of a single diet – the effect of grinding and pelleting (Blaxter & Graham, 1956) and the frequency with which it was given (Blaxter *et al.*, 1956*a*, *b*). Armstrong then embarked on a major undertaking to describe the utilization of the energy of a series of grasses cut and artificially dried at different stages of maturity (Armstrong, 1964). In this work the methods devised by Dr Waite in the Technical Chemistry Department of the Institute for the comprehensive analysis of grasses (Waite & Gorrod, 1959) were applied, thus permitting the first estimates to be made of the relationship between nutritive value and a chemical composition determined other than by the antiquated Weende methods (Armstrong *et al.*, 1964; Waite *et al.*, 1964). When Dr Bateman from Costa Rica visited the Institute he undertook similar studies with lucerne, which showed that legumes did not depart from the general relationship established between lignin content and net energies determined above and below maintenance (Bateman & Blaxter, 1964). The work was then extended to fresh herbage by comparing energy utilization of grass given fresh or artificially dried to the same animal. A. Ekern from Norway worked on this with me, and we found that fresh grass was invariably associated with higher increments of heat compared with those found when precisely the same amounts of identical grass were given in the dry form. Drying increased the net energy of herbage by about 5% (Ekern *et al.*, 1965*a*). A smaller difference in favour of dried material was found when grass was frozen before feeding (Ekern *et al.*, 1965*b*). The reasons for this curious difference, though investigated, were never found. Work was then commenced by J. L. Clapperton on silages, a problem fraught with considerable technical difficulty, but the results were not published during the period under review.

The new system of feeding that had been devised, usually called the 'ARC System' or the 'metabolizable energy system', classified feeds in feed tables in terms of their metabolizable energy per unit weight. To determine this attribute entails measuring energy losses in faeces, in urine and as methane. The first is a usual measurement in many animal science laboratories, the second is not, while the third requires respiration equipment. We therefore decided to examine our accumulated data with the purpose of devising methods for predicting methane and urinary energy losses so that those laboratories with digestion facilities only could estimate metabolizable energy knowing only feed composition, faecal energy loss and possibly urinary N loss. The results of 2500 24-hour experiments on methane production and 500 determinations of mean daily energy loss in urine (Blaxter & Clapperton, 1965; Blaxter *et al.*, 1966*a*) provided such factors and most tabulated values in official tables of feed values appear to have been based upon them.

The energy expenditures of ruminants

The energy required to maintain weight in a ruminant reflects its expenditure of energy, the minimal level of which is its basal metabolism. In most of our experiments fasting metabolism was measured and these data were used to provide predictions of the expenditures under conditions in which the animals were not under any stress and undertook minimal movement (Blaxter, 1962*b*; Blaxter & Wainman, 1966). These studies showed declines in metabolism with age in sheep and real differences due to breed in the metabolism of cattle. What was remarkable was that sheep generally had a low metabolism, as judged by the interspecies generalization that the metabolic rate of all homeotherms is 290 kJ/kg body weight raised to the power 0·75. Work with six breeds of sheep also revealed breed differences in metabolism (Blaxter *et al.*, 1966*b*). It was shown that metabolism was a highly individual characteristic in this work and that it was remarkably stable in animals given different amounts of feed for long periods (Clapperton & Blaxter, 1965). These experiments provided the experimental basis for estimating the energy needs of stall-fed cattle and penned sheep used in the ARC scheme for estimating energy requirements.

The energy expenditure in exercise was studied by Clapperton using the treadmill in one of the older chambers and a particularly co-operative sheep, Hank. Clapperton measured the energy cost of horizontal and uphill locomotion showing that the sheep expended similar amounts of energy per kg weight to those expended by man in both horizontal and grade walking (Clapperton, 1964*a*). The most important point established, however, was that when standard amounts of muscular work were imposed the effects on energy retention were those to be expected on the assumption that the utilization of the energy of food in meeting the demands of physical work were the same as those noted in maintenance and those were independent of feeding level (Clapperton *et al.*, 1960; Clapperton, 1964*b*). Similar work with cattle was started by A. J. F. Webster who came to the

Department in 1963. This showed that the cost of horizontal locomotion in cattle per kg weight was less than that in sheep. This work was not, however, entirely satisfactory; it was carried out in the 18th-century horse mill house at the Institute (since demolished) and the major difficulty was to ascertain precisely the distance walked when an ox was walking in a circle of relatively small radius. Was distance to be computed from the radius of the outer or inner leg or the mean?

Environmental effects

Studies of the effects of the physical environment on the energy exchange were commenced with the express objective of ascertaining the increase in feed requirements which occurred when ruminants are kept in conditions which might be expected to elicit a change in their heat production and how these responses might be affected by the amount of feed given. The first experiments we made were with clipped sheep given three different amounts of feed and kept at seven temperatures ranging from 5 to 35°C for periods of about a week (Graham *et al.*, 1959). This work showed that the critical temperature varied with feeding level, that the whole of the metabolizable energy of feed was available to keep the animal warm in cold conditions and that the reduction in energy retention which occurred in the cold was specifically at the expense of body fat. At very high temperatures there was an acceleration of metabolism which was proportional to the elevation of deep body temperature. This work which was all undertaken in the chambers over long periods, was extended to sheep with fleeces (Blaxter *et al.*, 1959*a*), to different breeds of sheep (Armstrong *et al.*, 1960*a*), to adult cattle (Blaxter & Wainman, 1961*a*, *b*) and to milk-fed calves (Gonzalez-Jimenez & Blaxter, 1962). This work provided estimates of the air temperatures below which feed would be diverted to the unproductive purpose of maintaining body temperature in these classes of stock.

It was obvious that in the United Kingdom environmental coldness was a far more important consideration than environmental warmth, and a large programme of work was commenced to extend the chamber observations to the outdoor conditions in which animals are normally kept, where besides air temperature, air velocity, the infra-red and solar radiation intensity and precipitation had to be considered. This work was developed within a framework of theory modified from the work of Burton (1934) and Burton & Edholm (1955) which entailed making a largely physical approach to the effects of the environment. The first application of the theory of heat flow to farm animals was made during the course of the first experiments on clipped and fleeced sheep in the respiration chambers to provide estimates of insulation terms – the tissue insulation, fleece insulation and the insulation provided at the interface between the animal and the environment (Blaxter *et al.*, 1959*a*, *b*). The first variable studied was air velocity. In the cattle chamber a small increase in air velocity could be invoked by judicious placing of fans to increase turbulence, and Wainman and I showed that even this small change increased heat production in store-fed cattle with thick coats kept in cold conditions by 6% (Blaxter & Wainman, 1964). Work with sheep commenced when J. P. Joyce from New Zealand joined the Department and most of the work was subsequently carried out using short-term measurements of the respiratory exchange carefully calibrated against the chambers. These studies demonstrated the effect of simulated winds on heat production and the destruction of coat insulation caused by wind (Blaxter *et al.*, 1963; Joyce & Blaxter, 1964*a*). Work on solar radiation followed, entailing measurements of the radiation profile of the sheep and the relating of incoming short-wave radiation to the heat load of the sheep. The latter involved measuring the albedo of the fleece, its permeability to solar radiation and the reflectance by the terrain (Clapperton *et al.*, 1965). The effect of rain was studied by simulation in the cattle respiration chamber so to ascertain the heat lost not only to the rain but also the energy cost of drying out the coat afterwards, and experiments of a short-term nature were made with sheep (Joyce, 1964), both sets showing further destruction of the insulation of the coat on wetting it. Infra-red radiation effects were more difficult to estimate. Work with infra-red heaters, though made, was hardly relevant and to obtain data within likely natural bounds the device was adopted of altering the emissivity of the chamber environment and thus the incoming infra-red radiation (Joyce *et al.*, 1966). This was effective in providing good estimates of the fundamental parameters involved.

Having obtained quantitative estimates of the effects of air temperature, air velocity, solar radiation intensity, the infra-red radiation environment and rain on the animal and the interaction of these variables with the coat, we then tested their value in predicting the effects of the natural outdoor environment. This entailed tethering a sheep out of doors and measuring both its respiratory exchange and the minute by minute changes in the weather. The results accorded extremely well with those predicted from a complex but none-the-less dimensionally correct equation derived from the experimental results (Joyce *et al.*, 1966).

Other questions then arose. Firstly, the work with calves and cattle had shown considerable differences in the insulating values of their coats, and in sheep the fleece cover of head and legs varies appreciably with breed. A large experiment with six breeds of sheep showed marked differences in their resistance to cold; Hampshires were the most resistant to cold wind and Scottish Blackfaced to cold rain (Blaxter *et al.*, 1966*b*). Secondly, virtually all the work in cold environments had been made at environmental temperatures above freezing point. Although the temperature of the extremities fell, the environmental stress was never sufficient to bring into play defence mechanisms to prevent tissue damage. The constants of the equations used to predict heat production and hence food requirements assumed that tissue insulation was constant in the vasoconstricted animal and that cutaneous and respiratory loss of water vapour was also constant and unaffected by subzero temperatures. Large increases in respiratory minute volume were associated with responses to cold (Joyce & Blaxter, 1964*b*), implying an increased loss of moisture, whereas exposure of sheep to very low temperatures led to vasodilatation and to decreases in the insulation of the tissues (Webster & Blaxter, 1966). Thirdly, since it appeared that in assessing the thermal demand of the environment using the living animal we were largely dealing with matters of physics, complex though they might be, and not with biological responses, it followed that an integrating probe could be built which could add up the effects of the insulating terms and allow for the interactions between them. This was the origin of the artificial sheep, which was basically a right cylinder covered with fleece heated to maintain constant internal temperature but with no heat capacity (Blaxter & Joyce, 1963*a*). A small flock of comparable 'animals' was used to measure and integrate the infra-red and solar radiation impinging on a cylinder the same size as a sheep, and the fleeced artificial sheep gave estimates of the overall effect of climate with little time lag. Agreement with the living animal was excellent. The way in which the real animal established equilibrium on changing the environment, that is the capacitance of the system, was studied by Webster & Blaxter (1966) who showed that patterns of time response varied with the depth of the fleece, raising intriguing questions about the nature of the temperature receptors involved.

All of this work on cold environments, later extended to conditions of much more severe cold by Webster when he moved to Canada, was concerned with the quantitation of the metabolic response to the environment so to enable feed requirements to be estimated. We avoided or side-stepped many interesting problems in the thermal physiology of the cold ruminant. At the end of the period we could predict the demands that any British environment made on cattle and sheep and the outcome in terms of either the effect on productivity or in terms of the additional food energy needed to combat it.

The voluntary intake of feed

Some earlier work (Blaxter, 1950*b*) had suggested that the voluntary intake of feed by cattle was related in some way to the quality of the feed, and studies of the rate of passage of feed residues through the digestive tracts of sheep further suggested that digestive tract distension, which is a function of both the digestibility of feed and its rate of passage through the gut, was the determining factor (Blaxter *et al.*, 1956*a*). Attempts were made to quantify these relationships (Blaxter *et al.*, 1961). This work showed that the maximal voluntary intake of sheep was proportional to the apparent digestibility of the feed, and that the estimated dry matter content of the tract was the same at 100 g/kg $W^{0 \cdot 73}$ irrespective of the quality of the diet. These studies with sheep were repeated with cattle and similar results were obtained (Blaxter & Wilson, 1962). They were fully confirmed by others, and direct test of the hypothesis showed that ruminants given *ad libitum* access to feed do in fact eat to constant fill of their rumens irrespective of the quality of the diet and that little regulation of voluntary intake can be attributed to distension of the hind gut (Ulyatt *et al.*, 1967).

When the sheep and cattle trials were compared it appeared that per unit of metabolic body size ($W^{0 \cdot 73}$) cattle consumed only slightly more feed than sheep over the range of feeds studied. Sheep had been shown to have a lower maintenance requirement per unit metabolic body size than cattle and superficially it seemed that sheep might well be the more efficient converters. The comparison was not a direct one; different feeds were given and to confirm or deny the apparent finding direct experiments were made in which both species were given the same feeds to ascertain whether the intakes of feed relative to maintenance needs were the same for both species. The results showed they were, but sheep ate more of a specific feed relative to their maintenance needs and digested it less well than cattle. The apparent higher efficiency of sheep was indeed an artefact (Blaxter *et al.*, 1966*d*).

During this work studies were made of mixtures of roughages of different qualities and of concentrates with roughages. With roughages there was direct proportionality of substitution. With high quality hay (9 % crude protein) and concentrates (21 % crude protein) there was similar proportionality, but with low quality roughages addition of concentrates initially increased voluntary intake.

The relevance of this work to farm practice was then examined (Blaxter & Wilson, 1963) in the specific context of optimizing the time at which hay is cut. The later hay is cut, the higher the yield and the poorer its quality. Optimal cutting time is clearly that at which yield of animal product per acre is maximal, not necessarily yield of dry matter or even of digested matter. This experiment showed that late cutting increased yield per acre by 57 %, increased metabolizable energy per acre by 38 %, but because of lower intake and the lower efficiency of utilization of the metabolizable energy of late-cut hay, the energy retained in the bodies of sheep that could be kept on an acre of late-cut hay was 78 % less than that for early hay. This work, combining energy metabolism, voluntary intake and yield data showed for the first time the futility of measuring the yields of crops destined to be eaten by animals as dry matter, or digested matter or net energy; voluntary intake was a major determinant.

Studies with milking cows

Throughout the period under review work took place with milking cows. Some of it has been referred to when discussing magnesium. Much was related to effects of levels of feeding of the cow on her production and some early work was published by Rook (1953). The main studies arose from suggestions made by the late Professor R. A. Fisher, that it might be possible to select for efficiency of conversion of feed to milk, but that to do so would entail estimating parameters characteristic of the individual cow which could be used to distinguish her in breeding programmes. H. Ruben, a mathematical statistician, spent periods at the Institute attempting to devise experimental designs to measure partial efficiency, and a number of models were tested, all to little avail. Eventually a mathematical model of lactation was constructed which appeared to agree with what had emerged from many abortive tests made at the Institute, namely that the cow senses her current and previous production continuously and that the differential response to feed is not constant. This model, which was complex, was fitted by hand (this was before the computer age) to two lactations in which there had been sequential changes in feed with reasonable success, but apart from reporting the results to the ARC, no publication was made of this phase of the work which was terminated.

A systematic and more formal approach was then made to the problem of how best to feed dairy cows using a variety of complex statistical designs, the most successful of which was an orthogonal sequence of treatments in time which enabled direct and carry-over effects of diet to be estimated for individual animals. These were applied to a variety of cows of different apparent intrinsic lactational ability, including Ayrshires, Jerseys, Herefords and Blue Grey Shorthorns. A preliminary account of this work was published (Blaxter, 1956, 1958), but the main results were only published some years later (Blaxter, 1966) in summary form. In fact, the main findings that cows are grossly underfed in early lactation when fed according to the then current standards and that allowances for milk production should not be proportional to yield were adopted by many farmers during the period and subjected to experimental test (Griffiths, 1965) and verification, largely as a result of the initial accounts of the work and a number of talks given to the Advisory Services about the work. This work on level of feeding and lactation has been continued by W. H. Broster at the National Institute for Research in Dairying, Reading, with considerable success and he has always acknowledged the pioneering work done in this field at the Hannah Research Institute.

CONCLUSIONS

A number of other minor investigations were made during the period and several reviews were prepared. Generally, however, the efforts of the Nutrition Department were wholly devoted to the two main themes, the description of nutritional deficiency states and the quantitation of ruminant response to diet. At the time the more spectacular aspects of deficiency disease in animals attracted most attention from those who visited the the Institute, but in the longer term, the slow accretion of information about the effects on animals of changes in the quality and amount of their normal diets undoubtedly constituted the Department's more important contribution.

REFERENCES

Agricultural Research Council (1965) *The Nutrient Requirements of Farm Livestock, No. 2, Ruminants*, Agricultural Research Council, London

Armstrong, D. G. (1964) *J. Agric. Sci.* **62**, 399–416

Armstrong, D. G. & Blaxter, K. L. (1957*a*) *Br. J. Nutr.* **11**, 247–272

Armstrong, D. G. & Blaxter, K. L. (1957*b*) *Br. J. Nutr.* **11**, 413–425

Armstrong, D. G. & Blaxter, K. L. (1961) in *Proc. 2nd Symp. Energy Metabolism*, pp. 187–197

Armstrong, D. G. & Blaxter, K. L. (1965) in *Proc. 3rd Symp. Energy Metabolism*, pp. 59–72

Armstrong, D. G., Blaxter, K. L. & Graham, N. McC. (1957) *Br. J. Nutr.* **11**, 392–408

Armstrong, D. G., Blaxter, K. L., Graham, N. McC. & Wainman, F. W. (1958) *Br. J. Nutr.* **12**, 177–188

Armstrong, D. G., Blaxter, K. L., Clapperton, J. L., Graham, N. McC. & Wainman, F. W. (1960*a*) *J. Agric. Sci.* **55**, 395–401

Armstrong, D. G., Blaxter, K. L. & Graham, N. McC. (1960*b*) *Proc. Nutr. Soc.* **19**, xxxi

Armstrong, D. G., Blaxter, K. L. & Graham, N. McC. (1961) *Br. J. Nutr.* **15**, 169–175

Armstrong, D. G., Blaxter, K. L. & Waite, R. (1964) *J. Agric. Sci.* **62**, 417–424

Bateman, J. V. & Blaxter, K. L. (1964) *J. Agric. Sci.* **63**, 129–131

Bensadoun, A., Paladines, O. L. & Reid, J. T. (1962) *J. Dairy Sci.* **45**, 1203–1210

Blackburn, P. S., Blaxter, K. L. & Castle, E. J. (1957) *Proc. Nutr. Soc.* **16**, xvi

Blaxter, K. L. (1948) *J. Agric. Sci.* **38**, 207–215

Blaxter, K. L. (1950*a*) *Agric. Prog.* **25**, 1–9

Blaxter, K. L. (1950*b*) *Nutr. Abstr. Rev.* **20**, 1–21

Blaxter, K. L. (1952) *Br. J. Nutr.* **6**, 12–18

Blaxter, K. L. (1953*a*) *Vet. Rec.* **65**, 835–838

Blaxter, K. L. (1953*b*) *Proc. Br. Soc. Anim. Prod.*, 94–101

Blaxter, K. L. (1955) in *Proc. Ciba Conf. on Bone*, p. 117

Blaxter, K. L. (1956) *Proc. Br. Soc. Anim. Prod.*, 3–31

Blaxter, K. L. (1957) in *Proc. 75th Congr. Br. Vet. Assoc.*, pp. 3–8

Blaxter, K. L. (1958) in *Proc. Conf. Scientific Principles of Feeding Farm Livestock*, pp. 21–46

Blaxter, K. L. (1961*a*) in *Proc. 2nd Symp. Energy Metabolism*, pp. 211–224

Blaxter, K. L. (1961*b*) in *Digestive Physiology and Nutrition of the Ruminant* (Lewis, D. ed.), pp. 183–197, Butterworth, London

Blaxter, K. L. (1962*a*) *The Energy Metabolism of Ruminants*, Hutchison, London
Blaxter, K. L. (1962*b*) *Br. J. Nutr.* **16**, 615–626
Blaxter, K. L. (1963) *Br. J. Nutr.* **17**, 105–115
Blaxter, K. L. (1966) *George Scott Robertson Memorial Lecture*, Queen's University, Belfast
Blaxter, K. L. (1967) *Proc. Nutr. Soc.* **26**, 86–96
Blaxter, K. L. & Clapperton, J. L. (1965) *Br. J. Nutr.* **19**, 511–522
Blaxter, K. L. & Czerkawski, J. W. (1966) *J. Sci. Food Agric.* **17**, 417–421
Blaxter, K. L. & Graham, N. McC. (1955) *J. Agric. Sci.* **46**, 292–306
Blaxter, K. L. & Graham, N. McC. (1956) *J. Agric. Sci.* **47**, 207–217
Blaxter, K. L. & Joyce, J. P. (1963*a*) *Anim. Prod.* **5**, 216 (Abstr.)
Blaxter, K. L. & Joyce, J. P. (1963*b*) *Br. J. Nutr.* **17**, 523–537
Blaxter, K. L. & McGill, R. F. (1956) *Vet. Rev. Annot.* **2**, 35–55
Blaxter, K. L. & Martin, A. K. (1962) *Br. J. Nutr.* **16**, 397–407
Blaxter, K. L. & Rook, J. A. F. (1954) *J. Comp. Pathol. Ther.* **64**, 176–186
Blaxter, K. L. & Rook, J. A. F. (1955) *Br. J. Nutr.* **9**, 121–132
Blaxter, K. L. & Rook, J. A. F. (1956*a*) *J. Agric. Sci.* **48**, 194–209
Blaxter, K. L. & Rook, J. A. F. (1956*b*) *J. Agric. Sci.* **48**, 210–219
Blaxter, K. L. & Sharman, G. A. M. (1953) *Nature* (*Lond.*) **172**, 1006–1007
Blaxter, K. L. & Sharman, G. A. M. (1955) *Vet. Rec.* **67**, 108–115
Blaxter, K. L. & Wainman, F. W. (1961*a*) *J. Agric. Sci.* **56**, 81–90
Blaxter, K. L. & Wainman, F. W. (1961*b*) *J. Agric. Sci.* **57**, 419–425
Blaxter, K. L. & Wainman, F. W. (1964) *J. Agric. Sci.* **62**, 207–214
Blaxter, K. L. & Wainman, F. W. (1966) *Br. J. Nutr.* **20**, 103–111
Blaxter, K. L. & Wilson, R. S. (1962) *Anim. Prod.* **4**, 351–358
Blaxter, K. L. & Wilson, R. S. (1963) *Anim. Prod.* **5**, 27–42
Blaxter, K. L. & Wood, W. A. (1951*a*) *Br. J. Nutr.* **5**, 11–25
Blaxter, K. L. & Wood, W. A. (1951*b*) *Br. J. Nutr.* **5**, 29–55
Blaxter, K. L. & Wood, W. A. (1951*c*) *Br. J. Nutr.* **5**, 55–67
Blaxter, K. L. & Wood, W. A. (1952*a*) *Br. J. Nutr.* **6**, 56–70
Blaxter, K. L. & Wood, W. A. (1952*b*) *Br. J. Nutr.* **6**, 144–163
Blaxter, K. L. & Wood, W. A. (1953) *Vet. Rec.* **65**, 889–892
Blaxter, K. L., Watts, P. S. & Wood, W. A. (1952) *Br. J. Nutr.* **6**, 125–143
Blaxter, K. L., Brown, F. & MacDonald, A. M. (1953*a*) *Br. J. Nutr.* **7**, 105–123
Blaxter, K. L., Brown, F. & MacDonald, A. M. (1953*b*) *Br. J. Nutr.* **7**, 287–298
Blaxter, K. L., Brown, F., Wood, W. A. & MacDonald, A. M. (1953*c*) *Br. J. Nutr.* **7**, 337–349
Blaxter, K. L., Wood, W. A. & MacDonald, A. M. (1953*d*) *Br. J. Nutr.* **7**, 34–50
Blaxter, K. L., Graham, N. McC. & Rook, J. A. F. (1954*a*) *J. Agric. Sci.* **45**, 10–18
Blaxter, K. L., Rook, J. A. F. & MacDonald, A. M. (1954*b*) *J. Comp. Pathol. Ther.* **64**, 157–175
Blaxter, K. L., Graham, N. McC. & Wainman, F. W. (1956*a*) *Br. J. Nutr.* **10**, 69–91
Blaxter, K. L., Graham, N. McC. & Wainman, F. W. (1956*b*) *Proc. Nutr. Soc.* **15**, ii
Blaxter, K. L., Sharman, G. A. M. & MacDonald, A. M. (1957) *Br. J. Nutr.* **11**, 234–246
Blaxter, K. L., Graham, N. McC. & Wainman, F. W. (1959*a*) *J. Agric. Sci.* **52**, 41–49
Blaxter, K. L., Graham, N. McC., Wainman, F. W. & Armstrong, D. G. (1959*b*) *J. Agric. Sci.* **52**, 25–40
Blaxter, K. L., Cowlishaw, B. & Rook, J. A. F. (1960) *Anim. Prod.* **2**, 1–10
Blaxter, K. L., Wainman, F. W. & Wilson, R. S. (1961) *Anim. Prod.* **3**, 51–61
Blaxter, K. L., Joyce, J. P. & Wainman, F. W. (1963) *Nature* (*Lond.*) **198**, 1115–1116
Blaxter, K. L., Clapperton, J. L. & Martin, A. K. (1966*a*) *Br. J. Nutr.* **20**, 449–460
Blaxter, K. L., Clapperton, J. L. & Wainman, F. W. (1966*b*) *Br. J. Nutr.* **20**, 283–294

Blaxter, K. L., Clapperton, J. L. & Wainman, F. W. (1966*c*) *J. Agric. Sci.* **67**, 67–75
Blaxter, K. L., Wainman, F. W. & Davidson, J. L. (1966*d*) *Anim. Prod.* **8**, 75–83
Brown, F. (1952*a*) *Biochem. J.* **51**, 237–239
Brown, F. (1952*b*) *Biochem. J.* **52**, 523–526
Brown, F. (1953*a*) *J. Sci. Food Agric.* **4**, 161–165
Brown, F. (1953*b*) *Nature* (*Lond.*) **171**, 790–791
Brown, F. & Blaxter, K. L. (1951) *Chem. Ind.* (*Lond.*), 633–634
Burton, A. C. (1934) *J. Nutr.* **7**, 481–496
Burton, A. C. & Edholm, O. G. (1955) *Man in a Cold Environment*, Edward Arnold, London
Clapperton, J. L. (1964*a*) *Br. J. Nutr.* **18**, 47–54
Clapperton, J. L. (1964*b*) *Br. J. Nutr.* **18**, 39–46
Clapperton, J. L. & Blaxter, K. L. (1965) *Proc. Nutr. Soc.* **24**, xxxiii
Clapperton, J. L., Armstrong, D. G. & Blaxter, K. L. (1960) *Proc. Nutr. Soc.* **19**, xvii
Clapperton, J. L., Joyce, J. P. & Blaxter, K. L. (1965) *J. Agric. Sci.* **64**, 37–49
Cowlishaw, B. & Blaxter, K. L. (1955) *Proc. Nutr. Soc.* **14**, iii
Czerkawski, J. W. (1970) *World Rev. Nutr. Diet.* **11**, 240–282
Czerkawski, J. W., Blaxter, K. L. & Wainman, F. W. (1966*a*) *Br. J. Nutr.* **20**, 349–362
Czerkawski, J. W., Blaxter, K. L. & Wainman, F. W. (1966*b*) *Br. J. Nutr.* **20**, 485–494
Czerkawski, J. W., Blaxter, K. L. & Wainman, F. W. (1966*c*) *Br. J. Nutr.* **20**, 495–508
Ekern, A., Blaxter, K. L. & Sawers, D. (1965*a*) *Br. J. Nutr.* **19**, 417–434
Ekern, A., Blaxter, K. L. & Sawers, D. (1965*b*) in *Proc. 3rd Symp. Energy Metabolism*, pp. 217–224
Garton, G. A., Duncan, W. R. H., Blaxter, K. L., McGill, R. F., Sharman, G. A. M. & Hutcheson, M. K. (1956) *Nature* (*Lond.*) **177**, 792–793
Garton, G. A., Lough, A. K. & Vioque, E. (1961) *J. Gen. Microbiol.* **25**, 215–225
Goettsch, M. & Pappenheimer, A. M. (1931) *J. Exp. Med.* **54**, 145–165
Gonzalez-Jimenez, E. & Blaxter, K. L. (1962) *Br. J. Nutr.* **16**, 199–212
Graham, N. McC., Wainman, F. W., Blaxter, K. L. & Armstrong, D. G. (1959) *J. Agric. Sci.* **52**, 13–24
Griffiths, T. W. (1965) *Anim. Prod.* **7**, 263–272
Hjärre, A. & Lilleengen, K. (1936) *Virchows Arch.* **297**, 565–593
Huffman, C. F., Conley, C. L., Lightfoot, C. C. & Duncan, C. W. (1941) *J. Nutr.* **22**, 609–620
Joyce, J. P. (1964) PhD Thesis, University of Glasgow
Joyce, J. P. & Blaxter, K. L. (1964*a*) *Br. J. Nutr.* **18**, 5–27
Joyce, J. P. & Blaxter, K. L. (1964*b*) *Res. Vet. Sci.* **5**, 506–516
Joyce, J. P., Blaxter, K. L. & Park, C. (1966) *Res. Vet. Sci.* **7**, 342–359
MacDonald, A. M., Blaxter, K. L., Watts, P. S. & Wood, W. A. (1952) *Br. J. Nutr.* **6**, 164–169
MacDougall, D. B., Bremner, I. & Dalgarno, A. C. (1973) *J. Sci. Food Agric.* **24**, 1255–1263
Marr, T., Sharman, G. A. M. & Blaxter, K. L. (1956) *Vet. Rec.* **68**, 408–410
Martin, A. K. (1966*a*) *J. Sci. Food Agric.* **17**, 496–500
Martin, A. K. (1966*b*) *Br. J. Nutr.* **20**, 325–337
Martin, A. K. (1973) *Br. J. Nutr.* **30**, 251–267
Martin, A. K. & Blaxter, K. L. (1961) in *Proc. 2nd Symp. Energy Metabolism*, pp. 200–209
Martin, A. K. & Blaxter, K. L. (1963) *Proc. Nutr. Soc.* **22**, xiv–xv
Martin, A. K. & Blaxter, K. L. (1965) in *Proc. 3rd Symp. Energy Metabolism*, pp. 83–91
Ministry of Agriculture, Fisheries and Food, Department of Agriculture and Fisheries for Scotland & Department of Agriculture for Northern Ireland (1975) *Energy*

Allowances and Feeding Systems for Ruminants, Tech. Bull. no. 33, Her Majesty's Stationery Office, London
Moe, P. W., Tyrell, H. F. & Flatt, W. P. (1970) in *Proc. 5th Symp. Energy Metabolism*, pp. 65–68
Moore, L. A., Sholl, L. B. & Hallman, E. T. (1936) *J. Dairy Sci.* **19**, 441–442 (Abstr.)
Moore, L. A., Hallman, E. T. & Sholl, L. B. (1938) *Arch. Pathol.* **26**, 820–838
Muth, O. H., Oldfield, J. E., Remmert, L. F. & Schubert, J. R. (1958) *Science* (*Wash. D.C.*) **128**, 1090
Patterson, E. L., Milstrey, R. & Stokstad, E. L. R. (1957) *Proc. Soc. Exp. Biol. Med.* **95**, 617–620
Rook, J. A. F. (1953) *Proc. Nutr. Soc.* **12**, viii
Schwarz, K. & Foltz, C. M. (1957) *J. Am. Chem. Soc.* **79**, 3292–3293
Sharman, G. A. M. (1954) *Vet. Rec.* **66**, 275–279
Sharman, G. A. M., Blaxter, K. L. & Wilson, R. S. (1959) *Vet. Rec.* **71**, 536
Sojka, W. (1965) *Commonw. Bur. Anim. Health Rev. Ser.* no. 7
Stewart, A. B. & Holmes, W. (1953) *J. Sci. Food Agric.* **4**, 401–408
Ulyatt, M. J., Czerkawski, J. W. & Blaxter, K. L. (1966) *Proc. Nutr. Soc.* **25**, xviii
Ulyatt, M. J., Blaxter, K. L. & McDonald, I. (1967) *Anim. Prod.* **9**, 463–470
Vercoe, J. E. & Blaxter, K. L. (1965) *Br. J. Nutr.* **19**, 523–530
Wainman, F. W. & Blaxter, K. L. (1958*a*) in *Proc. 1st Symp. Energy Metabolism*, pp. 85–92
Wainman, F. W. & Blaxter, K. L. (1958*b*) in *Proc. 1st Symp. Energy Metabolism*, pp. 80–84
Wainman, F. W. & Blaxter, K. L. (1969) in *Proc. 4th Symp. Energy Metabolism*, pp. 429–433
Waite, R. & Gorrod, A. R. N. (1959) *J. Sci. Food Agric.* **10**, 308–317
Waite, R., Johnston, M. J. & Armstrong, D. G. (1964) *J. Agric. Sci.* **62**, 391–398
Webster, A. J. F. (1967) *Br. J. Nutr.* **21**, 769–785
Webster, A. J. F. & Blaxter, K. L. (1966) *Res. Vet. Sci.* **7**, 466–479

ENVIRONMENTAL AND THERMOREGULATORY PHYSIOLOGY OF FARM ANIMALS

J. BLIGH

INTRODUCTION

THE forms and functions of living things have been shaped by selective adaptations to ancestral and contemporary environments and the distribution of a species is limited to the range of ambient conditions in which successive generations can thrive and reproduce. An otherwise suitable environment will fail to support the species if reproduction alone is impaired, or if a physiologically intolerable condition prevails for only a very short time in the course of an annual cycle.

Man is a notable exception to this general principle; his inventiveness enables him to create tolerable local environments for himself in otherwise intolerable climates. Like all other permanently thermostable and thermoregulating mammals, man can tolerate the climatic extremes of his distribution only so long as he can obtain food throughout the year. The domesticated herbivores serve this need in two ways: they act as converters of inedible cellulose-bound plant matter into edible matter; and they buffer man against famine, for even when underfed, underwatered, and slowly deteriorating in condition, they continue to be a source of food.

PROGRESSIVE AND TRADITIONAL ANIMAL HUSBANDRY

Since animals were first domesticated, a vast difference has arisen between the history of animal husbandry in areas of progressive and of traditional agriculture. In the former areas improvements in animals and their husbandry are of four kinds: (i) the deliberate selection to create breeds capable of greater productivity; (ii) the provision of artificial shelter from the elements; (iii) the harvesting of animal fodder during seasons of plenty, for use during seasons of scarcity, and (iv) pasture management to improve the quantity and quality of animal fodder. By a combination of these improvements dietary standards have risen in some parts of the world despite sharp increases in populations. By contrast, wherever agriculture and animal husbandry have remained traditional, productivity has failed to match population increases largely brought about by improved medicine, and there has been little or no improvement in dietary standards. These latter areas tend to coincide with those that have tropical or subtropical environments. The cattle of these areas have become adapted, largely by processes of natural selection, to the poor quality of the vegetation, to endemic diseases, and perhaps also to such local seasonal stresses as high solar radiation and drought, but there was, until relatively recently, little improvement in pastures, a failure to harvest, and practically no deliberate selection of animals capable of improved productivity. In such circumstances populations reach equilibrium states, but the quality of life for both man and beast may be very low.

Attempts to improve food production in many tropical areas have encountered many difficulties. These include the traditional measure of social prestige in terms of number of livestock rather than in terms of quality and productivity, and a profound lack of basic information about everything from pasture management to animal physiology. Without such basic information, the local efforts to improve meat and milk production could be only marginally effective. Efforts to close the gap between the agriculturally developed and agriculturally undeveloped regions were largely negated by widespread use of insecticides such as DDT, and of antibiotics, which resulted in explosive rises in the population in many parts of the world after the Second World War. These advances in the control of human diseases changed the balance between human population and food supplies, particularly in the agriculturally undeveloped parts of the world, and increased the incidence of other less readily cured diseases related directly or indirectly to malnutrition.

Chronic malnutrition shortens average life expectancy largely by increasing infant mortality. The pot-bellies of children, resulting from protein deficiency, provide an immediate index to the nutritional status of a community. This syndrome of infant malnutrition is called kwashiorkor in West Africa, and means ‘the disease of the displaced child’ since the symptoms arise when a child is displaced from the breast on the arrival of its successor. Since a continuing supply of mother’s milk would have protected the child from this condition, cows’ milk might be expected to afford a similar protection from the trauma of being weaned on to a plant diet deficient in first-class protein.

THE INITIATIVE OF NORMAN C. WRIGHT AND THE INVOLVEMENT OF THE HANNAH RESEARCH INSTITUTE

The Director of the Hannah Research Institute at the time of the Second World War was Dr (later Sir) Norman C. Wright. Not only was he involved in the problems of feeding Britain during those difficult years but he had also, from 1938 onwards, paid visits to India, Ceylon, Greece and some fifteen Middle Eastern territories to advise on the development of animal husbandry and dairy industries. These experiences enhanced his awareness of the more chronic plight of other parts of the world. He may also have sensed that neglect of this situation by the relatively privileged, and agriculturally more advanced, parts of the world would ultimately represent a politically and morally unacceptable division of the world's population. Thus the Hannah Research Institute, although created to investigate means of improving the dairy industry at a more local level, came to devote a tithe of its efforts to a problem which was totally outside the experience of its staff.

How, Dr Wright wanted to know, could a substantial increase in the local production of dairy produce in tropical and subtropical parts of the world be brought about? Clearly this was a multifactorial problem involving the need to improve grassland management; to develop systems of harvesting; to establish means of collecting and distributing dairy supplies; to overcome the traditional estimation of wealth in terms of numbers of cattle regardless of quality and productivity, and so on. Most of these technical and social problems were not matters with which the Hannah Research Institute could be concerned directly. Its contribution, if it was to make one at all, had to relate simply to one aspect of the overall problem: the identification of the physiological qualities which enable cattle to thrive in hot, and possibly also arid, environments, and the provision of a basis for selection of breeds, or individual animals within a breed, which would be most effective as producers of dairy produce in hot and humid or hot and arid environments.

THE CREATION OF A DEPARTMENT OF PHYSIOLOGY AT THE HANNAH RESEARCH INSTITUTE

At that time there was no Department of Physiology at the Institute but the late Dr J. D. Findlay, although primarily a chemist, had studied physiology under Professor R. C. Garry. Thus, in 1945 he was given the daunting task of trying to find out how quickly the quality of dairy livestock in tropical environments could be improved. He set to work by recruiting a staff consisting of a histologist (Miss A. M. Goodall) to investigate the structures of the skin related to channels and processes of heat loss; a physicist (W. R. Beakley) to tackle the dual task of studying heat fluxes between animal and environment and, with the help of an instrument maker and engineer (W. Nisbet), of designing a climatic chamber for the investigation of the capacity of cattle to lose heat in different environmental conditions. Dr Findlay undertook the formidable task of reading and digesting everything that was then known about the relations between cattle and their environments. He also started, with the aid of his first research student (S. H. Yang) to investigate the skin structures of cattle.

THE FIRST RESEARCHES

The first research paper (Findlay & Yang, 1948) was on the capillary distribution in cow skin, and this was followed by a paper by Findlay *et al.* (1950) on the sweat glands in the helix of the cow's ear. This paper touched on the persisting problem of finding an instant predictive index of potential milk yield: it reported that, contrary to the claims of others, the sweat gland density could not be related to milk yield. Findlay (1950) then summarized the existing knowledge of the effects of temperature, humidity, air movement and radiation on the behaviour and physiology of cattle and other farm animals. He dealt with the means of measuring both environmental and physiological quantities, and with the effects of environmental factors on the different physiological factors such as body temperature, heart rate, respiratory rate, and blood and milk constituents that might indicate the ability of the animals to withstand environmental stress. This work served a great need and probably did more than anything else to establish the Hannah Institute as a centre of research in the field of cattle and climate. The immediate and sustained effect was that people started coming to the Institute to get a firm grounding in the problems of dairy productivity in adverse climates. At the same time, Dr Findlay was asked to go and advise in several such climatically harsh areas of the world. The establishment of a Veterinary Institute of Tropical and High Altitude Research in Peru, financed by the Food and Agriculture Organization (FAO), was largely on Findlay's recommendation after his visit to Peru on behalf of FAO. He also went on similar fact-finding missions to East Africa and elsewhere.

By about 1950, a small climatic chamber had been constructed and commissioned by W. R. Beakley and W. Nisbet. The first study, made on Ayrshire male calves, was to determine whether, within a breed, individual differences could be found in the ability to equate heat production and heat loss during thermal loading. The result of this study was published as a series of papers (Beakley & Findlay, 1955*a*, *b*, *c*, *d*, *e*, *f*). Although not stated explicitly, this study also touched on the nub of the matter of selection for survival and sustained productivity in stressful climates. All genetic differences arise first as individual mutations. The natural or artificial selection of such mutations gives rise to strains or breeds with some common characteristics. The possibility of further selection from within a breed of animals better suited to a particular climatic pattern depends on the existence and size of a gene pool within a population. This is just another way of saying that there must be differences between individuals within a breed if a progeny testing programme is to yield marked improvements in performance.

THE MATCHING OF ANIMALS TO ENVIRONMENTS OR ENVIRONMENTS TO ANIMALS

To increase farm animal productivity one can do two things: engineer improvements in the performance of the animal or engineer improvements in the environment of the animal. Improvements in the performance of the animal are mostly genetic and achieved by selective breeding, but there can also be improvements in phenotypic processes of adaptation and these include immunological protection against the environmental hazards of pathogenic organisms. Improvements in the environment include disease protection by good hygiene and improvement in the quality and quantity of foodstuffs as well as physical protection from adverse meteorological conditions. The role of the physiologist is confined largely to defining the qualitative and quantitative differences in physiological processes that allow one breed, rather than another, to function productively in a particular environment, and of defining the optimum environmental conditions for a particular breed. The assumed value of these definitions is that they will provide breeders with guides to the physiological qualities for which they should select, and provide environmental engineers with guides to the ambient conditions which they should endeavour to provide.

It may be doubted how far selective breeding can still substantially influence European animal husbandry, if only because a consequence of the long-establishment of breeds is the considerable reduction in the magnitude of the gene pools (individual genetic variations) from which further selections can be made. On the other hand the structure of European agriculture is such that economic benefit can accrue from the provision of suitable local environments. Thus locally orientated environmental physiology of farm animals might be expected to relate more to environmental engineering than to selective breeding.

By contrast, artificial selection of breeds has not featured greatly in the tropical and subtropical areas where human malnutrition is most serious. Thus the size of the gene pools within most breeds of *Bos indicus* may be supposed to be greater than in most breeds of *Bos taurus*. They may, therefore, provide greater scope for the selection of strains of cattle (and other species) more suited physiologically to quite different environmental conditions. On the other hand, anything but the simplest and cheapest of environmental engineering is unlikely to be generally applicable or generally economically acceptable. Thus, the emphasis of the research effort orientated towards animal productivity in tropical and subtropical environments must be on the definitions of physiological quantities which might form the basis for selection of new breeds rather than on the definitions of environments for established breeds.

THE MEASURE OF PRODUCTIVITY AND THE SEARCH FOR AN INDEX

Agricultural productivity is easier to define than to measure. Overall productivity relates to both the *quantity* and *quality* of the food produced for *human consumption* from a unit area of land per year and *which is sustainable indefinitely*. This last proviso, although imperative for the future of man, renders the measurement of productivity almost impossible. The measure of the suitability of a breed or species must be narrower and relate to its *potential* contribution to productivity rather than to an *actual* and demonstrable contribution. This is essentially what progeny testing involves. Assuming a strong genetic basis for the measured weight gain, carcass quality or lactation yield, it can be supposed that the performance will be repeatable through successive generations in a similar environment. But measures such as that require great effort. There is a need for a readily measurable tell-tale index which correlates with the productivity of the individual or breed, and which can be used as the basis for a preliminary selection of those animals most likely to be productive.

As has been indicated already, meteorological stresses are not the only ones to which the animal may be exposed – disease, famine and drought may be quite as debilitating – but the environmental physiologist tends to focus his attentions on the meteorological stresses, their effects on the animal and the means of ameliorating them. Environmental temperature and humidity are important because they determine the rate at which heat can flow from the animal to the environment. When this rate cannot be made equal to the rate of heat production, heat will be stored in the body and body temperature will rise to a new equilibrium level. Most mammals are considered to be homeothermic – i.e. they stabilize body temperature at a particular species–specific level so long as they are physiologically capable of doing so. It has long been supposed that the stability or otherwise of body temperature can be used as an index of whether or not cattle can withstand the thermal load imposed on them. It has also been supposed that heat-stressed cattle will seek shade to reduce the radiation load, and will decrease feed intake to lower the rate of heat production. These responses alone, quite apart from any other effects of the environment, will lower productivity. The effect on rectal temperature of a standardized period of exposure to thermal stress, such as in the Iberian heat tolerance test (Rhoad, 1944), is still used extensively in field studies of 'heat tolerance' but a relation between rectal temperature and productivity has never been demonstrated clearly, and the Iberian heat tolerance test was soon abandoned at the research station at which it originated (Vernon *et al.*, 1959).

In seeking the needed predictive index Findlay clearly appreciated the need to face up to the profound ignorance of the basic environmental physiology upon which the heat-tolerance or intolerance of cattle must depend. The researches in the Department of Physiology of the Hannah Institute were thus directed towards a better understanding of underlying mechanisms, rather than towards the search for the much-needed index.

ORGANISM–ENVIRONMENT INTERACTIONS

As was implied at the beginning of this essay, all physiology relates directly or indirectly to the sustaining of life in the context of a continuously varying environment (see Bligh *et al.*, 1976). Every process by which an animal reacts appropriately to a change in its external or internal environment must be activated by a sensor of the disturbance, which sends a signal to the central nervous system. This then generates and emits appropriate response signals which act on effector processes. Thermoregulation, with which we are especially concerned here, is an example of this disturbance–response relation. In mammals the thermosensors both at the periphery and in the deeper tissues send neural signals to the hypothalamic area of the brain whence efferent signals pass to the effector mechanisms which vary the rates of both heat production and heat loss. The efferent signals are of two fundamentally different kinds: neurally transmitted signals which cause immediate responses to disturbances, and humorally (hormonally) transmitted signals which seem to act more slowly because their effects are generally to evoke structural and functional changes in the effector processes so as to modify their activities in response to neural signals. These more slowly developing hormonally induced changes are an important component of the processes of acclimatization by which phenotypic adaptations occur in response to both seasonal and man-imposed changes in the environment.

As is discussed below, the studies made in the Institute's Department of Physiology have straddled this whole pattern of life–environment relations, with particular reference to thermoregulation of cattle.

Sensors

My first assignment when I joined the department in 1952 was to assess the significance of peripheral thermosensitivity in the control of thermoregulation in cattle. By monitoring the temperature of blood passing up the carotid arteries to the brain, we demonstrated that the onset of panting in response to a high ambient temperature was effected with no change in the temperature of the blood supplying the brain, or of any change in body temperature generally (Bligh, 1957*b*). This indicated that the thermoregulatory response to a thermal disturbance was activated by the high ambient temperature acting on peripherally located thermosensors.

This result led to doubts about the generally accepted dominance of the direct thermosensitivity of the brain in mammalian thermoregulation. Subsequent studies at the Institute involving local heating and cooling of the hypothalamic region of the brain (Ingram *et al.*, 1963; Calvert *et al.*, 1972*a*, *b*; Calvert, 1975), and of the spinal cord (Jessen *et al.*, 1972; Clough *et al.*, 1973), showed that the temperature of the deep structures can influence thermoregulatory effector activities, but this was not evidence of the overriding importance of particular core sensors. These findings added weight to the evidence from other laboratories of some sort of integration of the neural information from peripheral, spinal cord, hypothalamic, and possibly other thermosensitive structures, in the activation of thermoregulatory effector functions. This line of investigation, which started with a simple question of possible immediate practical significance, has led to fundamental questions being asked about the nature of the central nervous control of the internal environment of the body.

The central controller

Exactly how the brain relates thermal disturbances to thermoregulatory responses remains largely unknown. That these functions reside in the complex interrelations between millions of neurons, and that these neurons communicate with each other through the release of chemical transmitters, is beyond doubt. The specificity of the thermoregulatory effects observed when certain putative transmitter substances are injected into local areas of the brain, has opened up the possibility of understanding the central nervous processes of thermoregulation in terms of neurons and synapses. The early results of this type of experiment caused some consternation because of the differences reported within, as well as between, species. The observation of Findlay & Thompson (1968) that the thermoregulatory effects of many of these substances are dependent on the prevailing environmental temperature, has greatly eased the burden of finding patterns in both structure and function. The full implication of this finding is only just reaching the pharmacologists who continue to observe drug effects at 'room temperature' irrespective of the species and their differing zones of thermoneutrality. This observation, however, had a profound influence on studies in my own laboratory (Bligh *et al.*, 1971) and on those of many others working on the central nervous control of body temperature. Other Hannah Institute contributions to this topic include studies on the hypothalamic control of plasma unesterified fatty acids (Clough & Thompson, 1972) and of metabolism (Darling *et al.*, 1974) in sheep. The Institute's Department of Physiology has thus made an important contribution to the current exciting attempts to understand a regulatory function wholly in terms of brain cells and their interrelations.

Effectors

There are four kinds of thermoregulatory effector functions: (i) shivering and non-shivering processes of heat production; (ii) changes in peripheral vasomotor tone, by which means heat loss through the outer surfaces of the body can be varied; (iii) evaporative heat loss by panting and sweating, and (iv) behavioural activities.

Heat production

Heat production did not feature in the Institute's earlier researches because the basic interest was in heat stress, not cold stress. High productivity inevitably involves high heat production so although animals with low heat production may be better regulators of body temperature they are unlikely to be good producers of meat and milk. The more important consideration relative to tropical environments is the ability to shed heat despite a diminished thermal gradient from the animal to the environment.

Now that there is need to relate agricultural research in the United Kingdom to domestic production (see below), cold tolerance is claiming the Institute's attention. Recent studies of the metabolic contributions of muscular tissue during cold stress (Bell *et al.*, 1974*b*, 1975); of the effects of cold exposure on plasma and muscle lipids (Thompson & Clough, 1972; Bell *et al.*, 1974*a* Bell & Thompson, 1974) and the effects of acute cold exposure on circulation (Bell & Thompson, 1974), are evidence of this shift of interest. Although this development deviates from the original objectives, there can be no doubt that as Britain seeks to increase its own meat production from marginal grasslands, such studies will become of great importance.

Peripheral vasomotor tone

The skin is the principal interface between animal and environment and at least some of the heat being produced in the body must pass through the skin to the environment. Skin temperature depends largely on the extent of its blood supply and on the ambient temperature, while the rate of heat flow into or out of an animal depends on the temperature difference between the skin and the surrounding air and surfaces. Findlay's first research reports (Findlay & Yang, 1948; Findlay *et al.*, 1950) were on the structure of bovine skin, about which little was known then. This led to interesting observations on the presence of arterio-venous anastomoses, and the consideration of the role which these 'short-circuits' may play in the control of skin surface temperature. This work could and should have been followed up by dynamic observations on living tissues. An attempt was made to use small 'windows' in the ears of calves to observe the circumstances in which the arterio-venous anastomoses of the ear open and close. This project, however, was not carried through to a conclusion. Presumably it was a casualty of an overstretched research programme.

Evaporative heat loss

A possible reason why there has been relatively little work on peripheral vasomotor tone compared with that on sweating and panting is that variations in peripheral circulation of blood are of significance in controlling the flow of heat between organism and environment only when there is a thermal gradient from organism to environment down which heat can flow. When the radiant heat transfer is from the environment to the organism, or when air temperature approximates to or exceeds that of the skin, the animal must rely on adequate heat loss by the evaporation of fluid from external surfaces to maintain thermal stability. Excepting the processes of wallowing and saliva-spreading, which are partly behavioural and which are of little significance in cattle, the principal processes of evaporative heat loss are by panting and sweating. European cattle pant when in a thermally stressful environment, but Indian breeds of cattle which are, presumably, better adapted to heat stress are said not to. Since, however, evaporation must be the means of heat loss in Indian cattle also, they must be capable of a considerable degree of sweating.

Evaporative heat loss by sweating

There has been conflicting evidence and opinion on whether or not all breeds of cattle sweat when exposed to heat. Because the sweat glands of cattle 'have a very slow cycle of secretion which is ill-adapted for the production of large quantities of fluid, and a very poor blood supply which would scarcely supply enough fluid for the glands to be of much use' the early judgement of Findlay (1953) was that cattle have only a very limited capacity to sweat. However, subsequent work in his Department (see review by Jenkinson in this volume, p.198), has demonstrated sweat glands in cattle, shown the importance of sweating for heat loss, and played an important part in the development of current theories on sweat gland control.

Evaporative heat loss by panting

The paradox of cattle is that they both sweat and pant, and an important question is whether in all breeds and in all environmental conditions one of the processes of evaporative heat loss is more effective than the other. Despite the evidence of McLean (1963) that sweating is more important than panting during heat stress, earlier work in the Department of Physiology had shown that when respiratory evaporation is the only channel of heat loss available to an ox calf, it can still maintain a stable body temperature

(Bligh, 1957*a*). It was also demonstrated at that time that there is no change in the temperature of the blood as it passes through the lungs from the right side of the heart to the left. This was evidence that evaporative cooling is restricted to the upper respiratory passages, and does not occur at all at the alveolar level.

There can be no doubt that the respiratory physiology of the panting animal is still a somewhat neglected subject. Panting is treated as something additional to the 'prime' function of ventilation, which is the control of the blood gases. However, most mammals pant during heat stress, and the non-panting animals are in the minority. It is therefore likely that in mammalian respiration in general there is an interplay between a process which regulates blood gases and one which regulates body temperature. The few species (e.g. man) in which respiration is concerned only with the control of blood gases, have apparently lost the capacity to pant (Bligh & Allen, 1970).

Acclimatization

Processes of acclimatization involve hormonally induced improvements in the capacity of effector processes to respond to a neural stimulus. Much is known in man and in laboratory animals of the changes in heat production and in evaporative heat loss occurring in response to experimentally imposed thermal stresses, to transportation to a different location, and to seasonal changes. It is now realized that the processes of acclimatization relate not just to the effector functions concerned with obvious responses to external stresses and that virtually every tissue of the body can be shown to change with the seasons, or in response to such unnatural stresses as may be imposed on domesticated species.

In view of the obvious importance of the effects of genetically fixed variations on the extent to which phenotypic adaptations can occur, it may seem surprising that this is an area of research with which the Institute's Department of Physiology has never become heavily involved. Review papers emanating from the department (e.g. Bianca, 1965) show that this was not due to oversight, and Bianca (1959) reported on a study of acclimatization of calves to a hot, dry environment. The simple fact is that breed variations and individual variations, in processes of acclimatization, cannot be readily investigated in simulated environments within laboratories because of the extensive and expensive facilities that would be required. It was thus sound judgement not to concentrate on this aspect of thermal physiology. However, now that the research emphasis has moved from problems of heat stress abroad to domestic problems of cold stress, attention must be paid to the processes of acclimatization, if only because they constitute a shifting physiological background against which all other investigations have to be conducted.

COMMENT

The aim of the work in the Institute's Department of Physiology, formulated by the late Sir Norman Wright, and faithfully preserved by the late J. D. Findlay throughout the 25 years of his headship of the department, was to provide basic information about the thermal physiology of dairy cattle which would assist field workers in their attempts to increase the productivity of these animals in tropical environments where protein deficiency is a major problem.

Because these researches had to be done far away from the tropics, it was never likely that they would provide information of direct and immediate application, and the value of the Institute's researches cannot be measured by the extent to which they have contributed directly and demonstrably to the achievement of the ultimate objective.

Medicine has long been aided by the massive support of detailed studies of human physiology but agriculturally applied animal sciences have lacked a similar fund of fundamental knowledge. No one can deny that the most prevalent disease in the world today is malnutrition, that there can be no therapeutic pill and that the only cure is the adequacy of food supplies and their distribution. It can be fairly argued that the best hope of a solution is in a vast increase in the availability of plant matter of a type or form directly consumable by man. However, many tropical areas are unsuitable for arable farming, and their agricultural value depends on the continued use of herbivorous animals to render the sparse and poor vegetation edible by man. To achieve the maximum productivity of such marginal lands we need to know how these animals interact with their environments. The modest, yet complex and difficult, task of Findlay and his colleagues was to provide an understanding of the basic environmental and thermal physiology of dairy cattle that did not exist and that, but for the work of this group, would still not exist.

The work of Findlay's group on farm animals has also added much to the understanding of the thermoregulatory processes of mammals in general. Four books have appeared in recent years – on environmental physiology, edited by Robertshaw (1974), on comparative thermoregulation, edited by Whittow (1970-73), a single-author work on thermoregulation on mammals and other vertebrates by Bligh (1973) and a book on animals in the heat by Ingram & Mount (1975). D. L. Ingram, D. Robertshaw, G. C. Whittow and myself were all led into this subject at the Hannah Institute. The formative influence of the Department of Physiology and of J. D. Findlay, in the production of so much consolidating literature is self-evident. The conclusion must be that the department has earned its distinction not only in the field of environmental physiology of farm animals, but also in the environmental and regulatory physiology of animals in general. The consequences of such a broad influence on biology are quite unpredictable.

POSTSCRIPT

In recent years the ultimate practical object of the study of environmental physiology at the Hannah Institute has been away from the international problem of hunger and towards the national problem of equating food production and population. This shift in objectives is well recorded in the opening paragraphs of the sections on the Department of Physiology in some of the Annual Reports of the Institute. In 1965 there was no doubt about the objective of increasing milk production in tropical and subtropical areas. In 1968 it can be read that 'It is well known that the productivity of animals in this country and throughout the world . . . is markedly affected by climatic conditions'. The 1971 Report contains no reference to conditions outside the United Kingdom, and in 1974 the redirection of the work of the department is stated with sad but commendable frankness: 'In recent years priorities in the Agricultural Service have altered so as to favour research topics related to agriculture in the United Kingdom rather than overseas. This has resulted in a change in emphasis in the work of the department towards examination of heat exchange problems in housed animals; the protective role of the skin glands against disease rather than against heat, and the effects of cold rather than warm environments.'

In many ways this change in motivation for the study of the environmental physiology of farm animals must be a matter for regret. Food supplies and population remain unequated in many of those parts of the world which were the concern of Norman Wright, and there is a continuing need to study the basic relations between organisms and their environments which applied workers in the areas of agriculture and medicine alike need to know in order to manipulate these relations to the benefit of man. However, despite the high level of productivity of British agriculture in terms of the sustained yield of food for human consumption per hectare per year we still need to import nearly half of the food we consume. Were there a strictly equitable distribution of the world's scant food reserves, we could not buy-in so much of our dietary needs. Domesticated animals may no longer be strictly essential for the production of a well-balanced diet and modern practices of intensive indoor husbandry squander grain that is directly consumable by man. The ruminant, however, will long remain the harvester and converter of vegetation on lands unsuitable for arable cropping, and of arable crops that cannot be consumed directly by man. Thus, while need of a better understanding of the influences of farm animal–environment interrelations on food supplies in

Britain is clearly evident, our persisting interest in food production elsewhere in the world is equally evident. This continuing need for research in environmental physiology of farm animals initiated at the Hannah Institute by Norman Wright should ensure the continuation of the high standard of research in the Department of Physiology which was the hallmark of J. D. Findlay's stewardship.

REFERENCES

Beakley, W. R. & Findlay, J. D. (1955*a*) *J. Agric. Sci.* **45**, 339–352
Beakley, W. R. & Findlay, J. D. (1955*b*) *J. Agric. Sci.* **45**, 353–364
Beakley, W. R. & Findlay, J. D. (1955*c*) *J. Agric. Sci.* **45**, 365–372
Beakley, W. R. & Findlay, J. D. (1955*d*) *J. Agric. Sci.* **45**, 373–379
Beakley, W. R. & Findlay, J. D. (1955*e*) *J. Agric. Sci.* **45**, 452–460
Beakley, W. R. & Findlay, J. D. (1955*f*) *J. Agric. Sci.* **45**, 461–468
Bell, A. W. & Thompson, G. E. (1974) *Res. Vet. Sci.* **17**, 265–267
Bell, A. W., Gardner, J. W. & Thompson, G. E. (1974*a*) *Br. J. Nutr.* **32**, 471–477
Bell, A. W., Thompson, G. E. & Findlay, J. D. (1974*b*) *Pfluegers Arch. Eur. J. Physiol.* **346**, 341–350
Bell, A. W., Gardner, J. W., Manson, W. & Thompson, G. E. (1975) *Br. J. Nutr.* **33**, 207–217
Bianca, W. (1959) *J. Agric. Sci.* **52**, 296–304
Bianca, W. (1965) *J. Dairy Res.* **32**, 291–345
Bligh, J. (1957*a*) *J. Physiol.* (*Lond.*) **136**, 404–412
Bligh, J. (1957*b*) *J. Physiol.* (*Lond.*) **136**, 413–419
Bligh, J. (1973) *Temperature Regulation in Mammals and Other Vertebrates*, North Holland, Amsterdam
Bligh, J. & Allen, T. E. (1970) in *Physiological and Behavioral Temperature Regulation*, (Hardy, J. D., Gagge, A. P. & Stolwijk, J. A. J., eds), p. 97, C. C. Thomas, Springfield, Illinois
Bligh, J., Cottle, W. H. & Maskrey, M. (1971) *J. Physiol.* (*Lond.*) **212**, 377–392
Bligh, J., Cloudsley-Thompson, J. L. & Macdonald, A. (1976) *Environmental Physiology of Animals*, Blackwell, Oxford
Calvert, D. T. (1975) *Bibl. Radiol.* **6**, 224–230
Calvert, D. T., Clough, D. P., Findlay, J. D. & Thompson, G. E. (1972*a*) *Life Sci.* **11**, 223–229
Calvert, D. T., Findlay, J. D. & McLean, J. A. (1972*b*) *Biometeorology* **5**, 40–41
Clough, D. P. & Thompson, G. E. (1972) *Neuroendocrinology* **9**, 365–374
Clough, D. P., Darling, K. F., Findlay, J. D. & Thompson, G. E. (1973) *Pfluegers Arch. Eur. J. Physiol.* **342**, 137–144
Darling, K. F., Findlay, J. D. & Thompson, G. E. (1974) *Pfluegers Arch. Eur. J. Physiol.* **349**, 235–245
Findlay, J. D. (1950) *Hannah Dairy Res. Inst. Bull.* no. 9
Findlay, J. D. (1953) *Br. Agric. Bull.* **6**, 212–216
Findlay, J. D. & Thompson, G. E. (1968) *J. Physiol.* (*Lond.*) **194**, 809–816
Findlay, J. D. & Yang, S. H. (1948) *Nature* (*Lond.*) **161**, 1012–1013
Findlay, J. D., Goodall, A. M. & Yang, S. H. (1950) *J. Dairy Res.* **17**, 22–36
Ingram, D. L. & Mount, L. E. (1975) *Man and Animals in Hot Environments*, Springer Verlag, New York
Ingram, D. L., McLean, J. A. & Whittow, G. C. (1963) *J. Physiol.* (*Lond.*) **169**, 394–403

Jessen, C., McLean, J. A., Calvert, D. T. & Findlay, J. D. (1972) *Am. J. Physiol.* **222**, 1343–1347

McLean, J. A. (1963) *J. Physiol. (Lond.)* **167**, 427–447

Rhoad, A. O. (1944) *Trop. Agric.* **21**, 162

Robertshaw, D. (1974) *Environmental Physiology*, Butterworth, London

Thompson, G. E. & Clough, D. P. (1972) *Q. J. Exp. Physiol. Cogn. Med. Sci.* **57**, 192–198

Vernon, E. H., Damon, R. A., Harvey, W. R., Warwick, E. J. & Kincaid, C. M. (1959) *J. Anim. Sci.* **18**, 91–94

Whittow, G. C. (1970–73) *Comparative Physiology of Thermoregulation*, vols 1–3, Academic Press, New York

THE METABOLISM OF VITAMINS AND MINERALS IN THE RUMINANT

E. C. OWEN

FOR nutritional studies of ruminants, special diets that differ considerably from those needed by non-ruminants have to be designed. Thus I have studied diets free from carotenes, diets poor in water-soluble vitamins and diets deficient in certain amino acids needed by rats, chickens, pigs and man. In all such diets there is a danger that not all the minerals essential for health of the ruminant may be present; for protein-rich natural foods contain elements essential for the growth of the plants from which they come, so that diets which are poor in whole seeds may lack sufficient phosphorus or other minerals that are stored in seeds in the form of phytate. It was considerations of this kind that caused me to design an up-to-date salt mixture for addition to experimetal diets (Owen, 1967*a*, *b*) which were used to study the synthetic powers of the rumen micro-organisms.

Many trace elements are toxic in small excess of requirements. For example, copper, manganese, molybdenum, selenium, iron and cobalt all exhibit very small ratios of the dose which is toxic to the dose which is essential or beneficial to the animal. By adding new trace elements to salt mixtures so that the ratio in the mixture is comparable with the ratio found by R. L. Mitchell of the Macaulay Institute for Soil Research, in healthy pastures, the danger of over-dosage can be avoided. In feeding ruminants we need to remember that it is important to keep the fermentation processes in the rumen normal. Normality of rumen processes is a prerequisite for the normality of the host animal. Thus the ruminant's requirement of cobalt per unit of its body weight is greater than the corresponding requirement for the non-ruminant be it rat, pig, chicken or man. The reason for this is that the non-ruminant requires salts of cobalt mainly in the form of vitamin B_{12} (hydroxocobalamin) while the ruminants depend on bacteria in the rumen to make the vitamin B_{12} for them. In utilizing cobalt, only some of the rumen bacteria make vitamin B_{12}. Other rumen bacteria make other cobalt compounds, which bypass much of the incoming cobalt into forms which the ruminant's tissues cannot themselves use (Owen, 1959*a*).

Many of the minor elements form compounds with specific proteins or with other organic bodily constituents. Examples are the long-known association of the iron-containing haematoporphyrin with globin to form haemoglobin, the much more recently discovered occurrence of copper as an essential constituent of desmosine in elastic tissues, the association of manganese with bone formation and of zinc with the formation of collagen. Rumen fermentations depend on iron, for McNaught *et al.* (1950) found that growth *in vitro* of the rumen flora could be inhibited by iron-chelators such as *o*-phenanthroline. The precipitation of barium as its sulphates has shown that the sulphate also is essential (E. C. Owen, unpublished work). Zinc occurs in corticotrophin of the pituitary gland and is associated with the production of enzymes and hormones in the pancreas. Zinc also forms part of the enzyme retinol (vitamin A alcohol) dehydrogenase which by turning retinol into retinaldehyde potentiates the formation of the photosensitive pigments that are the basis of sight throughout the animal kingdom (Owen, 1965*a*). An injection of the zinc chelator, dithizone, can cause blindness in the dog and produce diabetes in the rat. I have referred to such biochemical functions of minor elements in symposia (Owen 1959*a*, 1960) but I wish to discuss in some detail the metallo-enzyme xanthine oxidase (XO), also called Schardinger's enzyme, in milk. The inactivation of XO was once proposed as a test for pasteurization of cow's milk but had to be dropped because boiling does not completely inactivate this enzyme. Being an enzyme, XO is a protein but its molecule also contains iron, molybdenum, riboflavin and phosphate. The riboflavin (vitamin B_2) and the phosphate are present as flavinadenine dinucleotide (FAD) which is riboflavin linked to adenine by means of a pyrophosphate bridge.

We now know that, by and large, the bodies of mammals work in comparable ways and that their requirements of vitamins, amino acids, minerals, proteins, fats and carbohydrates are, at the cellular level, very much the same. There are, of course, some glaring exceptions such as the need of primates for vitamin C (ascorbic acid) and the need of the cat for preformed vitamin A, which in the cat's diet is not replaceable by carotenes. Non-primate mammals synthesize vitamin C in their own tissues and MacPherson *et al.* (1976) have recently shown that, in the sheep, cobalt is needed for this synthesis. Ruminants are also peculiar in shifting at the onset of rumination from glucose and fructose as the main source of energy circulating in the blood to acetic acid of which the rumen produces so much. In some of our experiments interactions between vitamins and hormones were studied in cows and goats. The hormones we used were thyroxine, its antagonist thiouracil, growth hormone, thyrotropic hormone and the female-sex-hormone substitute, stilboestrol. The hormones we used had no effect on the total secretion in milk of any water-soluble vitamin but we discovered effects of thyroxine and thiouracil on the oxidative state of ascorbic acid, on the partition of vitamin B_1, between its free and its phosphorylated forms, and on the level of ester-phosphorus in milk. The ratio of phosphorus to nitrogen in casein was also changed. Hormonal effects were also found on the extent of digestion of β-carotene and on the ratio of free to esterified vitamin A in the milk of the cow. In one set of experiments I designed a zein-based diet which was free from the amino acids lysine and tryptophan and from the water-soluble vitamins (Edwards & Darroch, 1956; Crossland *et al.*, 1958). By giving this diet to goats that were in positive N balance we were able to demonstrate the synthesis of 40 g of protein and 5 to 6 mg riboflavin in the rumen per goat per day, whereas the level of XO in their milk did not change. Samples taken, by a pump of my own design (Crossland *et al.*, 1958), showed that the rumen was the site of these syntheses.

STUDIES OF XANTHINE OXIDASE

The literature on XO (E.C. 1.2.3.2) is very extensive for the enzyme itself was known for a long time before anyone knew that its molecule contained (see earlier) two metals, vitamin B_2 and phosphorus or that it is widely distributed throughout living things (Modi *et al.*, 1959*a*). In our experiments (Hart *et al.*, 1967) we found a close correlation between molybdenum and XO in cows' milk, but not in the milk of the goat. For cows we found:

$$XO = 171\ Mo + 44$$

The correlation between XO and molybdenum in the milk was large ($r = 0{\cdot}94$) and significant ($P < 0{\cdot}001$). This equation implies that all the molybdenum in cows' milk is probably there as a constituent of the XO enzyme.

It had long been known that both XO and molybdenum rise with the cream of cows' milk, and more recently R. K. Morton showed that XO of cows' milk is present in microsomes among the cellular debris from the mammary epithelium. Morton's work also supplies the first rational explanation of the phenomenon which had long been known as, 'the revelation of xanthine oxidase in cows' milk'. This is the name given to the observation that the XO titre of cows' milk is much increased if the milk is agitated before the XO is measured, whether by Warburg manometry (Hart *et al.*, 1967) or in Thunberg tubes by triphenyltetrazolium chloride (Owen, 1971*a*, *b*). There is an extensive literature on this phenomenon, which the work of R. K. Morton clearly shows to be due to the prepackage by the mammary tissue of XO in microsomal membrane. Many other instances of such prepacked enzymes are now known.

Attempts to correlate the XO in cows' milk with the molybdenum of its food have not proved altogether rewarding, either in my own experience or in that of Kiermeier & Capellari (1958). In the rat, provided that (1) the amount and availability of molybdenum or iron in the food are not limiting, that (2) the amount of vitamin B_2 in the diet is adequate and that

(3) the biological value of the protein in the diet is not limiting, the level of XO in the rat's liver is a reflexion of the biological value of the dietary protein. In addition, copper and sulphate levels in the rat's diet also have to be optimal. It is obviously much easier to handle the interactions of all these variables in the diet of a 250 g rat than in the diet of a cow weighing 400kg, and, need I say it, much less expensive in food and other overheads! Leaving out the dependence of the biological value of the protein in the cow's food on the state of its rumen organisms (McNaught *et al.*, 1954) much more food has to be mixed with molybdenum, iron, copper and so forth for one cow for a single day. It is thoughts like this that make my heart leap for joy when I behold a large small-animals house in the Hannah Research Institute grounds!

Notwithstanding these daunting facts, I tried to affect, by various means, the XO in ruminant milk. In some of the diets the salt mixture was changed to that of Miller *et al.* (1956), in which the sulphate was entirely replaced by chloride. We knew that in chronic molybdenosis of cattle, scouring is the most prominent symptom. To our surprise neither dosing with Na_2MoO_4 nor dosing with $(K,Na)_2SO_4$ caused any scouring; but when both were given together scouring promptly began. Evidently the hind-gut of the cow is sensitive to the MoO_4^{2-}/SO_4^{2-} ratio in the food rather than to the level of either.

In the same series of experiments we measured the effect of dietary molybdate on the XO in the milk of the cow and the goat but only minimal effects were found. In fungi, in bacteria, in root-nodules of leguminous plants and in the chicken, the essential role of molybdenum in potentiating the formation of XO (or of xanthine dehydrogenase) is inhibited by sodium tungstate. In both the cow and goat, milk XO was considerably depressed by oral tungstate (Owen & Proudfoot, 1968). In goats, the milk XO was likewise depressed when tungstate was injected subcutaneously (Owen & Dundas, 1969). Ingestion of tungstate by kids for 3 months led to a diminution of the level of XO in their livers as compared with the livers of undosed control kids (Owen & Proudfoot, 1968). In none of these experiments was there any evidence of a deleterious effect of the tungstate. Tests *in vitro* showed that tungstate added to milk did not affect the XO test nor did milk containing no active XO affect the XO value of normal milk when added to it *in vitro*.

PARTITION OF RIBOFLAVIN IN MILK

There are two commonly used ways of measuring vitamin B_2 in milk, the fluorimetric and the microbiological. For this work I had to design my own fluorimeter using a war-surplus photomultiplier tube (Crossland *et al.*, 1958) and I am very grateful to W. Nisbet of the Institute's staff who showed me how to make the instrument electrically stable. Each method for vitamin B_2 has its own difficulties but each can be checked against the other. The microbiological assay is carried out with *Lactobacillus casei*, a well known rumen organism that contributes to the characteristic odour of butter and cheese by producing diacetyl and acetylmethylcarbinol.

Almost invariably the fluorimetric result exceeds the microbiological result because of fluorescent impurities in food, urine, milk or faeces. The tests agree if extracts are chromatographically purified before microbiological assay. By chromatographic and other methods we were able to characterize three of these fluorescent but microbiologically inactive congeners of vitamin B_2 as catabolites of riboflavin, new to animal science, and to show that they occurred not only in goats but in cows, sheep, rabbits and man himself (West & Owen, 1973*a*) but not in the dog (Owen & Dzialoszynski, 1965). In animal tissues riboflavin can be found in a number of forms, all of which can be hydrolysed to microbiologically active riboflavin. These can appropriately be called anabolites of riboflavin. Milks of various species of mammals can contain FAD or riboflavin either free or bound to protein. If they are free they can be ultrafiltered out, but if bound to protein they cannot. Thus cows' milk contains FAD as an integral part of XO and in addition contains free riboflavin along with various catabolites of riboflavin and sometimes other, as yet unidentified, yellow-fluorescent substances as well. Unlike sows' milk, the milk of the cow does not contain free FAD, and sows' milk contains no XO (Modi *et al.*, 1959*a*, *b*). Manson & Modi (1957) showed that FAD is rapidly hydrolysed to free vitamin B_2 by fresh cows' milk, but that destroying the phosphatase, but not presumably the pyrophosphatase activity, in cows' milk by pasteurization leads to the appearance of riboflavin phosphate (FMN, flavinmonoucleotide) in the milk. Thus conceivably the free riboflavin in cows' milk is secreted into the milk as FMN.

In our earlier experiments (Crossland *et al.*, 1958) for microbiological assays of riboflavin we used the medium of Barton-Wright & Booth (1943) which is based on skimmed cows' milk. We found that at infrequent intervals a batch of tests failed because *L. casei* neither grew nor produced its characteristic 'buttery-cheesy' odour. This phenomenon had nothing to do with the technique or with the lack of care in collection of milk from the Institute herd, for it could occur suddenly when there was no change of personnel, supervision or manipulation. Indeed, while we were worrying about this phenomenon, Miss Gregory, who was using this same assay at the National Institute for Research in Dairying, Reading, asked me if I had encountered this very difficulty. We discussed the matter but my suggestions at that time were no more enlightening than hers. Nowhere in the literature was this difficulty mentioned until Lambooy (1966) detected in his *L. casei* cultures a less exacting mutant. Such a mutant would not explain the failure of visible growth that we observed. About that time factory-made media became available and the problem was no longer the researchers' worry. Nevertheless, I kept the problem in mind. Later the extreme variability of XO in milk (Kiermeier & Capellari, 1958; Owen & Proudfoot, 1968) and in particular its paucity near the termination of lactation made me wonder whether *L. casei* needed molybdenum as a growth factor. I began by testing *L. casei* itself to see whether it could produce XO as many other bacteria were reported to be able to do. There was a large non-specific reduction of triphenyltetrazolium by the bacterium but no evidence whatever of increased reduction in the presence of xanthine. Dikstein *et al.* (1957) had, likewise, failed to find XO in *L. casei*. We therefore grew *L. casei* in the presence of various concentrations of tungstate which in many organisms requiring molybdate as a specific inhibitor of it (Owen & Pringle, 1970). From the results I concluded that *L. casei* needs molybdenum in the medium and that the sudden failures of growth of *L. casei* could be due to milk of very low XO titre because (see earlier in this review) a low XO in cows' milk indicates a low molybdenum as well. Another possible contributing factor to such a failure could be an insufficient 'revelation' of XO in the milk which could possibly result in too much XO and therefore too much molybdenum being discarded with the cream.

XO in cows' milk, and in goats' milk, has aldehyde oxidase activity and when samples of cows' milk are tested on successive days from the same animal with xanthine and formaldehyde as alternative substrates in separate tests the variations from day to day are strikingly parallel (E. C. Owen, unpublished work) as would be expected from Schardinger's original work. He used formaldehyde in the presence of methylene blue in his tests; triphenyltetrazolium was a much later innovation.

Owen & Proudfoot (1965) found an interesting correlation between the percentage of fat and the concentration of riboflavin (vitamin B_2) in cows' milk. The data were from the same experiment as was later more fully described in the paper by Hart *et al.* (1967). This correlation was expressed by the equation:

$$\text{Fat } (\%) = 0{\cdot}94 \text{ vitamin } B_2 \text{ (mg/100 ml)} + 2{\cdot}5$$

The coefficient of correlation of fat percentage with vitamin B_2 in cows' milks, was $+0{\cdot}73$ and it was statistically significant ($P<0{\cdot}001$). This correlation suggests that in metabolism, just as vitamin B_1 (thiamin) is required chiefly in the metabolism of carbohydrate, riboflavin is required in the metabolism of fat (Owen, 1967*a*, *b*). Observations of others serve to confirm this hypothesis. Thus Virtanen (1967), in his studies of milk produced by cows fed on a diet in which protein was entirely replaced by urea plus ammonium salts, found abnormally large percentages of both fat and vitamin B_2 in their milk but nevertheless three of his four results fell very near the straight line represented by the above equation. Thyroxine injections, in our experiments on cows (Chanda *et al.*, 1952; Modi *et al.*, 1959*b*), produced, as expected, abnormally large at percentages but no change in vitamin B_2. Conversely, the antithyroid drug, thiouracil, depressed the fat percentage in the milk but still had no significant effect on the vitamin B_2 cencentrations. These hormonal effects are thus of a different nature from the symbiotic effects of the rumen bacteria: for, in my opinion, it is the fermentative and the synthetic powers of the rumen bacteria which cause the positive correlation between the concentrations of fat and riboflavin in the milk of the cow. Thyroxine has a direct effect on the translocation of fat from the depots to the milk but has no detectable effect on the rumen bacteria.

Several animal species are known to have a vitamin B_2 requirement that increases when their fat intake is increased, e.g. cats (Gershoff *et al.*, 1959), guinea-pigs (Hara, 1960) and rats (Koyanagi & Oikawa, 1965). Using rats, Kim & Lambooy (1967) found that increasing the fat content of an ulcerogenic diet increased the incidence of peptic ulcers, unless the vitamin B_2 intake was increased at the same time. Superimposition of stress, by means of a mechanism which caused the bottom of the rat's cage to see-saw for 2 hours each day, caused a re-appearance of ulceration which could be prevented by increasing the riboflavin still further. It is known that eating fried foods predisposes people to ulceration of the stomach (Meulengracht, 1939) and that a stressful occupation has an aggravating effect. Probably it is the riboflavin in cows' milk that makes it so effective in the treatment of peptic ulcers by nasal drip in hospital patients, but doubtless the great buffering power of milk helps by decreasing the pH of the stomach contents. Doubtless also the vitamin A in the milk helps to stimulate (see Owen, 1965*a*) the synthesis of mucosal chondroitin and mucoitin sulphates in the wall of the stomach.

RIBOFLAVIN METABOLITES

Crossland *et al.* (1958) were the first to try to find out the nature of the yellow-fluorescent substances that accompanied riboflavin in milk and urine of animals. These became evident when goats on our vitamin-B-free diet were given riboflavin in gram quantities by mouth. Such ingestion produced yellow milk and yellow-fluorescent urine. We found that the increased fluorescence was in part due to increased riboflavin but that the main increase was attributable to newly discovered metabolites of riboflavin, three of which were subsequently isolated and characterized as hydroxyethylflavin (Owen, 1962*a*, *b*, *c*; 1965*b*; Owen & West, 1971*a*, *b*), formylmethylflavin (Owen & West, 1971*a*, *b*; West & Owen, 1973*a*) and flavin-10-acetic acid (West & Owen, 1973*b*). Thus:

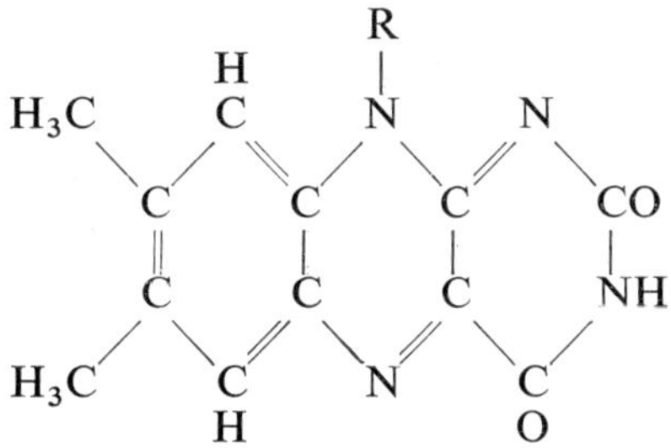

Ethanolflavin	R is CH_2CH_2OH
Formylmethylflavin	R is CH_2CHO
Flavinacetic acid	R is CH_2COOH

In the identification of these metabolites a variety of techniques was used. The very first isolation (Owen, 1962*a*, *b*, *c*; 1965*b*; 1971*a*, *b*) was much facilitated by my personal use of the mineralogical microscope; for isoalloxazines are both birefringent and pleochroic (Owen, 1971*a*, *b*) but lumichrome is an alloxazine and is not pleochroic (E. C. Owen, unpublished work). These properties made it very easy to detect the first and chief metabolite, ethanolflavin, in several hundred times its weight of a colourless material that was on the same place on the paper chromatograms from which the first isolations from goat urine were made. The colourless material was also well crystallized and birefringent but had much lower orders of polarization colours (Owen *et al.*, 1961; Owen, 1962*a*, *b*, *c*; 1965*b*; 1971*a*, *b*), a much lower melting point of 123°C and a u.v. spectrum that changed when the pH of its solution was changed but it had an isosbestic point, i.e. a point where all the graphs of different pH intersected and so had equal extinctions, at 230 mμ. In p-toluidine the colourless substance had a molecular weight of 268; the value was three times bigger by Rast's camphor method presumably because in camphor solution the form is trimolecular. The molecular extinction at 230 mμ was 561. It was not a metabolite of vitamin B_2 for it occurred in urine extracts before dosage of the animals with vitamin B_2 had begun and was isolated by putting spots of the then unidentified hydroxyethylflavin on the linear origin of wide sheets of filter-paper when chromatographing control urine. Many other colourless substances of various R_f values also occurred on the papers. They were readily crystallizable and none of them contained phosphate when tested colorimetrically by my own micro-modification of the method of Martin & Doty (1949). Further studies of these interesting natural constituents of normal urine from goats were not made.

R. I. Reed of the Chemistry Department of the University of Glasgow recorded some preliminary positive mass spectra of our metabolites but later, because they had already studied riboflavin, we contacted the

Forschungsinstitut Manfred von Ardenne, Dresden-Weisser-Hirsch, who recorded both positive and negative mass spectra for us (Tümmler *et al.*, 1971).

K. Overton of the Chemistry Department of the University of Glasgow recorded n.m.r. (nuclear magnetic resonance) spectra of our metabolites. These elegantly confirmed by a non-destructive technique, the structures we had assigned to our metabolites on the basis of their chemical reactions and on the basis of comparisons with genuine synthetic samples.

Such synthetic samples were prepared by degradation of riboflavin by purely chemical means but, when working with the first metabolite, I was provided with a sample prepared by a new synthetic route by H. Wood of the University of Strathclyde (Owen, 1962*c*). Synthetic and metabolic isoalloxazines were compared in various chromatographic systems both on paper and on preparative layers of silica gel and by electrophoretic methods in phosphate or borate buffers. Such tests were also made on acetylation products of the metabolites and their synthetic counterparts.

A problem in obtaining the n.m.r. spectra arose from the low solubility of the isoalloxazines in commonly used solvents such as chloroform. Dr Scheidegger, of Varian Laboratories, Switzerland, however, circumvented this difficulty by the use of computer-averaged transient scanning to overcome the very low signal to noise ratio of the weak solutions of our isoalloxazine.

Professor Overton, besides having a much more sophisticated n.m.r. machine (it cost over half-a-million pounds), also used a new solvent. It was dideuteriodimethylsulphoxide – the deuterium serving to obviate confusion of solvent signals with those from the methyl groups of the isoalloxazines themselves.

An important variable as between the parent compound and the various metabolites is the number of oxygen atoms in the molecule, e.g. riboflavin 6, hydroxyethylflavin 3, lumiflavin 2, flavin-10-acetaldehyde-ethylhemialcoholate 4. T. Reichstein of the Organische Chemisches Institut, Basel, Switzerland, kindly provided us with elementary analyses with directly determined oxygen figures which removed the doubts that arise when oxygen is determined by difference between the weight of the compound and the sum of all the elements other than oxygen.

Professor Dzialoszynski of the University of Poznan, Poland, visited my laboratory under the auspices of the British Council, and collaborated with me on the problem of the origin of the riboflavin metabolites. We found no metabolites in the urine of a dog to which a large dose of the vitamin had been fed. Moreover, when incubated *in vitro* with riboflavin, neither fresh ruminant kidney nor fresh ruminant liver slices from the local slaughterhouse produced any ethanolflavin. These results made us think that the rumen must be the site of metabolite production, but when we pumped samples from the rumen of goats that had had an oral dose of riboflavin half-an-hour earlier, no metabolites were found until the rumen samples were incubated *in vitro*. Rumen samples obtained by fistula from either cows or goats also produced metabolites when incubated *in vitro*. Later experiments, by D. W. West and R. Proudfoot, showed that half-an-hour was too soon after a vitamin dose for sampling of the rumen and that the best time was just before the next feed. At this preprandial time the pH of the rumen had reverted to nearer neutrality or had become slightly alkaline. Evidently organisms disfavoured by too low a pH were responsible for the metabolites.

P. N. Hobson, of the Rowett Research Institute, Aberdeen, had immunologically identified many different species of rumen bacteria from sheep and cattle and had also perfected a method of mass incubation of pure cultures of rumen bacteria of many different species, both from sheep and cattle. However, when Hobson made such incubations in the presence of added riboflavin and sent the resulting incubates to us at the Hannah Research Institute we could not detect any of our metabolites on paper chromatograms of any of the cultures. There were some yellow-fluorescent products but none with the R_f values of our metabolites (Hobson *et al.*, 1969). We therefore concluded that the bacterium we were seeking was not any of the species isolated by Hobson and his colleagues. We then collaborated with Margaret Taylor who was in charge of the Bacteriology Department of the Hannah Institute. Eventually several varieties of a new bacterium that could produce hydroxyethylflavin were isolated. Of these, two varieties studied by West and myself were readily distinguishable under the microscope, both from other rumen micro-organisms and from one another. One looked like short chains of sausages for they were linked end to end and sometimes branched. The other species formed much longer chains which made it much easier to precipitate by centrifugation.

Taylor (1972) identified these varieties of bacteria by plating them out on agar and showed them to have a hitherto unrecorded facies of cultural properties. She therefore reported them as a new species of the genus *Eubacterium* under the specific name *fissicatena*, which referred to their power to split a three-carbon fragment from the ribityl side-chain of riboflavin. She deposited cultures of these bacteria with the National Collection of Industrial Bacteria (Taylor, 1972).

I have already referred to our experiments (Owen & Pringle, 1970) on the relation of the rumen bacterium *L. casei* to the concentrations of molybdenum and tungsten in the medium. This is different from the reaction to molybdenum reported by McKillen & Spencer (1970) for *Salmonella typhimurium*, which reaction we have since confirmed (E. C. Owen & S. A. R. Dempster, unpublished work); Owen & Dempster (1970) grew *Eubacterium fissicatena* in two different media A and B. In seven triplicated cultures in each medium *E. fissicatena* averaged a turbidity of $0{\cdot}478 \pm 0{\cdot}043$ (S.D.) in A and $0{\cdot}791 \pm 0{\cdot}090$ in B. Either molybdate or tungstate stimulated growth of *E. fissicatena* in A and inhibited growth in B, the results being $0{\cdot}762 \pm 0{\cdot}061$ in A and $0{\cdot}619 \pm 0{\cdot}046$ in B. Evidently there is scope for a greater use of molybdenum and tungsten in cultural methods of bacterial classification, particularly when it is recalled that the molybdenum requirement of the bacterium *Escherichia coli* is increased if the medium contains either nitrate or nitrite, that the fungus *Neurospora crassa* exhibits a competition between molybdenum and tungsten when ammonium in the medium is replaced by nitrate and that analogous effects occur in the bacterium *Pseudomonas aeruginosa* and in the N-fixing bacteria of plants (for references see Owen & Pringle, 1970). Even in plants that do not fix nitrogen, the molybdenum requirement increases when nitrate replaces ammonium in the soil.

When I suspected that the riboflavin metabolites were of symbiotic origin I at once wondered whether the rabbit could produce them in its large caecum. The rabbit empties its caecum at night but the faecal pellets so produced are pink in colour and soft, quite unlike the daytime faecal pellets. These caecal pellets never touch the ground for the rabbit eats them as they emerge from the anus. They provide the rabbit with all its requirements of B-vitamins and doubtless many essential amino acids as well. Their pink colour is due to vitamin B_{12} and other 'cobamides'. When the rabbit's urine was tested for hydroxyethylflavin I was not surprised to see a positive result (Owen & West, 1970). Later when we knew what our metabolites were, and that they were produced in both the rumen and caecum of the cow, goat and sheep but not in the newborn kid or newborn calf, we were pleased that Marie E. Coates of the National Institute for Research in Dairying, Reading, who was already experimenting with rabbits in her germ-free animal apparatus, could arrange for us to investigate the riboflavin metabolism of these nutritionally intriguing animals. We found (Owen *et al.*, 1970) that, as we had expected, the germ-free rabbits' caecal contents contained no hydroxyethylflavin but did contain 27 to 46 p.p.m. riboflavin. By contrast, the contents of the caeca of conventional rabbits contained only a trace of riboflavin together with 30 to 35 p.p.m. of hydroxyethylflavin. Conventional rabbits kept on a diet free from riboflavin, contained no trace of either the vitamin or its hydroxyethyl metabolite.

Before the establishment of rumination neither kids nor calves could produce metabolites from ingested riboflavin, but when they drank the milk of lactating goats which were secreting metabolites, these metabolites appeared in their urine. When riboflavin was subcutaneously injected into goats it caused a prolonged secretion of riboflavin in both milk and urine, but did not give rise to metabolites. Such injected riboflavin readily reached the bile thus further confirming that the site of metabolite production is the rumen and that absorption from the caecum is a highly selective process. Formylmethylflavin (flavin-10-acetaldehyde), when injected, was secreted in the milk and excreted in the urine partly unchanged and partly reduced to hydroxyethylflavin (ethanolflavin). This reduction readily occurs in the liver of ruminants and certain other species and the kidney is also active in this respect.

All our experiments (Owen & West, 1970; 1971*a*, *b*) led to the conclusion that all three metabolites arise from rumen bacteria (West & Owen, 1973*a*, *b*).

THE PRODUCTION OF SWAYBACK IN THE GOAT

An observation which arose incidentally from the studies of the partition of riboflavin in milk was the demonstration that swayback, a disease common in lambs, can occur in the kid. The disease affects only the newborn lamb but is prevented by antenatal administration of copper to the ewe. An experimental diet which I designed in an attempt to lower the XO titre of milk, when fed to goats throughout pregnancy, resulted in the birth of kids with the symptoms and lesions of swayback (Barlow *et al.*, 1962).

The diet was therefore used in further studies (Owen *et al.*, 1965) of the biochemical and histological lesions in collaboration with the West of Scotland Agricultural College and the Animal Diseases Research Association, Edinburgh. Significant positive correlations were found between the blood copper of the kids and that of their dams and both the liver copper and brain copper of the kids were positively correlated with their blood copper. However, to quote from Owen *et al.*, (1965) 'A new biochemical feature of the present work is the finding of an inverse relationship between the levels of copper in whole blood and vitamin-A in the serum. The swayback kids of 1962 were among the group which showed abnormally high concentrations of vitamin-A in the blood serum. It is a reasonable hypothesis that vitamin-A, having found its way into the blood of the kid from its dam's milk, was inhibited from being stored in the liver when the blood contained insufficient copper. Supporting this hypothesis are the low levels of vitamin-A in adult livers and colostrum and the experiments of Šimek *et al.* (1961) and Moore *et al.* (1964)'. Šimek worked with pigs and Moore with rats.

Sulphate was shown by C. F. Mills and his colleagues (Fell *et al.*, 1961) to aggravate swayback in lambs and the normal metabolism of sulphur compounds is known to need vitamin A (Owen, 1965*a*). It is therefore of interest that the non-sulphate sulphur exceeded the sulphate sulphur in both concentrates and hay in our swayback diet (Owen *et al.*, 1965; Owen, 1965*c*).

As we said in the main paper (Owen *et al.*, 1965),' . . . the present work appears to be the first in which copper and vitamin-A have been studied at the same time and in the same animal in relation to the integrity of the central nervous system'.

REFERENCES

Barlow, R. M., Robertson, J. M., Owen, E. C. & Proudfoot, R. (1962) *Vet. Rec.* **74**, 737–739

Barton-Wright, E. C. & Booth, R. G. (1943) *Biochem. J.* **37**, 25–30

Chanda, R., McNaught, M. L. & Owen, E. C. (1952) *Biochem. J.* **51**, 543–552

Crossland, A., Owen, E. C. & Proudfoot, R. (1958) *Br. J. Nutr.* **12**, 312–329

Dikstein, S., Bergman, F. & Henis, Y. (1957) *J. Biol. Chem.* **224**, 67–77

Edwards, D. C. & Darroch, R. A. (1956) *Br. J. Nutr.* **10**, 286–292

Fell, B. F., Williams, R. B. & Mills, C. F. (1961) *Proc. Nutr. Soc.* **20**, xxvii–xxviii

Gershoff, S. N., Andrus, S. B. & Hegsted, D. M. (1959) *J. Nutr.* **68**, 75–88

Hara, H. (1960) *J. Vitaminol.* (*Kyoto*) **6**, 24–42

Hart, L. I., Owen, E. C. & Proudfoot, R. (1967) *Br. J. Nutr.* **21**, 617–630

Hobson, P. N., Summers, R., Owen, E. C., Spencer, J. C. & West, D. W. (1969) *Proc. Nutr. Soc.* **28**, 53A–54A

Kiermeier, F. & Capellari, K. (1958) *Biochem. Z.* **330**, 160–168

Kim, Y. S. & Lambooy, J. P. (1967) *J. Nutr.* **91**, 183–188

Koyanagi, T. & Oikawa, K. (1965) *Tohoku J. Exp. Med.* **86**, 19–22

Lambooy, J. P. (1966) *Arch. Biochem. Biophys.* **117**, 120–124

McNaught, M. L., Owen, E. C. & Smith, J. A. B. (1950) *Biochem. J.* **46**, 36–43

McNaught, M. L., Owen, E. C., Henry, K. M. & Kon, S. K. (1954) *Biochem. J.* **56**, 151–156

McKillen, M. N. & Spencer, B. (1970) *Biochem. J.* **118**, 27P

MacPherson, A., Moon, F. E. & Voss, R. C. (1976) *Br. Vet. J.* **132**, 294–308

Manson, W. & Modi, V. V. (1957) *Biochim. Biophys. Acta* **24**, 423–425

Martin, J. B. & Doty, D. M. (1949) *Anal. Chem.* **21**, 965–967

Meulengracht, E. (1939) in *Proc. Br. Med. Assoc. Conf.*

Miller, R. F., Price, N. O. & Engel, R. W. (1956) *J. Nutr.* **60**, 539–547

Modi, V. V., Owen, E. C. & Darroch, R. A. (1959*a*) *J. Dairy Res.* **26**, 277–280

Modi, V. V., Owen, E. C. & Proudfoot, R. (1959*b*) *Proc. Nutr. Soc.* **18**, i

Moore, T., Constable, B. J., Day, K. C., Impey, S. G. & Symonds, K. R. (1964) *Br. J. Nutr.* **18**, 135–146

Owen, E. C. (1959*a*) *Rev. Pathol. Gen. Physiol. Clin.* **59**, 224–243

Owen, E. C. (1959*b*) *Vet. Rec.* **71**, 1114–1127

Owen, E. C. (1960) *Proc. Nutr. Soc.* **19**, 154–162

Owen, E. C. (1962*a*) *Biochem. J.* **82**, 7P–8P

Owen, E. C. (1962*b*) *Proc. Nutr. Soc.* **21**, xvii–xviii

Owen, E. C. (1962*c*) *Biochem. J.* **84**, 96P

Owen, E. C. (1965*a*) *World Rev. Nutr. Diet.* **5**, 132–208

Owen, E. C. (1965*b*) *Proc. 1st Int. Congr. Food Sci. Technol.* vol. III, pp. 669–678

Owen, E. C. (1965*c*) *Food & Cosmet. Toxicol.* **3**, 701–706

Owen, E. C. (1967*a*) in *Urea as a Protein Supplement* (Briggs, H. M., ed.), pp. 329–370, Pergamon Press, London

Owen, E. C. (1967*b*) *Br. Goat Soc. Monthly J.* **60**, 48–50
Owen, E. C. (1971*a*) *Methods Enzymol.* **18**, 296–305
Owen, E. C. (1971*b*) *Methods Enzymol.* **18**, 590–592
Owen, E. C. & Dempster, S. A. R. (1970) *Biochem. J.* **119**, 24P
Owen, E. C. & Dundas, I. (1969) *Proc. Nutr. Soc.* **28**, 59A
Owen, E. C. & Dzialoszynksi, L. (1965) *Proc. Nutr. Soc.* **24**, ix–x
Owen, E. C. & Pringle, J. (1970) *Lab. Pract.* **19**, 591–592
Owen, E. C. & Proudfoot, R. (1965) *Proc. Nutr. Soc.* **24**, xxii–xxiii
Owen, E. C. & Proudfoot, R. (1968) *Br. J. Nutr.* **22**, 331–340
Owen, E. C. & West, D. W. (1970) *Br. J. Nutr.* **24**, 45–60
Owen, E. C. & West, D. W. (1971*a*) *Methods Enzymol.* **18**, 574–579
Owen, E. C. & West, D. W. (1971*b*) *Methods Enzymol.* **18**, 579–581
Owen, E. C., Montgomery, J. P. & Proudfoot, R. (1961) *Biochem. J.* **82**, 8P
Owen, E. C., Proudfoot, R., Robertson, J. M., Barlow, R. M., Butler, E. J. & Smith, B. S. W. (1965) *J. Comp. Pathol.* **75**, 241–251
Owen, E. C., West, D. W. & Coates, M. E. (1970) *Br. J. Nutr.* **24**, 259–267
Šimek, L., Mandel, L., Trávníček, J. & Syřínek, F. (1961) *Zivocisna Vyroba* **6**, 427–434
Taylor, M. M. (1972) *J. Gen. Microbiol.* **71**, 457–463
Tümmler, R., Steinfelder, K., Owen, E. C. & West, D. W. (1971) *Org. Mass Spectrom.* **5**, 41–52
Virtanen, A. I. (1967) in *Urea as a Protein Supplement* (Briggs, M. H., ed.), pp. 185–212, Pergamon Press, London
West, D. W. & Owen, E. C. (1973*a*) *Br. J. Nutr.* **29**, 33–41
West, D. W. & Owen, E. C. (1973*b*) *Br. J. Nutr.* **29**, 43–50

CURRENT RESEARCH TOPICS

GRASSLAND

M. E. Castle and D. Reid

The early research work on grassland described by W. Holmes (see this volume, p. 15) was the foundation on which many of the future investigations at the Hannah Research Institute were based. The search for a greater measure of self-sufficiency on the dairy farm demonstrated the importance of grassland, which was, and still is, the crop occupying the largest area in south-west Scotland. Thus, from 1950 onwards, increasing emphasis was placed on research involving grass and the dairy cow. In 1950, the potential value of grassland was appreciated fully by relatively few enthusiasts, and its contribution to livestock feeding was low. To augment the available knowledge concerning grassland two main lines of research work were initiated at the Institute and are still progressing. First, the work on grassland production was done mainly on small-plot experiments, but was augmented by further investigation of the more promising results on a field scale. Secondly, grassland utilization studies included both grazing experiments with dairy cows in the summer and feeding experiments with housed cattle in the winter. For many years, the main purpose of these grassland experiments was to explore and develop techniques that would allow the dairy farmer to make maximum use of grassland for milk production.

HERBAGE PRODUCTION

Studies on herbage production in small-plot experiments at the Institute before 1950 (e.g. Holmes & MacLusky, 1955) showed clearly the superiority of grassland over other crops in producing animal feed per hectare. In addition, the results showed that nitrogen and mineral fertilizers could be used to maintain high production. Subsequent plot experiments at the Institute have developed this work in three main aspects:

(1) The effects of cutting management;

(2) The response of herbage yields to nitrogen fertilizer;

(3) The role of legumes in herbage production.

Cutting management

In the early experiments, the cutting management consisted usually of four or five cuts per year, each taken when the herbage reached the 'long leafy' stage of growth. The mower was a reciprocating-blade type, which cut to 6 cm from ground level. To investigate the effects of frequency and closeness of cutting, and of the type of mower on herbage production, further experiments began in 1954.

One of the most striking and consistent effects was the large increase in herbage yield obtained by cutting closer to ground level. Thus the total dry-matter yield from a perennial ryegrass–white clover sward was increased by between 39 and 49% when cut to 2·5 cm from ground level rather than to 6 cm (Reid, 1959). The yield of crude protein and the white clover content of the herbage was also increased by close cutting. Similar but slightly smaller effects were obtained on a timothy–white clover sward (Reid, 1962), but on a cocksfoot–white clover sward (Reid, 1968) close cutting increased yield by only 10% in the first year, and had no effect in later years.

There was no interaction between closeness and frequency of cutting, but on average the yield of dry matter increased when the herbage was cut less often. On the ryegrass sward (Reid, 1959) cutting three to four times at the 'early flower' stage increased yield by 28% compared with cutting five to six times at the 'leafy' stage, and on the cocksfoot sward (Reid, 1968) the increase was 31%.

The effects of various mowers on yield were investigated also. Cutting a perennial ryegrass sward (Reid & MacLusky, 1960) to 2·5 cm from ground level with a cylinder mower gave an 8% greater yield of dry matter than cutting to the same height with a reciprocating-blade mower. An opposite, but slight, effect was noted on a cocksfoot sward (Reid, 1968). A flail mower cutting to 2·5 cm from ground level or to 6 cm on a perennial ryegrass sward gave a 16 to 18% greater yield than a reciprocating-blade mower. Part of the increased yield probably came from the more efficient collection of the cut herbage.

Response to nitrogenous fertilizer

Before 1963, grassland manuring experiments (e. g. Holmes & MacLusky 1955; Reid & Castle, 1965*b*) included only a few rates of nitrogen application covering the range 0 to 350 kg nitrogen/ha. The relationship between herbage yield and rate of nitrogen application was invariably linear with occasional evidence of slight curvilinearity at the highest rate of application. To study this relationship more closely, particularly at the higher rates of nitrogen application, another series of experiments was begun.

In the first experiment (Reid, 1966) thirty-three rates of application varying between 0 and 900 kg nitrogen/ha were applied on a perennial ryegrass sward for one growing season. The results from this experiment and from later ones (Reid, 1970, 1972) agreed closely and can be summarized jointly. The relationship between total yield of both dry matter and crude protein, and amount of nitrogen applied could be described by a model of the form

$$y = a - b \exp(-cx^d)$$

where y is the estimated yield, x the rate of nitrogen application and a, b, c and d are constants. The response of total dry-matter yield was almost linear with the amount of nitrogen applied between 0 and 336 kg/ha but declined with further increases in rate of nitrogen application to reach insignificant values above the 560 kg/ha rate. Between 0 and 336 kg nitrogen/ha the mean response was about 27 kg dry matter/kg nitrogen applied, and this value varied little either between years or between experiments. However, the yield at a particular level of nitrogen application varied slightly between years depending on the environmental conditions. The response of crude-protein yields was linear over a range of nitrogen applications up to 500 to 600 kg/ha.

Many factors influence the pattern of herbage-yield response to nitrogen described above. In certain areas of Britain the most important is moisture supply, but the rainfall at the Institute is relatively high and well distributed over the season. Experiments showed that although irrigation increased herbage yields slightly in the early summer of some years, it had no effect on the response of yield to nitrogen application (Reid & Castle, 1965*a*).

Experiments on the effects of seed mixture on the yield response to the application of nitrogenous fertilizer showed that the presence or absence of white clover was a more important factor than the species or variety of grass. The clover effects are discussed later. In comparisons of grass varieties the response patterns for S23, S24 and Barvestra perennial ryegrass were similar over a range of nitrogen application from 0 to 900 kg/ha whereas the responses for S37 cocksfoot and S53 meadow fescue were poorer than for ryegrass at levels of nitrogen application over 300 kg/ha.

The influence of frequency of defoliation on yield response to nitrogen was investigated on a perennial ryegrass sward when the five-cut system used previously was compared with either a three-cut or a ten-cut system. Three cuts gave slightly greater yields than did five cuts, but with three cuts the response to nitrogen was almost linear up to only 280 kg nitrogen/ha, and then decreased rapidly to insignificant values at 448 kg nitrogen/ha. In contrast, ten cuts gave much smaller yields than did five cuts, but the response was linear up to 500 kg nitrogen/ha, and decreased much less rapidly at the higher levels of nitrogen application. The patterns of response of nitrogen yield to nitrogen application for the three cutting frequencies were almost identical, which suggests that the rate of absorption of nitrogen by the grass is greater than the rate of utilization of nitrogen for protein synthesis.

The frequency of nitrogen application and the amounts applied at each dressing also affect the total yield of herbage (Castle & Reid, 1968). The most important effects, however, were on the pattern of herbage production over the season. More recent experiments have provided further information on these effects, and mathematical models have been constructed.

The importance of adequate dressings of potassium fertilizers to maintain the yield of grass swards receiving nitrogen fertilizer was shown in early work at the Institute (Castle & Holmes, 1960), and was confirmed in wider co-operative experiments (Reith *et al.*, 1961). More recent work has covered a greater range of nitrogen application. Potassium fertilizer had little effect on the yield response to applications of nitrogen varying from 0 to 112 kg/ha but from 112 to 336 kg nitrogen/ha about 0·8 kg of potassium was required for every 1 kg of nitrogen applied. Further increases in the level of nitrogen applied required little or no increase in potassium rate.

In all the above experiments a solid fertilizer was used with ammonium nitrate as the nitrogen source. When applied at equivalent amounts of nitrogen, fertilizers containing ammonium sulphate gave the same yield response (Reid & Castle, 1970*b*), but those containing urea gave poorer responses. Anhydrous ammonia injected as a gas also gave smaller yields and a poorer seasonal distribution of yield (Reid & Castle, 1970*a*). The major disadvantage was the delayed response of herbage yields to ammonia injection in the spring (Reid & Castle, 1970*b*).

Another important, but under-utilized source of nitrogen and potassium is the urine from housed livestock, and the effects of applying this in mixtures with water have been studied. Where equivalent amounts of nitrogen and potassium were applied either as a solid fertilizer or as diluted urine the yields of herbage dry matter were almost identical (Drysdale, 1965). On the farm, slurry contains urine and faeces, and experiments showed that when slurries were applied to grassland the urine was the important constituent (Castle & Drysdale, 1966), since the response of herbage yield decreased as the ratio of faeces to urine increased.

The role of legumes

The most valuable legume for forage production at the Institute is undoubtedly white clover. The yield of the best lucerne variety tested was 13 tonnes of dry matter per hectare in the first harvest year, but only about 6 tonnes/ha in the second and third years (Castle & Reid, 1960). Similar results were noted with red clover (Castle & Watson, 1974), but with a suitable management white clover has remained vigorous for many years. White clover studies at the Institute have been reviewed recently (Reid, 1976), and the findings are summarized below.

White clover can be used as the sole nitrogen source for herbage growth from grassland. Thus the average yield from a perennial ryegrass sward, with neither clover nor applied nitrogen, was 2 tonnes dry matter per hectare annually, which was increased to 5 to 8 tonnes/ha simply by sowing a medium-leaved white clover, S100, with the grass. By replacing the S100 with a large-leaved variety, Blanca, the yield was further increased to 11 to 12 tonnes/ha. To obtain this yield from a pure-grass sward would require the application of 350 to 400 kg nitrogen/ha annually.

Where nitrogen was applied on grass–clover swards, greater yields were obtained than from pure-grass swards at all levels of nitrogen application up to 350 kg/ha with S100, and 450 kg/ha with Blanca. However, the yield response to nitrogen on these swards was about half that on pure-grass swards. Taking account of the relative costs of fertilizer and clover seed, the optimum level of nitrogen application on the grass–clover swards was estimated to be at 200 to 250 kg/ha annually. Predictions from these results suggest that sufficient herbage could be produced to graze five dairy cows per hectare by applying 400 kg nitrogen/ha on a pure perennial ryegrass sward, 320 kg/ha on a ryegrass–S100 sward, or 200 kg/ha on a ryegrass–Blanca sward. These represent reductions in the level of nitrogen application of 20 and 50% with S100 and Blanca respectively. White clover therefore offers a considerable potential for saving nitrogen fertilizer on grassland.

HERBAGE UTILIZATION

Grazing

The importance of the efficient utilization of herbage was stressed by Holmes (1954) who stated that the guiding principles were – grow the grass to the leafy stage, defoliate it rapidly, and then rest the sward for the next growth.

Although research before 1954 had demonstrated clearly that grass was a relatively cheap source of nutrients, and that well-managed pasture was capable of sustaining high-yielding dairy cows, a high proportion of dairy herds was still given supplementary concentrates during the grazing season. An early investigation showed that dietary concentrates did not affect herbage intake if pasture quality was high and that the increase in milk yield was uneconomic (MacLusky, 1955). In later grazing experiments, the concentrates gave some increases in milk yield, but again at an uneconomic level (Castle *et al.*, 1960). The feeding periods in these experiments were only 2 to 3 weeks, but in further work the supplements were given continuously over the whole grazing season. However, feeding 2·1 kg concentrates/10 kg milk during the summer period (Castle *et al.*, 1964) had no significant effect on milk yield. In these experiments, the cows always had access to leafy herbage, and the concentrates in the diet made a relatively small contribution to the total nutrient intake. The effects of feeding supplements where herbage intake was restricted was investigated in two further grazing experiments with various densities of stocking (Castle *et al.*, 1968). As the rate of stocking was increased the response to feeding extra concentrates increased also, but an increase in stocking rate was found to be more profitable (Castle *et al.*, 1968). The importance of stocking rate in achieving a high output of milk per hectare was confirmed in a further experiment comparing diploid and tetraploid varieties of perennial ryegrass at different rates of stocking (Castle & Watson, 1971).

The supplements given in the above experiments consisted mainly of barley. However, because milk yields from cows on high-digestibility silage were increased by giving a high-protein supplement (Castle & Watson, 1976), preliminary studies have started on the effects of feeding a similar supplement to cows on high-quality pasture.

Since 1970, a series of experiments comparing various systems of grazing has been conducted. Over a 2-year period, a system using twenty-eight separate 1-day paddocks gave almost identical outputs of milk per hectare as did the Wye College system in which four paddocks were grazed on a predetermined rotation (Castle & Watson, 1973). The advantages of the Wye College system were: simplicity, ease of layout and operation, and the need for fewer management decisions. Further comparisons confirmed the effectiveness of the Wye College system, and indicated some merits in set stocking, *i.e.* a system in which the cattle are grazed on a single paddock throughout the whole grazing season (Castle & Watson, 1975*a*). Because of its simplicity, set-stocking was investigated for a further 2 years and was found to be as effective as other systems in the first half of the grazing season only. After mid-July, with set-stocking the daily milk production declined as the amount of herbage declined. At this time a larger area of grass must be made available to cows or a supplement must be offered.

The published results show that the output of milk per hectare in experiments has increased over the years. In the 1949 experiments the yield of milk was 4 to 6·5 tonnes/ha. The maximum output was almost 8 tonnes/ha in 1964–65 and 9 tonnes/ha in 1969. Since 1970, yields of 12 to 15 tonnes/ha have been reached. This increase is attributed mainly to more productive swards, greater use of fertilizer nitrogen and increases in the rate of stocking. The factors affecting milk production were broadly confirmed in a study of farm output at the Institute between 1951 and 1970 (Castle *et al.*, 1972). The most important of these were: the rate of stocking, milk yield per cow and the proportion of grass on the farm.

Conservation

The study of conservation crops has been an important part of the research programme at the Institute since pre-war years. In the early experiments, the main emphasis was on artificially dried grass, and the value of this high-quality food as a replacement for concentrates was shown (Reid & Holmes, 1956; Castle *et al.*, 1958). Limited work was conducted on hay, and a feeding experiment showed clearly that higher milk yields could be obtained by early cutting of the hay, and that ryegrass hay was superior to cocksfoot hay (Castle *et al.*, 1962). The differences in milk yield were attributed to differences in the digestibility of the hays and focused attention on this factor. Other experiments included both hay and silage until about 1957 when emphasis was placed on silage for future experimental work. A review of the silage experiments described recent findings and posed problems requiring investigation (Castle, 1975). These silage experiments showed that the major factor influencing milk production was the content of digestible organic matter in the dry matter (D-value) of the silage. A highly significant, positive correlation ($r = +0{\cdot}777$) was found between the D-value of the silage and the mean daily yield of milk. Regression analysis showed that an increase of one unit in the silage D-value gave an increase of 0·23 kg in the mean daily yield of milk.

The main factors influencing the D-value of the silage were the date the herbage was cut, the type of sward, and the use of a silage additive (Castle, 1975). In one experiment, silages made from a tetraploid ryegrass, Reveille, produced significantly higher yields of milk than silages made from a diploid ryegrass, S24. Similar results were obtained when the two ryegrasses were compared in a grazing experiment (Castle & Watson, 1971). In four comparisons with dairy cows, silages made in the spring gave significantly higher yields of milk than silages made in the autumn from the same swards. Five comparisons of silages with and without an additive made at the Institute have been summarized (Castle, 1975). The advantages attributed to the use of 2·2 litres of 85% formic acid per tonne were listed as follows: (1) silage temperature was reduced by 8°C, (2) pH was lowered by 0·7 units, and (3) D-value was increased by 3·1 units. As a result, silage intake was higher with the treated than with the untreated silages, and hence milk yield was increased significantly. It was stressed that the use of additives was only one factor in the process of making satisfactory silage and could not rectify a deficiency in another factor.

In further experiments, nine silages made from tetraploid red clover were investigated (Castle & Watson, 1974), since high yields of organic matter and crude protein can be obtained from this crop without the use of nitrogen fertilizer. The dry-matter content of red clover was only 12·5% when cut, and wilting for 1 to 2 days was vital to obtain a satisfactory silage fermentation. An application of 2·2 litres formic acid/tonne was considered essential. The red-clover silages had a mean D-value of 52, but the dry-matter intakes of cows fed on it were high at 10 to 11 kg/cow per day. This supports the evidence that the intake of legumes is 20 to 40% higher than that of grasses of the same digestibility. Although the crude protein content of the red-clover silage was high at 19·6% the milk yield of the cows was increased significantly by increasing the crude-protein content of the supplement.

The object of recent investigations was the making of a highly digestible silage that was acceptable in large amounts to dairy cows. Grass silages with D-values of 68 to 70 can be made successfully on a farm scale in ordinary clamp silos using normal farm equipment and good conventional techniques. To obtain this high D-value, early cutting is essential but precision chopping, the use of an additive, and wilting, are also important (Castle & Watson, 1976). In three experiments with dairy cows receiving silage only, the mean daily intake of silage with D-values from 68·4 to 70·4 was 11·2 kg/cow. This was equivalent to 2·34% of the live weight, and the mean daily milk yield was 14·6 kg/cow (Castle *et al.*, 1977). Adequate supplementary feeding is obviously required particularly in early lactation.

Suitable supplements for silage of high D-value have been investigated recently. A supplement of high D-value dried grass was superior to one of barley, and a ration of high quality silage and dried grass supported daily yields of 20·2 kg milk/cow. Barley was not an ideal supplement as it depressed the intake of silage and gave little increase in milk production (Castle & Watson, 1975*b*). In another experiment with high D-value silage, a supplement of groundnut cake did not depress the intake of silage and was a better supplement than barley (Castle & Watson, 1976). The value of groundnut was confirmed in a further experiment which indicated that the optimal feeding rate was about 1·4 kg/10 kg milk (Castle *et al.*, 1977). The effect of protein in increasing milk yield agrees with earlier work on both grass silage and red-clover silage, and indicates a deficiency in silage not previously recognized.

CONCLUSIONS

There has been a growing awareness of the value of grass in the feeding of the dairy cow over many years, and research work at the Institute has contributed significantly to this. For example, research work on higher levels of fertilizer nitrogen to increase herbage production has always been at the forefront of the programme, and there is now a slow but steady increase in the mean levels of nitrogen application to grassland on dairy farms. Intensive systems of grazing are now practised on one dairy farm in two in Scotland, and our research programme on grazing systems has helped to supply much of the basic and practical knowledge required on this subject by the grassland farmer. Similarly, work at the Institute on grassland conservation has demonstrated the importance of high digestibility silage made with a minimum of loss, and has contributed considerably to the knowledge of this topic. Throughout the years, our grassland research programme has kept in close touch with both the economic realities of grassland husbandry at a practical level, and with fundamental investigation at the research stage. With this blend of interests, many important advances have been made in a subject of vital consequence to the dairy farmer and the nation.

REFERENCES

Castle, M. E. (1975) *Agric. Prog.* **50**, 53–60
Castle, M. E. & Drysdale, A. D. (1966) *J. Agric. Sci.* **67**, 397–404
Castle, M. E. & Holmes, W. (1960) *J. Agric. Sci.* **55**, 251–260
Castle, M. E. & Reid, D. (1960) *J. Br. Grassl. Soc.* **15**, 281–286
Castle, M. E. & Reid, D. (1968) *J. Agric. Sci.* **70**, 383–389
Castle, M. E. & Watson, J. N. (1971) *J. Agric. Sci.* **77**, 69–76
Castle, M. E. & Watson, J. N. (1973) *J. Br. Grassl. Soc.* **28**, 7–11
Castle, M. E. & Watson, J. N. (1974) *J. Br. Grassl. Soc.* **29**, 101–108
Castle, M. E. & Watson, J. N. (1975*a*) *J. Br. Grassl. Soc.* **30**, 1–6
Castle, M. E. & Watson, J. N. (1975*b*) *J. Br. Grassl. Soc.* **30**, 217–222
Castle, M. E. & Watson, J. N. (1976) *J. Br. Grassl. Soc.* **31**, 191–195
Castle, M. E., MacLusky, D. S., Waite, R. & Watson, J. N. (1958) *J. Dairy Res.* **25**, 365–372
Castle, M. E., Drysdale, A. D. & Watson, J. N. (1960) *J. Dairy Res.* **27**, 419–426
Castle, M. E., Drysdale, A. D. & Watson, J. N. (1962) *J. Dairy Res.* **29**, 199–206
Castle, M. E., Drysdale, A. D. & Watson, J. N. (1964) *J. Br. Grassl. Soc.* **19**, 381–386
Castle, M. E., Drysdale, A. D. & Watson, J. N. (1968) *J. Br. Grassl. Soc.* **23**, 137–143
Castle, M. E., MacDaid, E. & Watson, J. N. (1972) *J. Br. Grassl. Soc.* **27**, 87–92
Castle, M. E., Retter, W. C., Watson, J. N. & Zewdie, E. (1977) *J. Br. Grassl. Soc.* **32**, 43–48
Drysdale, A. D. (1965) *J. Agric. Sci.* **65**, 333–340
Holmes, W. (1954) *Proc. Nutr. Soc.* **13**, 19–22
Holmes, W. & MacLusky, D. S. (1955) *J. Agric. Sci.* **46**, 267–286
MacLusky, D. S. (1955) *Proc. Br. Soc. Anim. Prod.*, 45–51
Reid, D. (1959) *J. Agric. Sci.* **53**, 299–312
Reid, D. (1962) *J. Agric. Sci.* **59**, 359–368
Reid, D. (1966) *Proc. 10th Int. Grassl. Congr.*, pp. 209–213
Reid, D. (1968) *J. Agric. Sci.* **70**, 59–64
Reid, D. (1970) *J. Agric. Sci.* **74**, 227–240
Reid, D. (1972) *J. Agric. Sci.* **79**, 291–301
Reid, D. (1976) *Hannah Res. Inst. Rep.*, 66–69
Reid, D. & Castle, M. E. (1965*a*) *J. Agric. Sci.* **64**, 185–194
Reid, D. & Castle, M. E. (1965*b*) *J. Agric. Sci.* **65**, 109–119
Reid, D. & Castle, M. E. (1970*a*) *J. Agric. Sci.* **75**, 347–353
Reid, D. & Castle, M. E. (1970*b*) *J. Agric. Sci.* **75**, 523–532
Reid, D. & Holmes, W. (1956) *J. Dairy Res.* **23**, 159–168
Reid, D. & MacLusky, D. S. (1960) *J. Agric. Sci.* **54**, 158–165
Reith, J. W. S., Inkson, R. H. E., Stewart, A. B., Holmes, W., MacLusky, D. S., Reid, D., Heddle, R. G., Clouston, D. & Copeman, G. J. F. (1961) *J. Agric. Sci.* **56**, 17–29

STRUCTURAL COMPONENTS OF GRASSES

I. M. MORRISON

DURING the 1940s, much of the effort in the Crop and Stock Husbandry Department of the Hannah Research Institute was aimed at improving the yield and nutritive value of the grass grown on the farm. It was realized that generally the yield of grass from a given area was inversely related to its nutritional value but that a fuller knowledge of the chemical composition was required for the reliable assessment of nutritive value. The idea of forage analysis was not new. Henneberg and Stohmann (1860/65) had made their research institute at Weende famous by their pioneering work on the subject. Their system of analysis, however, did not determine discrete chemical components. For example, the fraction referred to as 'crude fibre' was that portion of the forage which was insoluble in dilute acid and dilute alkali and the fraction termed 'nitrogen-free extractives' was estimated by difference. Although the Weende system was an important start to plant analysis, more precise chemical methods for the measurement of plant components were required.

At the University of Edinburgh, in the late 1940s and early 1950s, a large group of research workers was being formed under the late Sir Edmund Hirst, to study various aspects of carbohydrate chemistry. One of the major interests of the group was plant carbohydrates, with strong emphasis being placed on the hemicelluloses. The group's interests were mainly in the isolation, characterization and structural composition of these polysaccharides. The progress made by this group and the recognition of the practical importance of a better knowledge of the structural changes that take place when herbage matures stimulated Dr R. Waite and his colleagues to develop improved methods of herbage analysis. Emphasis was given to the various carbohydrate components as they are the major source of energy to the ruminant. The early part of the programme dealt with the quantities of soluble carbohydrates present in grasses (Waite & Boyd, 1953) but this essay will consider only the later work involving the structural components of the cell wall. In this, one must consider not only the structural carbohydrates but also the effect of increasing lignification on the utilization of these structural carbohydrates.

The initial aim of the programme was achieved with the publication by Waite & Gorrod (1959*a*, *b*) of two papers, the first on the structural carbohydrates of grasses and the second on the comprehensive analysis of grasses. This second paper showed that, by using sequential extraction procedures involving organic and aqueous solvents, 97 to 98 % of the grass dry matter could be accounted for. This was a notable achievement since earlier attempts by other workers had accounted for not more than 81% of the dry matter. The various components measured by the Waite and Gorrod scheme were fats, waxes, organic acids, water-soluble carbohydrates, phenolic compounds, crude protein, pectin, hemicellulose, cellulose, lignin, acetyl groups and ash. This formidable list of constituents meant that the method, as a whole, was too laborious to be used as a routine procedure but was ideal as a guide to other methods.

In particular it was the approach to the analysis of the structural carbohydrates that made this method so much superior to other methods. Up to that time, a reasonably accurate method for cellulose determination was available and there were a number of methods for the determination of lignin. The other structural carbohydrates could then be measured by difference. The Waite and Gorrod procedure was able to measure pectin and lignin directly, and, following hydrolysis of the holocellulose, both cellulose and hemicellulose could be determined. Cellulose was calculated from the glucose present in the hydrolysate whereas hemicellulose was the sum of arabinose, xylose, galactose and uronic acid. This method also indicated the composition of the hemicellulosic fraction. The separation of these sugars involved the use of paper chromatography which was a recently developed technique in sugar analysis at that time. Waite & Gorrod (1959*a*) showed how the analytical scheme for the structural components of the cell wall could be applied to different varieties of grasses of widely different maturity. It showed the steady increase in lignin content of all varieties with increasing maturity. The cellulose and total hemicellulose contents increased also, hemicellulose increasing at a faster rate than cellulose, but in the hemicellulose fraction only the proportion of xylose residues present seemed to increase with increasing maturity.

Although the procedure was too time-consuming to be used *in toto* in routine analyses, its usefulness in the understanding of how the plant cell wall was degraded in the gastrointestinal tract of the ruminant became evident in the evaluation of artificially dried grasses as a source of energy for sheep (Armstrong *et al.*, 1964; Waite *et al.*, 1964). This collaborative programme between the Nutrition and Technical Chemistry Departments was concerned with the determination by calorimetric methods of the nutritive value of artificially dried grasses. By involving the method for grass analysis described above, the composition of the grass was established. By applying the method to faecal samples obtained from digestibility trials, the indigestible components were estimated. The important conclusions to this work are best described from actual results obtained. When the digestibility of the organic matter of S23 ryegrass decreased from 86 to 62 units, i.e. in proceeding from a young leafy state to a mature stemmy state, the apparent digestibility coefficients of the cellulose, hemicellulose and lignin fractions decreased from 92, 93 and 23 to 73, 56 and 0 respectively. In other words, the digestibility of cellulose had decreased by about 25 %, that of the hemicellulose by about 50 % while the digestibility of lignin was negligible (there is some doubt whether lignin from immature plant tissue is partially digestible or not). During this period of plant growth, there is no change in the chemical structure of the cellulose and the changes occurring in the structure of the hemicelluloses are ones which make the polysaccharide chains less branched and which therefore ought to make them more easily degraded. As lignin is the only component of the cell wall which shows a marked increase in concentration during this period, it was thought that lignin, itself indigestible, might also influence the digestibility of other cell wall components.

The high negative correlation between lignin content and digestibility obtained by Waite *et al.* (1964) had been noted before (Phillips & Loughlin, 1949) and there was much speculation about the mechanism of this effect. The two points of view, which do not necessarily conflict, are that lignin protects the cell wall carbohydrates from bacterial attack in the rumen merely by its physical presence, or that lignin and some of the cell wall carbohydrates may be combined chemically. The first theory considers lignin as a coating agent forming a protective crust on the cell wall carbohydrates which would otherwise be available for use by the ruminant as an energy source. The second considers that actual chemical bonds exist between lignin and cell wall carbohydrates. These bonds would be stable to cleavage by enzymes in the rumen and they may also stabilize neighbouring bonds in the carbohydrate chains which otherwise might be cleaved by micro-organisms.

Evidence for the first theory is based on a combination of histological results and chemical analysis (Johnston & Waite, 1965). The histology of both leaf and stem tissue of grasses had been well documented (Esau, 1960) but there was little known about changes occurring with increasing maturity. To assess these changes, a series of ryegrass and cocksfoot samples of increasing maturity was analysed for lignin content and digestibility *in vitro*. At the same time, sections of tissue were stained with a lignin-specific stain to determine the position of lignin in the cell and the intensity of stain produced. There was little fibrous tissue in either leaf or leaf sheath and what was there was associated with the vascular bundles. Lignification of such tissue increased only slightly with increasing maturity. The thickened cells of the pericycle of the stem constituted the major lignified region and lignification here increased up to anthesis. After this, the larger cells of the tissue connecting the vascular bundles also became lignified. This work also confirmed the strong inverse relationship between lignin content and digestibility *in vitro*.

From a knowledge of the distribution of lignin in the tissue along with the digestibility of such tissue, it was clear that lignin did form a coat on the cell wall carbohydrates. As such a coat is indigestible, the carbohydrates would be protected from enzymic attack in the rumen.

The possiblity of lignin being bonded covalently to carbohydrates in the cell wall of grasses was investigated next. There were reports, exclusively from wood research, that lignin-carbohydrate bonds were present in a number of woods (Merewether, 1957). Lignins from grasses, however, have compositions and structural units different from those in wood lignins and there was no guarantee that lignin-carbohydrate bonds were present in the cell walls of members of the *Gramineae* family.

It was shown that lignin-carbohydrate complexes could be extracted by a variety of solvents from grass cell walls (Morrison, 1973*a*). The yields of the complexes were not high and considerable quantities of lignin were retained in the unextracted carbohydrate-rich residue. Each solvent examined gave a different yield of product and the most successful were dimethylsulphoxide, which afforded a lignin-carbohydrate complex, and dilute alkali, which gave a lignin-hemicellulose complex. More important than the extractant, however, was the fact that the grass samples had to be ground very finely in a ball mill before the neutral aprotic solvents had much effect.

Further analysis of leaf blade, leaf sheath and stem tissue from four varieties of grass using the same solvents (Morrison, 1974*a*) showed that lignin-carbohydrate and lignin-hemicellulose complexes could be obtained from each type of tissue. From the analyses, the leaf sheath and stem tissue gave complexes very similar to each other in terms of total yield, lignin content, nitrogen (hence protein) content and sugar composition. The leaf blade complexes were very different when the results were considered quantitatively. Further differences were noted when the full ultraviolet spectra of the complexes were determined. As well as lignin being present, these spectra showed considerable absorption, particularly with the material obtained from leaf sheath and stem tissue, in the area where ester bonds are known to absorb. Another feature of the lignin-carbohydrate complexes was the large proportion of glucose residues present. This, and the question of ester bonds, required further examination.

Larger quantities of the lignin-carbohydrate complexes from each type of S23 ryegrass tissue were prepared (Morrison, 1974*b*) and fractionation experiments indicated that the complexes were not homogeneous. However, heterogeneity seemed to be confined to the ratio of the individual sugar residues. Lignin was always present along with the carbohydrate. The complexes had high molecular weights ($>150\,000$) and although chemical treatment such as alkaline extraction and borohydride reduction altered the amount of lignin-like material present, the molecular weight of the complexes remained high. Low molecular weight fragments released from the complexes with alkali included acetic, ferulic, and *p*-coumaric acids. Other phenolic acids of higher molecular weight were present also. Without being able to define the actual chemical bonds in the complexes, it was postulated from the stability under different chemical conditions that at least three types were present. They were defined as bonds cleaved by borohydride reduction, bonds stable to borohydride reduction but labile under alkaline conditions and bonds resistant to both borohydride reduction and alkaline extraction. The carbohydrate portion of the complexes was composed of β-(1→4)-linked D-glucose residues (cellulose) and β-(1→4)-linked D-xylose residues to which side chains of arabinose and galactose residues were linked at C-3 of some of the xylose residues. How the components of the complexes were held together was not established but it was postulated that the phenolic acids, being bifunctional units, might act as cross-linking agents.

An interesting side to this work arose from the need for a small-scale method to measure the lignin content of these lignin-carbohydrate complexes. Most lignin methods in use were gravimetric procedures and since the yields obtained with some of the complexes were very small, some other method was required. The method developed was a spectrophotometric one whereby the plant cell wall was dissolved in acetyl bromide solution and the absorption at 280 nm (the wavelength of maximum absorption of the phenolic groups) was taken as a measure of the lignin content; such a method was very sensitive and could be used with very small amounts of sample. Proteins, which are present in most preparations of plant cell walls and also absorb at 280 nm, were insoluble in the reagent and caused no interference. One drawback to the method was that no pure grass lignin (i.e. uncontaminated with other cell wall components) could be obtained and therefore the absorbance had to be related to another method for determining lignin (Morrison, 1972*a*).

As has already been stated, lignin shows a high negative correlation with digestibility and it was found that this new lignin method could be used to predict the digestibility not only of grasses but also of legumes (Morrison, 1972*b*). It has many advantages over other methods for predicting digestigility. It is very rapid in that results can be obtained within one working day. Many samples can be analysed in a specific time without requiring any special equipment. It requires only milligrams of plant tissue and there is no need for animals to provide rumen liquor. Over the past 10 years there has been considerable interest in laboratory methods for predicting digestibilty and this method must be considered as one of the most promising.

The term 'hemicellulose' was first used to describe the polysaccharides in the plant cell wall that were similar to cellulose but could be extracted under alkaline conditions. Much has been discovered about the hemicelluloses and it is well established that those of the *Gramineae* are based mainly on β-(1→4)-linked D-xylose residues. Although the hemicellulose of a particular plant species is not a single polysaccharide, like cellulose, it had been assumed that there was always a major component to the hemicellulosic complex. Recent work on other *Gramineae* cell walls (Wilkie, 1977) has led to the conclusion that the hemicellulosic complex contains an arabinoxylan which has relatively few side chains, a galactoarabinoxylan which is more highly branched, and a glucan which contains both β-(1→3) and β-(1→4) bonds.

The presence of arabinoxylan and galactoarabinoxylan in S23 ryegrass was confirmed and the presence of a glucan which contained both β-(1→3) and β-(1→4) linkages was implied although it could not be obtained in pure form (Morrison, 1974*c*), The arabinoxylan concentration decreased very slightly during growth as a proportion of the total hemicellulose but the ratio of xylose to arabinose in the polysaccharide rose considerably with increasing maturity. The galactoarabinoxylan did not alter greatly in composition during ageing but as a percentage of the total hemicellulose it increased. Little can be said about the glucan because of the failure to purify it, but as a proportion of the total hemicellulose complex it decreased markedly with increasing maturity.

One of the major questions which remains unsolved in cell wall analysis is whether protein is an integral component of the secondary cell wall. To isolate the polysaccharides in the above study, it was necessary to have as little protein present as possible. Use of the proteolytic enzyme complex, pronase, gave cell wall preparations which contained less protein than any of the other methods tried (Morrison, 1973*b*) and since it is used at neutral pH, there was no effect on any other cell wall component.

To isolate pure hemicellulose from plant tissue, it is also necessary to delignify the tissue. Most work on delignification has come from wood chemistry and since woods have two to three times the lignin content of even the most mature grass, the methods used for delignification, particularly with regard to length of reaction and strength of reagents, may not be applicable. It is known, again from wood analyses, that delignification over an extended period causes dissolution of hemicelluloses. Before isolating the total hemicelluloses in the above study, the rate of delignification at each stage of maturity was determined so that the optimum conditions could be used (Morrison, 1975*a*). From these results, it was clear that the time of the delignification reaction was critical and that very short reaction times were required, particularly for immature herbage.

Since the purpose of this research programme is to improve the utilization of plant cell wall carbohydrates by ruminants, further studies were initiated on how the plant cell wall is degraded *in vivo*. It is well known that ruminants can degrade plant cell wall carbohydrates, and Dehority (1973), with co-workers, has isolated various rumen micro-organisms that are able to degrade these polysaccharides. It is still not known how the walls are degraded and so how one can improve their availability to rumen micro-organisms. Our studies at the Institute have started with what we consider to be the simplest situation, namely, how are the pure hemicelluloses degraded in the rumen? Our aim is to build up these studies to see how lignin-hemicellulose complexes, then lignin-carbohydrate complexes and finally cell walls themselves are broken down. These studies of a chemical and biochemical nature are supplemented by microbiological studies on the production of cell wall degrading enzymes from rumen micro-organisms.

Pure hemicelluloses are too large to be transported across the bacterial membrane but they are degraded by enzymes present in cell-free rumen fluid (Morrison, 1976). The levels of enzyme activity in the rumen are dependent on the time of sampling during the feeding cycle. They possess an optimum pH and temperature but the active ranges of both pH and temperature are quite wide since a mixture of enzymes is probably present. Many reagents, particularly heavy metal ions and phenols, inhibit the activity but reducing agents, providing an environment more like that of the rumen, seem to enhance the activity. The enzyme complex was not stable because proteolytic enzymes were present also, but purification did give more stable products.

When the rate of degradation was examined (Morrison, 1975*b*), the rate curve was consistent with only one type of enzyme being present. The products of degradation were a series of oligosaccharides with degrees of polymerization mostly from two to seven. There were no free monosaccharides produced by the enzymic complex. From analysis of the oligosaccharides by a new method of molecular size and end-group determination (Morrison, 1975*c*) it was concluded that the enzymes are endo-enzymes attacking the main chain in a fairly random manner but only doing so when two unsubstituted xylose residues are adjacent to each other. The oligosaccharides so produced would then be small enough to be transported across the bacterial membrane of either the xylan-degrading organism or any other organism. These cell-free enzymes are capable of degrading not only pure hemicelluloses but also lignin-hemicellulose and lignin-carbohydrate complexes. They do not attack intact cell walls or delignified cell walls.

Over the years that investigations into the structural components of grasses have been made, it can be claimed that considerable progress has taken place. However, our final aim of improving the availability of these carbohydrates for the ruminant is a continuing role and it will take some time yet to solve all the problems involved.

REFERENCES

Armstrong, D. G., Blaxter, K. L. & Waite, R. (1964) *J. Agric. Sci.* **62**, 417–424

Dehority, B. A. (1973) *Fed. Proc.* **32**, 1819–1825

Esau, K. (1960) *Plant Anatomy*, Wiley, New York

Henneberg, W. & Stohmann, F. (1860/65) *Beiträge zur Bergründung einer rationellen Fütterung der Wiederkäuer*, 2 vols, Schwetschke, Braunschweig

Johnston, M. J. & Waite, R. (1965) *J. Agric. Sci.* **64**, 211–219

Merewether, J. W. T. (1957) *Holzforschung* **11**, 65–80

Morrison, I. M. (1972*a*) *J. Sci. Food Agric.* **23**, 455–463

Morrison, I. M. (1972*b*) *J. Sci. Food Agric.* **23**, 1463–1469

Morrison, I. M. (1973*a*) *Phytochemistry* (*Oxf.*) **12**, 2979–2984

Morrison, I. M. (1973*b*) *J. Agric. Sci.* **80**, 407–410

Morrison, I. M. (1974*a*) *Phytochemistry* (*Oxf.*) **13**, 1161–1165

Morrison, I. M. (1974*b*) *Biochem. J.* **139**, 197–204

Morrison, I. M. (1974*c*) *Carbohydr. Res.* **36**, 45–51

Morrison, I. M. (1975*a*) *Phytochemistry* (*Oxf.*) **14**, 505–508

Morrison, I. M. (1975*b*) *Biochem. Soc. Trans.* **3**, 992–994

Morrison, I. M. (1975*c*) *J. Chromatogr.* **108**, 361–364

Morrison, I. M. (1976) *Carbohydr. Res.* **47**, 129–139

Phillips, T. G. & Loughlin, M. E. (1949) *J. Agric. Res.* **78**, 389–395

Waite, R. & Boyd, J. (1953) *J. Sci. Food Agric.* **4**, 197–204

Waite, R. & Gorrod, A. R. N. (1959*a*) *J. Sci. Food Agric.* **10**, 308–317

Waite, R. & Gorrod, A. R. N. (1959*b*) *J. Sci. Food Agric.* **10**, 317–326

Waite, R., Johnston, M. J. & Armstrong, D. G. (1964) *J. Agric. Sci.* **62**, 391–398

Wilkie, K. C. B. (1977) *Adv. Carbohydr. Chem. Biochem.* in press

TRANSFER OF METABOLIC HYDROGEN IN THE RUMEN

J. W. Czerkawski

INTRODUCTION

Rumen fermentation has been studied at the Hannah Research Institute for over 30 years. Starting with pioneering work on the metabolism of urea and synthesis of microbial protein (Pearson & Smith, 1943), research continued for a number of years, showing the importance of microbial matter in the rumen (McNaught *et al.*, 1950*a*, *b*). The experimental procedures used were simple but even in those days it became apparent that particular aspects of rumen fermentation could not be studied in isolation. Although J. A. B. Smith and his co-workers were interested mainly in nitrogen metabolism in the rumen, they soon realized that energy had to be supplied and that the extent of protein synthesis depended on the nature and amount of carbohydrate present. These studies were comprehensive, and measured the production of volatile fatty acids and even gases (McNaught, 1951).

For about 5 years, little or no work on rumen fermentation was done at the Institute, but interest was renewed when various workers began to consider the contribution of end-products of fermentation to the energy metabolism of ruminant animals. The products that received most attention were the volatile fatty acids (Armstrong & Blaxter, 1957) but in general, the rumen was treated as a 'black box' where the inputs and outputs were emphasized rather than what was going on inside it.

The work on methane (see reviews: Blaxter & Czerkawski, 1966; Czerkawski, 1969) began in 1964 with a reasonable, but naïve premise that since some micro-organisms in the rumen use hydrogen to hydrogenate unsaturated fatty acids, and other micro-organisms use hydrogen to form methane from carbon dioxide, the addition of unsaturated fatty acids to the diet should result in competition for available hydrogen and in inhibition of methane production. We found that methane production in sheep could be inhibited in this manner (Czerkawski *et al.*, 1966*a*) and we began to be interested in reactions within the rumen (Czerkawski, 1966), but the emphasis was still on energy metabolism as a whole. Nevertheless, it was a step in the right direction, because like Smith and co-workers some 20 years before, we considered more than one process at the same time. In addition we began to appreciate the importance of metabolic hydrogen in the rumen.

Although methanogenesis remained the central topic during subsequent research, the scope increased considerably. An artificial rumen was developed (Czerkawski & Breckenridge, 1969*a*; Czerkawski, 1976*c*) and was used successfully to study the reactions involved in degradations of various substrates, such as carbohydrates (Czerkawski & Breckenridge, 1969*c*) and simpler compounds (Czerkawski & Breckenridge, 1972). It was possible to trace out metabolic pathways (e.g. Czerkawski & Breckenridge, 1973) and to define conditions for work *in vitro* (Clapperton & Czerkawski, 1969; Czerkawski, 1973*a*; Czerkawski *et al.*, 1975). We have learned a great deal during this consolidating period and we are convinced that in spite of its complexity, the microbial system of the rumen must be studied as a whole. To make the task easier there should be a search for unifying principles. It appears that the production, transfer and utilization of metabolic hydrogen might be such a unifying principle. It might help in unravelling various interrelationships, and since the transfer of hydrogen is intimately connected with energy metabolism in the rumen, it may be pertinent to an intelligent manipulation of rumen fermentation (see Czerkawski, 1973*b*).

During the past few years we have tried to study rumen fermentation with particular emphasis on the involvement of metabolic hydrogen (Czerkawski, 1973*a*, *b*) and this is the subject of the present essay.

MICROBIAL FERMENTATION IN THE RUMEN: GENERATION OF REDUCING POWER

The ruminant animal consumes a predominantly fibrous diet that could not be digested without the intervention of the micro-organisms in the rumen. These micro-organisms must break down the food in order to grow and multiply; to be able to do this, they must carry out three processes. They must break down complex substances in food to small units; they must utilize these small units for the synthesis of their own cell material; and they must convert some of the substrate into end-products in such a way that useful energy is released. The main constituents of the ruminant diet are carbohydrates and it is the breakdown of these compounds that provides the energy and a large proportion of building material for growth. The ruminant diet also contains other components, such as proteins, lipids and nucleic acids. These too, are broken down and utilized by micro-organisms, partly to obtain energy, but mainly as nutrients necessary for growth.

There are about 200 known species of bacteria and fifteen species of protozoa in the rumen, with a density of 10^{10} bacteria and 10^6 protozoa/ml. Some bacteria in isolated cultures are very sensitive to even small traces of oxygen, whereas others not only tolerate oxygen, but can actually use it. There are two main groups of protozoa in the rumen – the holotrichs, which are mobile and use relatively simple sugars that may be converted to volatile fatty acids or may be stored in the form of polysaccharides, and the entodinia, which are usually more exacting in their nutritional requirements, ingesting starch and sometimes bacteria.

The rumen micro-organisms are often considered in the light of the particular food that they utilize. Some groups of micro-organisms attack cellulose; others utilize starch or maltose. There are micro-organisms that attack hemicellulose specifically whereas others must be provided with simple sugars. There are also some specialized micro-organisms in the rumen that can use the non-carbohydrate components of the diet; usually these micro-organisms have specific energy-generating mechanisms. Some micro-organisms, such as methanogens, rely on the end-products of metabolism of other micro-organisms.

Some examples of micro-organisms, their substrates and the end-products of their metabolism are given in Table 1. Several of the products listed (e.g. gaseous hydrogen and formic, lactic and succinic acids) do not normally accumulate in the rumen or are found in very small amounts. Formic acid and hydrogen can be used by methanogenic bacteria and there are micro-organisms that can degrade lactate instead of simple sugars. Even the volatile fatty acids (acetic, propionic and butyric) cannot be treated as true end-products of fermentation because there is inter-conversion, particularly between acetic and butyric acids.

A great deal of proteolysis occurs in the rumen and the resulting amino acids are broken down further to ammonia and various carbon compounds, including pyruvic acid. Other typical end-products of fermentation that are not given in Table 1 are branched chain fatty acids such as isobutyric and isovaleric acids, which are derived normally from the metabolism of branched chain amino acids. The small amounts of branched chain acids that occur in the rumen may be of considerable metabolic importance, since isobutyric and isovaleric acids are growth factors for cellulolytic bacteria and 2-methylbutyric acid has been reported as essential for methanogenic bacteria.

It can be seen from the simplified description above that the microbial system of the rumen is complex and that there is much interdependence between various microbial populations. We are helped in the studies of such a complex system in several ways. In general, the energy content of substances diminishes when they are progressively broken down and therefore the number of steps in a given substrate chain must be limited. Furthermore, many micro-organisms show a considerable substrate specificity and this will limit the substrate chain. We know much about the metabolic pathways involved and fortunately many of these pathways are common to the majority of micro-organisms, even though the substrates and the end-products may be different. The fermentation schemes that,

TABLE 1

Some substrates, micro-organisms and end-products of fermentation in the rumen

Substrate	Micro-organism	Acids						Gases		
		Acetic	Propionic	Butyric	Formic	Lactic	Succinic	H_2	CO_2	CH_4
Cellulose	*Bacteroides succinogenes*	+	...	...	+	...	+	...	...	...
	Ruminococcus species	+	...	...	+	+	+	+	...	...
Starch (or maltose)	*Bacteroides amylophilus*	+	..	...	...	+	+	...	...	...
	Succinomonas amylolytica	+	—	...	...	...	+	...	...	...
	Entodinium	+	—	+	+	...	...	+	+	...
Pectin (and hemicellulose)	*Lachnospira multiparus*	+	...	...	+	+	...	+	+	...
Simple sugars (or lactate)	*Eubacterium ruminantium*	+	...	+	+	...	...	+	+	...
	Peptostreptococcus elsdenii	+	+	+	...	...	...	+	+	...
	Holotrichs	+	+	+	...	+	...	+	+	...
CO_2+H_2 (or formate)	*Methanobacterium ruminantium*	...	...	...	...	...	...	...	...	+

we believe, account for most of the carbohydrate breakdown in the rumen have been published (see e.g. Czerkawski, 1969).

The reactions involved in the conversion of glucose to pyruvate, the associated changes in reducing power and the ATP yields are summarized in the first part of Table 2. In the scheme 1 mol of glyceraldehyde monophosphate gains another phosphate group and is oxidized to 1 mol diphosphoglyceric acid; at the same time 1 mol of NADH is formed from NAD. Thus indirectly, the *reduction* of NAD results in production of ATP, which on hydrolysis would release free energy of about 34 kJ/mol. On the other hand, under aerobic conditions, with flavo-proteins and cytochromes, the *oxidation* of 1 mol of NADH may result in the production of 3 mol ATP. Thus the energy exchanges that are associated with the transfer of reducing power depend very much on the reactions involved.

It can be seen in Table 2 that pyruvate is converted to acetyl CoA, which by phosphoroclastic reaction forms acetyl phosphate (a high energy compound). The hydrolysis of this compound to acetate results in the formation of 1 mol ATP/mol substrate. The overall change in free energy during the conversion of pyruvate to acetate and ATP is −21 kJ/mol substrate, but if the hydrogen gas could be used to reduce a nucleotide coenzyme, then the overall change in free energy would be −63 kJ/mol, which might be sufficient for the production of more ATP. Acetate is the main end-product of fermentation in the rumen and its formation provides most reducing power and yet we are not quite certain what is the correct energy yield from this reaction.

Propionate can be formed from pyruvate by the randomizing pathway involving succinate as an intermediate or by a direct reductive pathway involving the intermediary production of acrylate. If the former reaction terminates in succinate, as happens in some micro-organisms (Hopgood & Walker, 1967*a*, *b*), methylmalonyl-CoA is not involved and the starting compound is phosphoenol pyruvic acid. In the conversion of phosphoenol pyruvic acid to oxaloacetate and pyrophosphate one potential mol of ATP is lost but pyrophosphate is another high energy compound and it would be interesting to know whether useful energy may be obtained from this reaction. The conversion of fumarate to succinate is associated with a large change in free energy ($\triangle G = -84$ kJ/mol substrate or $\triangle G = -46$ kJ/mol substrate if the reaction involves a flavo-protein intermediate). This change in free energy would be sufficient to generate 1 mol ATP. If the conversion of methylmalonyl-CoA to propionyl-CoA is linked with the phosphorylation of ADP, more ATP may be generated. The randomizing pathway is interesting for two reasons – (a) isomerization involves a temporary cleavage of the C–C bond, which may explain some of the inconsistencies in the pathways of one-carbon compounds (see later), and

TABLE 2

Summary of reactions involved in the conversion of glucose to acetate, propionate and pyruvate and the amount of reducing power generated or utilized, the changes in free energy (ΔG) *and the ATP yields associated with each reaction*

ATP yields given in parenthesis are possible values that may include ATP production linked to electron transport and other sources

Reaction	Source of reducing power	Amount of reducing power generated or utilized (mol)	ΔG (kJ)	ATP (mol)
½ glucose + ½ ATP → ½ glucose-6-P + ½ ADP	...	...	−10	−0·5
½ glucose-6-P → ½ fructose-6-P	...	...	+1	...
½ fructose-6-P + ½ ATP → ½ fructose-1,6-diP + ½ ADP	...	...	−9	−0·5
½ fructose-1,6-diP → glyceraldehyde-3-P	...	...	+16	...
glyceraldehyde-3-P + ADP + NAD → 3-P-glycerate + NADH + ATP	NADH	1·0	−22	1·0
3-P-glycerate → phosphoenolpyruvate	...	...	+2	...
phosphoenolpyruvate + ADP → pyruvate + ATP	...	...	−24	1·0
Overall: ½ glucose → pyruvate	Total	1·0	−47	1·0
pyruvate + CoA → acetyl-CoA + CO_2 + H_2	H_2	1·0	−19	...
acetyl-CoA + P → acetyl-P + CoA	...	...	+11	...
acetyl-P + ADP → acetate + ATP	...	...	−13	1·0
Overall: pyruvate → acetate + CO_2 (if H_2 is used to reduce FAD)	Total	1·0	−21 (−63)	1·0 (2·0)
pyruvate + NADH → lactate + NAD	NADH	−1·0	−25	...
lactate + NADH → propionate + NAD + H_2O	NADH	−1·0	−68	(1·0)
Overall: pyruvate → propionate (through acrylate)	Total	−2·0	−93	(1·0)
pyruvate + CO_2 → oxaloacetate	...	...	+30	...
oxaloacetate + NADH → malate + NAD	NADH	−1·0	−29	...
malate → fumarate + H_2O	...	...	+4	...
fumarate + $FADH_2$ → succinate + FAD	$FADH_2$	−1·0	−45	1·0
succinate → methylmalonate → propionate + CO_2	...	...	−26	(1·0)
Overall: pyruvate → propionate (randomizing pathway)	Total	−2·0	−66	1·0 (2·0)
pyruvate + CoA → acetyl-CoA + CO_2 + H_2	H_2	1·0	−19	...
acetyl-CoA → ½ acetoacetyl-CoA + ½ CoA	...	...	?	...
½ acetoacetyl-CoA + ½ NADH → ½ β-OH-butyryl-CoA + NAD + ½ CoA	NADH	−0·5	−8	...
½ β-OH-butyryl-CoA → ½ crotonyl-CoA + ½ H_2O	...	...	?	...
½ crotonyl−CoA + ½ $FADH_2$ → ½ butyryl-CoA + ½ FAD	$FADH_2$	−0·5	−37	(1·0)
½ butyryl-CoA → ½ butyrate + ½ CoA	...	...	?	...
Overall: pyruvate → ½ butyrate + CO_2 (if H_2 is used to reduce NAD and FAD)	Total	0·0	−75 (−105)	(1·0)

(b) methylmalonate is involved in the synthesis of branched chain fatty acids, commonly found in the rumen (Duncan *et al.*, 1974).

In the direct reductive pathway (Ladd & Walker, 1959), which may account for about 30% of propionic acid formed, the reducing power is supplied by NADH and the change in free energy (−25 kJ/mol propionic acid) in the first part of the reaction is not sufficient for 1 mol ATP. The energetic yield may be adequate in the conversion of lactate to propionate (see Table 2).

The conversion of 2 mol of pyruvate to butyrate, CO_2 and HCO_3^- is associated with $\triangle G = -76$ kJ, and yet there is apparently no formation of ATP. The free energy change in conversion of 1 mol crotonyl-CoA to 1 mol butyryl-CoA, when the hydrogen donor is $FADH_2$ is −76 kJ and this might be sufficient for the formation of 2 mol ATP.

Clearly, our knowledge of energetic exchanges during typical rumen fermentation is incomplete, and the energy exchanges associated with hydrogen transfer contribute most to the uncertainty. If NADH could be re-oxidized by molecular oxygen, then the ATP yield during conversion of glucose to pyruvate would be 8 mol/mol glucose. Similarly, if the hydrogen produced during the formation of acetate could react with a suitable carrier, say NAD, and then with oxygen, the yield of ATP would increase to 10 mol/mol glucose. It can be calculated from Table 2 that when acetic, propionic and butyric acids are produced from glucose in a ratio 0·65 : 0·20 : 0·15, there would be a definite production of 3·5 mol ATP/mol glucose and a possible maximal production of 5·5 mol/mol. If the excess reducing power is used in a suitable electron transport system, then with oxygen as the final hydrogen acceptor, ATP production would be increased by a further 7·3 mol/mol glucose.

DISPOSAL OF EXCESS REDUCING POWER

Methanogenesis

Methanogenic bacteria use hydrogen and carbon dioxide to form methane. The amount of hydrogen used is 4 mol/mol methane and the change in free energy for this reaction is −134 kJ/mol methane. We have seen in the previous section that the conversion of pyruvate to acetate is associated with $\triangle G = -55$ kJ/mol substrate and that the reaction results in production of 1 mol ATP. Therefore, one would expect that methanogenesis with $\triangle G = -134$ kJ/mol CH_4 would result in a release of energy equivalent to at least 3 mol ATP/mol CH_4.

Let us assume that methane is formed from carbon dioxide, with formic acid, formaldehyde and methanol as intermediates as shown in equation 1. In equation 1 the values for each step refer to changes in free energy during the given step and are given in kJ. Clearly, only the last two steps will result in a release of energy.

$$CO_2 \xrightarrow[H_2]{+29} HCOOH \xrightarrow[H_2 \quad H_2O]{-8} HCHO \xrightarrow[H_2]{-46} CH_3OH \xrightarrow[H_2 \quad H_2O]{-109} CH_4 \qquad (1)$$

Our knowledge of the pathways involved in methanogenesis is far from complete, but it is increasing. Certainly most of the simple compounds in equation 1 above are not the real intermediates in methane production. The possible pathways are outlined in Fig. 1.

Pure cultures of methane bacteria can be grown with carbon dioxide and hydrogen as the sole source of energy, but it is unlikely that hydrogen gas is used as such. It is possible that the electrons from H_2 are transferred to CO_2 through ferredoxin and NADP by the reversal of formate dehydrogenase with an expenditure of 1 mol ATP/mol CO_2. It is usually assumed that the formic acid reacts with tetrahydrofolic acid, with an expenditure of a further 1 mol ATP/mol formic acid to give formyltetrahydrofolic acid. The methylene- and methyl-tetrahydrofolic acid intermediates correspond to the oxidation state of formaldehyde and methanol respectively. In the pathway on the left of Fig. 1 the first two steps result in a loss of water and the second two steps are reductions. The reducing power for these two steps is probably supplied by coenzymes which are regenerated during general metabolism within the cell. In the last two steps the methyl group is transfered to a reduced corrinoid compound and thence to coenzyme M and methane is formed from methyl-coenzyme M in the presence of reducing power (Taylor & Wolfe, 1974). With cell-free extracts this final reducing step requires ATP and ferredoxin but according to equation 1 this should be the step that yields most of the energy during methanogenesis.

There are several indications that the initial and final steps in the formation of CH_4 from CO_2 are as given in Fig. 1. For instance the specific radioactivity of CH_4 produced by mixed rumen micro-organisms in the presence of $[^{14}C]-CO_2$ and H_2 is the same as that of the CO_2 (Czerkawski & Breckenridge, 1975*a*) and CO_2+H_2 can be replaced by formate (Czerkawski *et al.*, 1972). Moreover, when methane production is inhibited, formate tends to accumulate. The powerful inhibition of methane production by various polyhalogen compounds (Czerkawski & Breckenridge, 1975*b*, *c*) and chloroform would implicate vitamin B_{12} or related corrinoid compounds because such compounds are known to react with polyhalogen methane analogues.

The inclusion of the tetrahydrofolic acid system in the pathway may not be justified even though it has been established that the methanogens have a functioning tetrahydrofolic acid system. When cell-free extracts are incubated with labelled formaldehyde or methanol, compounds such as histidine that are synthesized by pathways involving tetrahydrofolic acid are labelled but no label is found in the methane that is produced. There-

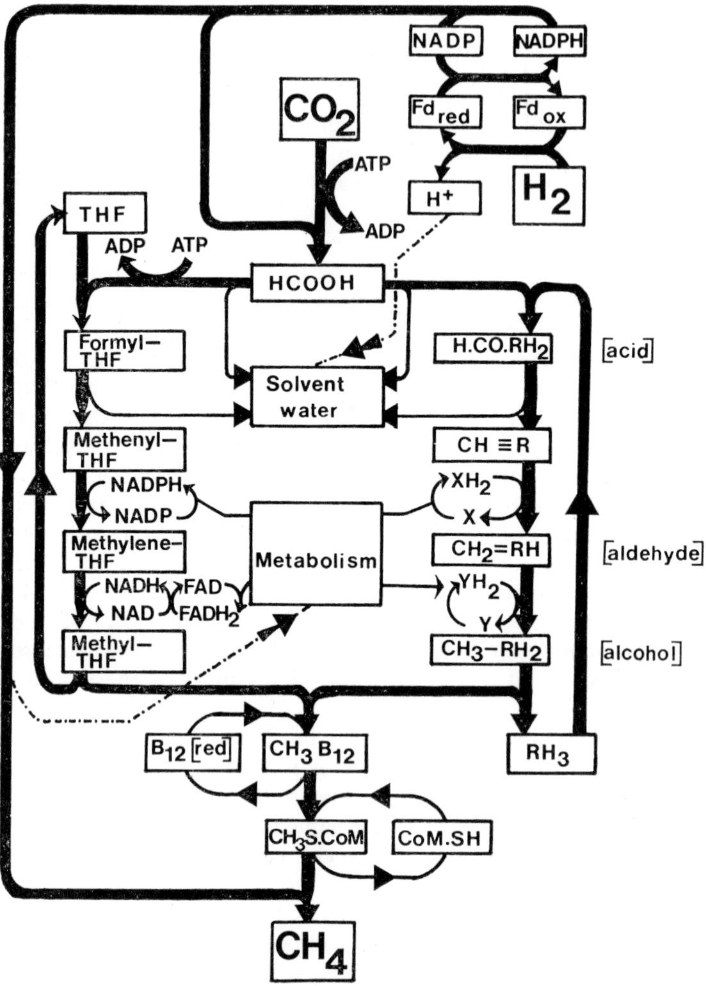

Fig. 1. *The pathways involved in the production of methane from carbon dioxide* (CO_2) *and hydrogen* (H_2). THF, tetrahydrofolic acid; CoM, coenzyme M, HS.$CH_2CH_2SO_3^-$; B_{12}, vitamin B_{12}; Fd, ferredoxin, a protein involved in electron transport; NAD, NADP, FAD, coenzymes involved in redox reactions; RH_3, a hypothetical compound.

fore, in Fig. 1, an alternative, hypothetical part of the pathway has been introduced, but it is difficult to suggest what the actual intermediates might be.

Methanogenic bacteria in pure culture use CO_2 as a source of carbon for the synthesis of cellular components. Therefore, they must be well equipped with enzymes that can bring about the fixation of CO_2. We have already referred to one of the reactions in which C_1 units are transferred (e.g. formation of propionate). Perhaps we should look there for intermediate pathways in methane production.

When mixed rumen micro-organisms are incubated under hydrogen gas, methane is formed at approximately 0·5 mol/mol hydrogen used (Czerkawski & Breckenridge, 1975*a*). If the hydrogen is labelled with tritium, radioactivity passes rapidly and quantitatively to water; the hydrogen pool in solvent water is very large and the methane that is formed is not labelled. On the other hand, when rumen contents are incubated under H_2 gas with tritium-labelled water in the reaction mixture, methane is labelled (but not H_2 gas) and the specific radioactivity of methane is precisely half of that in water (expressed per g atom of hydrogen). It would appear, therefore, that at least two hydrogen atoms in methane are derived from an intermediate that is in equilibrium with water. The results are consistent with the reaction $H_2 \rightarrow 2H^+ + 2e$ being mainly in the indicated direction, because there is virtually no labelling of hydrogen when rumen contents are incubated under hydrogen with tritiated water or glucose.

Alternative reactions that require hydrogen

Often the failure to detect an alternative hydrogen 'sink' is caused by a very close association of two or more microbial species. The association between different microbial species may be so close that they are considered as a single species. After many years of general acceptance of *Methanobacillus Omelianskii* as a micro-organism responsible for the 'oxidation' of ethanol to acetic acid and the stoichiometric formation of methane, it was shown elegantly by Bryant *et al.* (1967) that the overall reaction is carried out by two closely associated micro-organisms. One micro-organism converts ethanol to acetic acid and hydrogen gas, while the other is a true methanogen, using the resulting hydrogen and CO_2 to produce methane. The growth of the ethanol-utilizer is inhibited by hydrogen; this explains why the micro-organisms could not be grown unless special provisions were made for the continuous removal of hydrogen. We have shown that this type of reaction can also take place in the rumen (Czerkawski & Breckenridge, 1972).

An association of *Vibrio succinogenes* and *Ruminococcus albus* was studied in continuous culture by Iannotti *et al.* (1973). When grown alone, *R. albus* utilized glucose to form acetate, ethanol and hydrogen gas.

In a mixed culture of *R. albus* and *V. succinogenes* the products were acetate and succinate but no free hydrogen was produced. Clearly, *V. succinogenes* used the hydrogen to convert fumarate to succinate and obtain energy for growth (see Table 2). Yet, as we have seen above succinic acid does not accumulate during rumen fermentation and therefore there must be organisms that can utilize this acid. Incidentally, *V. succinogenes* is a cytochrome-producing anaerobe (Wolin *et al.*, 1961) and it would appear that the electrons from hydrogen pass through a specific cytochrome system.

When methane production is inhibited by adding a polyhalogen analogue of methane, such as chloroform, the rumen fermentation pattern changes. There is usually a decrease in the production of acetic acid and an increase in the production of propionic acid, but these changes do not reduce the output of metabolic hydrogen to any great extent (10 to 20% decrease). The inhibition of methane production is often accompanied by some accumulation of hydrogen gas, but in general, the accumulation is less than one would expect. Clearly the metabolic hydrogen is rechannelled from methanogenesis to other processes and we must consider what are the likely reactions that can use this excess of reducing power.

To illustrate some of the changes that take place when methane production is inhibited, we may consider the results obtained with a long-term artificial rumen (Czerkawski, 1976*a*; Czerkawski & Breckenridge, 1977). Some of these results are summarized in Table 3. When methane production was inhibited there was some accumulation of hydrogen gas, the production of acetic acid was depressed and the productions of propionic, butyric and isovaleric acids were increased. There was also an accumulation of formic acid. Using the values in Table 2, it was possible to calculate the net amounts of reduced cofactors and hydrogen produced during the formation of the listed products from glucose. In the control vessel the amount of hydrogen used for methane production was greater than the amount produced as hydrogen gas; therefore it must be concluded that some of the reducing power used in methanogenesis is derived from the reduced cofactors. In the presence of the inhibitor, the amount of hydrogen gas potentially produced was the same as in the control vessel, but the net amount of reduced cofactors was less than in the control vessel. If we assume that in the control vessel the reduced cofactors not used for methane production are used in cell synthesis (0·59 mol/kg food dry matter digested) and that this is proportional to cell mass, then the amount of cofactors used in the synthesis of cells during inhibition of methane production would be 0·57 mol/kg. Clearly, when methane production was inhibited about 1·5 mol hydrogen produced during fermentation could not be accounted for and must have been used elsewhere.

It was shown in this laboratory (Clapperton & Czerkawski, 1972; Czerkawski & Breckenridge, 1973) that mixed rumen micro-organisms dissimilate propanediol to propionic acid and *n*-propanol. The reaction may be summarized as follows:

$$2CH_3CH(OH)CH_2OH \rightarrow CH_3CH_2COOH + CH_3CH_2CH_2OH + H_2O \quad (2)$$

In this reaction, after the removal of water, some of the intermediate (*n*-propionaldehyde) is oxidized to propionic acid and some is reduced to *n*-propanol. In all our experiments there was more *n*-propanol than propionic acid in the products and therefore there was a net uptake of reducing power in the reaction, resulting in a small but consistent inhibition of methane production. We have then an example of micro-organisms competing for metabolic hydrogen with its main utilizers – methanogens. Admittedly, 1,2-propanediol is not a natural substrate for rumen micro-organisms, but there are other substances that might react in a similar manner. For example, if glycerol was attacked in this manner, the final reduced product might be 1,3-propanediol.

TABLE 3

Effect of trichloroethyladipate (TCE-A) on the fermentation of a mixture of hay, goat mix and sugar beet nuts (7, 1 and 2 g/day respectively) as assessed using the rumen simulation technique and an original inoculum of the rumen contents of sheep receiving the same diet

TCE-A was given with food. The results are the means for days 9–17, and are expressed as mol per kg dry matter digested. The outputs of microbial dry matter in the effluent, in the absence and in the presence of TCE-A were 112 and 109 g/kg food dry matter digested respectively

	Control	+TCE-A
Gas production (mol/kg)		
CH_4	1·70	0·21
H_2	0·00	1·61
CO_2	4·32	4·60
Short-chain fatty acid production (mol/kg)		
Acetic	3·65	2·91
Propionic	0·93	1·19
Butyric	0·68	0·88
Valeric	0·17	0·14
Isovaleric	0·15	0·48
Formic	0·01	1·29
Hydrogen balance (mol/kg)		
(a) Net reduced cofactors (NAD, FAD)	2·08	0·48
(b) H_2 produced	5·33	5·29
(c) H_2 used for CH_4, HCOOH or as gas	6·82	3·74
(d) Reduced cofactors for cells	0·59*	0·57
Balances ((a)+(b)−(c)−(d))	0·00	1·46

* Assumed.

Calculations show (Czerkawski, 1975) that the synthesis of amino acids from glucose and NH_3 might require variable amounts of reducing power. Also, since the ruminant diet contains long chain polyunsaturated fatty acids and since these acids are hydrogenated in the rumen (Czerkawski, 1967) some reducing power would be used in hydrogenation. In fact, it was shown in several experiments with sheep that polyunsaturated fatty acids, when added to the rations, are quite effective inhibitors of methane production (e.g. Czerkawski *et al.*, 1966*a*, *b*, *c*; Czerkawski, 1966).

A likely reaction that can act as a 'sink' for the excess of metabolic hydrogen is lipid synthesis. Let us return to the example in Table 3. The net production of acetic acid was depressed by 0·7 mol/kg dry matter digested, and it can be shown that the same quantity of acetic acid could be used in the synthesis of palmitic acid and would require 1·22 mol of metabolic hydrogen per kg dry matter digested (see the unaccounted hydrogen in Table 3). Unfortunately, this cannot be the correct explanation because when expressed in the same units, the production of an additional 0·7 mol acetate from glucose would also give 1·4 mol hydrogen. In other words the conversion of 4 mol glucose into 1 mol palmitic acid and 8 mol CO_2 with intermediary formation of acetate does not require any additional hydrogen. The conversions, however, need not be direct. For instance, consider the formation of glutamic acid from ketoglutaric acid, one of the most important anaplerotic compounds in the citric acid cycle. If this conversion involves the fixation of CO_2, a very prevalent reaction in the rumen (Sauer *et al.*, 1975), the overall reaction can be formulated as follows

$$\text{glucose} + CO_2 + NH_3 + H_2 \rightarrow \text{glutamic acid} + \text{acetic acid} + 2H_2O \quad (3)$$

Thus, acetate can be formed with an uptake of reducing power and if some of the acetate produced in this manner is used in lipid synthesis an uptake of reducing power will be observed.

The methanogenic bacteria in the rumen convert large amounts of CO_2 to CH_4 and there is no reason why they should not convert some of the CO_2 to polycarbon compounds; they certainly must do so in pure cultures. Certain micro-organisms, e.g. chlorobium, can use the citric acid cycle in reverse, using the energy from light to convert CO_2 to acetate.

$$2CO_2 + H_2 + 2NADH + FADH_2 \xrightarrow{Fd} CH_3COOH + 2NAD^+ + FAD + 2H_2O \quad (4)$$

If a similar process was carried out in the rumen (not necessarily in methane bacteria) we would have another reaction in which acetate is formed with an uptake of reducing power. Suppose the unaccounted hydrogen in Table 3 is used in both the fixation of CO_2 into acetic acid and in the synthesis of palmitic acid from acetic acid, then 1·46 mol of reducing power would result in the fixation of about 0·5 mol CO_2 and would give about 7 g palmitic acid/kg dry matter digested. The microbial lipid content is variable, 5 to 15% of dry matter (Czerkawski, 1976*b*), and therefore an increase of 7 g in 110 g dry matter is possible. It was shown previously (Czerkawski *et al.*, 1975) that when methane production was inhibited in sheep over extended periods, the capacity of rumen contents to incorporate [^{14}C] from acetate into lipid increased. The calculated additional lipid synthesized was of the same order as the value of 7 g calculated above.

In this connexion, Sauer *et al.* (1975) found that considerable amounts of [^{14}C]carbonate were incorporated into microbial amino acids during incubation of mixed rumen micro-organisms. The radioactivity was incorporated into most amino acids but the incorporation was somewhat greater in histidine, methionine, serine and aspartic acid. In alanine, ^{14}C from carbonate was incorporated exclusively into C-1 and when [$1-{}^{14}C$] acetate was used only C-2 was labelled. Since alanine is synthesized from pyruvate, the authors concluded that under the conditions used, synthesis of acetate from CO_2 *de novo* cannot be important. It would be interesting, however, to see whether this pathway is unimportant when methane production is inhibited and there is a large excess of reducing power.

The rumen is considered to be an anaerobic system, but it was shown in this Institute (Czerkawski & Breckenridge, 1969*b*) and elsewhere (e.g. Demeyer & van Nevel, 1975) that oxygen can be apparently utilized by rumen micro-organisms. Although the oxygen tension in the rumen is low, calculations show (Czerkawski, 1969) that large amounts of oxygen might diffuse into the rumen from the blood. Here we have a hydrogen 'sink' *par excellence* and an important subject for future research.

MANIPULATION OF RUMEN FERMENTATION

Several reactions have been listed in Table 4. The first summarizes the production of CH_4 from CO_2 and H_2 and has been discussed earlier.

The second reaction in Table 4 summarizes another complex multistage process, namely fixation of atmospheric nitrogen. It is usually accepted that fixation of N_2 requires ATP and yet the change in free energy associated with this reaction is negative and fairly large (-80 kJ/mol N_2). Is it possible that the nitrogen-fixing micro-organisms obtain energy for growth from the conversion of molecular N_2 to NH_3? It is well known that before sulphate can be used in microbial synthesis it must be reduced to sulphide. Thus we can write equations such as 3 or 4 in Table 4 and again we see that the free energy change is large and negative. We can take this a stage further and write equations describing the overall synthesis of acetate from CO_2 and palmitate from acetate (equations 5 and 6 in Table 4)

Again the change in free energy is large and negative. Thus, thermodynamically, all these reactions should result in the release of useful energy. We know that the synthesis of fatty acids from acetate is an expensive process, requiring a supply of ATP and reduced cofactors, particularly NADPH. The reduction of 1 mol of NADP requires at least 3 mol of ATP and this is the most energy demanding part of lipid synthesis. Under conditions where reduced cofactors are in excess, lipid synthesis need not be very costly.

TABLE 4

Free energy changes in a number of reducing reactions which occur within the rumen

Reaction	Free energy change for the reaction (kJ)	Free energy required to supply H_2 from water (MJ)
$CO_2+4H_2 \rightarrow CH_4+2H_2O$	−134	0·95
$N_2+3H_2 \rightarrow 2NH_3$	− 80	0·71
$SO_4^{=}+14H_2 \rightarrow H_5^{-}+OH^{-}+3H_2O$	−113	0·95
$H_2SO_4+4H_2 \rightarrow H_2S+4H_2O$	−155	0·95
$2CO_2+4H_2 \rightarrow CH_3COOH+2H_2O$	−101	0·95
$8CH_3COOH+14H_2 \rightarrow$ palmitic $+14H_2O$	−302	3·33
$O_2+2H_2 \rightarrow 2H_2O$	−474	...

Even though all the reactions in Table 4 are energy-producing, they all require hydrogen and it is obvious that some energy is required to form hydrogen from the most ubiquitous source – water. This is really the crux of the matter; methanogenesis is useful in the rumen because there is an ample supply of hydrogen produced in the course of fermentation. When methane production is inhibited specifically, there is no reason why other micro-organisms should not use some of the reactions listed in Table 4 to obtain useful energy and grow.

The inhibition of methane production and the rechannelling of metabolic hydrogen to less wasteful processes is only one example of the manipulation of rumen fermentation. The reactions in the rumen depend very much on the available substrates and therefore on the diet of the animal. For instance, a concentrate diet containing large porportions of readily available carbohydrate results in a vigorous fermentation and under these conditions the amount of methane produced per unit of dry matter digested is less than with a roughage diet. Such vigorous fermentation is associated with an increased production of propionic acid and a decreased production of acetic acid, and this would favour a decreased net production of reducing power and methane.

CONCLUSIONS

The microbial system of the rumen is remarkably stable and the fermentation pattern is determined largely by the type of substrates supplied. The choice of food is probably the best method of manipulating rumen fermentation. However, fermentation can be altered, albeit temporarily, by more artificial means, such as the use of dietary additives, but the system reverts rapidly to the conditions determined by the diet. The aim of our work at the Hannah Research Institute in recent years has been to show that in addition to the economically wasteful process of methanogenesis, there are other, more useful means of disposal of excess metabolic hydrogen. It will be a worthwhile task for the future to study these alternative processes in detail and to define the best conditions for the vigorous growth of the responsible micro-organisms. One can visualize the development of a microbial culture, which, when inoculated into the normal rumen and established there, would successfully compete with methanogens for metabolic hydrogen and would give rise to a product more useful to the host animal than methane; the efficiency of food conversion would thereby be increased.

REFERENCES

Armstrong, D. G. & Blaxter, K. L. (1957) *Br. J. Nutr.* **11**, 247–272

Blaxter, K. L. & Czerkawski, J. W. (1966) *J. Sci. Food Agric.* **17**, 417–421

Bryant, M. P., Wolin, E. A., Wolin, M. J. & Wolfe, R. S. (1967) *Arch. Microbiol.* **59**, 20–31

Clapperton, J. L. & Czerkawski, J. W. (1969) *Br. J. Nutr.* **23**, 813–826

Clapperton, J. L. & Czerkawski, J. W. (1972) *Br. J. Nutr.* **27**, 553–560

Czerkawski, J. W. (1966) *Br. J. Nutr.* **20**, 833–842

Czerkawski, J. W. (1967) *Br. J. Nutr.* **21**, 865–878

Czerkawski, J. W. (1969) *World Rev. Nutr. Diet.* **11**, 240–282

Czerkawski, J. W. (1973*a*) *J. Agric. Sci.* **81**, 517–531

Czerkawski, J. W. (1973*b*) *Process Biochem.* **8** (10), 25–30

Czerkawski, J. W. (1975) in *Tracer Studies on Non-Protein Nitrogen for Ruminants, II*, pp. 55–63, International Atomic Energy Agency, Vienna

Czerkawski, J. W. (1976*a*) in *Energy Metabolism of Farm Animals* (Vermorel, M., ed.), pp. 69–72, de Bussac, Clermont–Ferrand

Czerkawski, J. W. (1976*b*) *J. Sci. Food Agric.* **27**, 621–632

Czerkawski, J. W. (1976*c*) *Lab. Pract.* **25**, 15–20

Czerkawski, J. W. & Breckenridge, G. (1969*a*) *Br. J. Nutr.* **23**, 51–66

Czerkawski, J. W. & Breckenridge, G. (1969*b*) *Br. J. Nutr.* **23**, 67–80

Czerkawski, J. W. & Breckenridge, G. (1969*c*) *Br. J. Nutr.* **23**, 925–937

Czerkawski, J. W. & Breckenridge, G. (1972) *Br. J. Nutr.* **27**, 131–146

Czerkawski, J. W. & Breckenridge, G. (1973) *Br. J. Nutr.* **29**, 317–330

Czerkawski, J. W. & Breckenridge, G. (1975*a*) *Anal. Biochem.* **67**, 476–484

Czerkawski, J. W. & Breckenridge, G. (1975*b*) *Br. J. Nutr.* **34**, 429–446

Czerkawski, J. W. & Breckenridge, G. (1975*c*) *Br. J. Nutr.* **34**, 447–457

Czerkawski, J. W. & Breckenridge, G. (1977) *Br. J. Nutr.* **38**, 371–374
Czerkawski, J. W., Blaxter, K. L. & Wainman, F. W. (1966*a*) *Br. J. Nutr.* **20**, 349–362
Czerkawski, J. W., Blaxter, K. L. & Wainman, F. W. (1966*b*) *Br. J. Nutr.* **20**, 485–494
Czerkawski, J. W., Blaxter, K. L. & Wainman, F. W. (1966*c*) *Br. J. Nutr.* **20**, 495–508
Czerkawski, J. W., Harfoot, C. G. & Breckenridge, G. (1972) *J. Appl. Bacteriol.* **35**, 537–551
Czerkawski, J. W., Christie, W. W., Breckenridge, G. & Hunter, M. (1975) *Br. J. Nutr.* **34**, 25–44
Demeyer, D. I. & van Nevel, C. J. (1975) in *Digestion & Metabolism in the Ruminant* (McDonald, I. W. & Warner, A. C. I., eds), p. 366, University of New England Publishing Unit, Armidale, Australia
Duncan, W. R. H., Lough, A. K., Garton, G. A. & Brooks, P. (1974) *Lipids* **9**, 669–673
Hopgood, M. F. & Walker, D. J. (1967*a*) *Aust. J. Biol. Sci.* **20**, 165–182
Hopgood, M. F. & Walker, D. J. (1967*b*) *Aust. J. Biol. Sci.* **20**, 183–192
Iannotti, E. L., Kafkewitz, D., Wolin, M. J. & Bryant, M. P. (1973) *J. Bacteriol.* **114**, 1231–1240
Ladd, J. N. & Walker, D. J. (1959) *Biochem. J.* **71**, 365–373
McNaught, M. L. (1951) *Biochem. J.* **49**, 325–332
McNaught, M. L., Owen, E. C. & Smith, J. A. B. (1950*a*) *Biochem. J.* **46**, 36–43
McNaught, M. L., Smith, J. A. B., Henry, K. M. & Kon, S. K. (1950*b*) *Biochem. J.* **46**, 32–36
Pearson, R. M. & Smith, J. A. B. (1943) *Biochem. J.* **37**, 153–164
Sauer, F. D., Erfle, J. D. & Mahadevan, S. (1975) *Biochem. J.* **150**, 357–372
Taylor, C. D. & Wolfe, R. S. (1974) *J. Biol. Chem.* **249**, 4879–4885
Wolin, E. A., Wolin, M. J. & Jacobs, A. (1961) *J. Bacteriol.* **81**, 911–917

ENERGY AND NITROGEN METABOLISM IN THE RUMINANT

P. C. THOMAS

IN late 1971 a programme of work on the energy and nitrogen metabolism of ruminants was begun in the Applied Studies Department of the Hannah Research Institute. Some of the subjects investigated since that time are summarized briefly here.

MANIPULATION OF RUMEN FERMENTATION

By the early 1960s, K. L. Blaxter, D. G. Armstrong and their colleagues at the Hannah Research Institute and J. A. F. Rook and his colleagues at the National Institute for Research in Dairying, Shinfield, had provided evidence that the composition of the mixture of short-chain fatty acids in the rumen had an important influence on the efficiency of utilization of dietary energy for maintenance and fattening and on the yield and composition of milk (see Thomas & Rook, 1977). At that time, the mixture of acids was thought to depend closely on the composition of the diet and it was not until later that it became evident that with diets containing starchy concentrates the link was less close than had been supposed. Eadie *et al.* (1970) found that in steers given cubed barley at 80% of the limit of appetite there was a large ruminal population of ciliate protozoa and that the mixture of rumen acids was characterized by a high molar proportion of butyric acid. When the diet was given *ad libitum*, the numbers of protozoa were reduced, there was an increase in certain species of bacteria, and the fermentation was characterized by a high molar proportion of propionic acid. Ishaque *et al.* (1971), however, also observed differences in the composition of the mixture of rumen acids between animals, and in the same animal from time to time when sheep were offered a standard amount of a diet of ground barley, ground hay and flaked maize (56 : 24 : 20, w/w). They found that the efficiency of bacterial protein synthesis in the rumen, and thus the utilization of dietary nitrogen, was correlated with the composition of the mixture of fatty acids produced. These results suggested that in some circumstances, diet was not the major factor influencing the mixture of rumen acids and that it might be possible to manipulate metabolism in the rumen to obtain beneficial effects on the utilization of dietary energy and protein.

Control of rumen fermentation

At the Hannah Research Institute work was begun to investigate the metabolism in the rumen of sheep receiving a diet similar to that used by M. Ishaque and his colleagues. Initial results (Hodgson & Thomas, 1972) confirmed that the rumen fermentation pattern could be of either a high or a low propionic acid type but indicated that these two fermentation patterns were not characterized by differences in rumen pH or in the ruminal concentration of sodium, potassium, calcium, magnesium, phosphorus or chloride. The volume of liquid in the rumen and its rate of flow to the omasum varied between animals irrespective of the fermentation pattern. However, the clearance rate (the fraction of rumen volume flowing out per hour) differed significantly between the types of fermentation. Observations with other high-concentrate and highly digestible forage diets (Hodgson & Thomas, 1975) indicated that there was a widely applicable inverse relationship between the molar proportion of propionic acid in the rumen liquor and the clearance rate. Subsequent work with high concentrate diets indicated that the relationship was unaffected by variations in the level of feeding (Hodgson *et al.*, 1976) or by provision of the hay and concentrate parts of the diet in separate meals given several hours apart (M. R. Stokes, unpublished work). On the other hand, the relationship did not hold with diets consisting solely of moderate- or poor-quality forages (J. C. Hodgson, unpublished work). Attempts to influence clearance rate with intraruminal infusions of water were without success but in sheep receiving high concentrate diets clearance was altered by intraruminal infusion of a solution of a mixture of sodium chloride and potassium chloride or by intravenous injections of atropine sulphate, a drug that reduces salivation and gut motility. These induced changes in clearance rate caused alterations in the proportion of propionic acid in the rumen. In sheep receiving diets consisting of moderate quality forages, however, intravenous infusion of atropine sulphate reduced clearance rate but did not affect the proportion of propionic acid in the rumen (J. C. Hodgson, unpublished work).

These results suggested that with diets containing starchy concentrates and with high-quality dried forages, dietary treatments influencing the rumen clearance rate could influence the mixture of fatty acids in the rumen. This conclusion agreed closely with the findings at the Grassland Research Institute of Harrison *et al.* (1974) who altered the ruminal clearance rate and fermentation pattern through the intraruminal infusion of a variety of isotonic and hypertonic solutions. The question of the most effective dietary method of influencing clearance rate has yet to be resolved. Work at the Hannah Research Institute using additions of sodium chloride has had little success (M. R. Stokes, unpublished work) but Thomson *et al.* (1975) have obtained promising results with salt mixtures containing a large proportion of sodium bicarbonate.

Utilization of energy

The infusion studies of Blaxter and his colleagues in the fasting and fattening sheep (see Blaxter, 1962) indicated that the efficiency of utilization of the mixture of acids absorbed from the rumen had a negative correlation with the percentage of acetic acid absorbed and was lower than the efficiency for sugars, amino acids and long-chain fatty acids absorbed from the small intestine. More recent work also showed that acetic acid was utilized poorly during fasting (Holter *et al.*, 1970) but results for fattening animals were equivocal. Comparisons between the acids (Lawrence & Thomas, 1973) confirmed that acetic acid was less well utilized than propionic acid or butyric acid but the average efficiency of utilization for acetate was 70%. This value was consistent with a number of studies on acetate utilization (Bull *et al.*, 1970; Johnson, 1972; Robb & Reid, 1972) but higher than the 33% observed by Armstrong & Blaxter (1957). The explanation for this disagreement was, and still is, uncertain. Recent findings by Tyrell *et al.* (1976) suggest that it may lie in the interactions between the acetic acid infused and the substrates absorbed from the basal diet. Thus, although it was clear that different fermentations could occur in the rumen with a single diet, the effect of a particular type of fermentation on the utilization of energy by fattening animals was difficult to predict. Calorimetric studies were therefore undertaken in sheep offered a diet known to give wide variations in fermentation pattern. Initial results indicated that differences between sheep in fermentation were indeed associated with differences in energy utilization (Lawrence & Thomas, 1972), and in a more comprehensive study (M. R. Stokes, unpublished work) there was a significant inverse relationship between the efficiency of utilization of energy for fattening and the rate of production of acetic acid. These observations agree with those of Blaxter & Wainman (1964) who found in experiments with diets varying in the ratio of hay to cereals that the efficiency of utilization of metabolizable energy was related to the proportion of acetic acid in the mixture of acids in the rumen. However, as indicated by Thomas & Rook (1974), such correlations could result if acetic acid acted as an index of the mixture of nutrients absorbed from the gut as a whole.

To explore this possibility, four semipurified mixtures were formulated to represent diets varying progressively from a high-forage type, with low efficiency of energy utilization and giving rise to a high proportion of acetic acid in the rumen, to a high-concentrate type with high efficiency of energy utilization and resulting in a low proportion of acetic acid in the rumen (M. R. Stokes, unpublished work). The main ingredients in the diets were straw, maize starch and molasses; the proportions of straw and starch were varied so that straw provided 80, 60, 40 and 20 % of the mixture. The diets were isonitrogenous, containing 14 % crude protein, and the nitrogen was incorporated as soluble sodium caseinate and urea so that 60 % of the nitrogen in each diet was in non-protein form. The use of these semipurified diets ensured that the amounts of dietary long-chain fatty acids and undigested dietary protein flowing to the duodenum were the same for all the diets.

Calorimetric measurements confirmed that with these four diets the efficiency of utilization of metabolizable energy for fattening increased progressively as the proportion of starch in the diet increased. In parallel studies, sheep fitted with cannulae in the rumen, duodenum and ileum were used to determine the uptake of substrates from the stomach and intestines. As compared with the diets rich in straw, the diets rich in starch had a larger part of their digestible energy fermented but the volatile acids produced in the rumen contained a smaller proportion of acetic acid. However, the low proportion of acetic acid was linked with a high uptake of digested energy from the small intestine, mainly in the form of protein synthesized by rumen micro-organisms and in the form of α-linked glucose polymers derived from the diet.

Bacterial protein synthesis

By 1970 evidence was accumulating from studies with forage diets that the rate of bacterial protein synthesis in the rumen was relatively constant at about 23 g per 100 g of organic matter apparently digested, a figure in good agreement with that obtained *in vitro* (see Thomas, 1973). The results of Ishaque *et al.* (1971) were thus especially interesting since they suggested that with diets containing starchy concentrates the rate of bacterial protein synthesis could be low and variable.

The initial objective at the Hannah Institute was to determine the rate of protein synthesis in animals receiving a diet different from that used by Ishaque *et al.* (1971) but known to be associated with variation in the pattern of rumen fermentation between animals. A diet consisting of molassed sugar beet pulp and ground barley (60 : 40 w/w) met these requirements. But with this diet, bacterial protein synthesis in the rumen was constant and unrelated to the composition of the mixture of volatile acids produced in the rumen (Chamberlain *et al.*, 1976). However, the rate of synthesis of protein by rumen micro-organisms was low, on average 14·4 g per 100 g of organic matter apparently digested. This prompted a more extensive examination of ruminal protein synthesis with a range of diets varying in their ratio of hay to cereals (Chamberlain & Thomas, 1974). With diets containing more than 70 % hay, the rate of bacterial protein synthesis was high, approaching the value obtained *in vitro*, but with those diets containing smaller proportions of hay the rate decreased with the amount of hay in the diet. The reasons for this are still not fully understood but current evidence suggests that several factors are involved. In recent studies with a high cereal diet (D. G. Chamberlain, unpublished work) increased rates of bacterial protein synthesis have been observed when increases in the flow of liquid through the rumen were induced by intraruminal infusions of 'artificial saliva', and when increases in the supply of nitrogen to the rumen micro-organisms were brought about by intraruminal infusions of urea. Urea treatment produced rates of protein synthesis that were consistently higher than those in control animals but 'artificial saliva' produced higher rates in only some instances. When urea and 'saliva' were given together the effect was sometimes greater and sometimes only equal to that observed when urea was administered alone and the rates of protein synthesis were usually less than the theoretical maximum. The failure to achieve the theoretical maximum rate of protein synthesis may have been the result of a recycling of nitrogen within the rumen due to the lysis of rumen bacteria or to the engulfment of bacteria by protozoa (see Thomas, 1977).

THE NUTRITIVE VALUE OF SILAGE

Between 1967 and 1973 research at the institute led to the development of reliable farm-scale procedures for the production of highly digestible silages using formic acid as an additive (see the review by Castle & Reid in this volume, p. 105) and in 1974 work was begun to investigate the nutritional characteristics of materials ensiled in this way.

Typically the silages have a pH of 3·7 to 4·1 and contain 60 to 70 g of digestible organic matter per 100 g dry matter. The dry matter content varies from 18 to 30% and the dry matter contains 4 to 10% water-soluble carbohydrates, 5 to 13% lactic acid, 0·3 to 3% acetic acid and 0 to 0·5% butyric acid. Nitrogen contents are 1·9 to 3·2% and 40 to 60% of this is non-protein nitrogen of which 8 to 22% is ammonia. Judged in feeding trial terms the nutritional value of the materials is quite high (see review by Castle & Reid in this volume, p. 107) but the voluntary consumption of these silages by cattle is less than that for the original material ensiled and because of the content of fermentation end-products and non-protein nitrogen in the silages it is difficult to assess their feeding value.

Food intake

The factors influencing the intake of silages are poorly understood and vary from one type to another. Thus it was important to determine the factors operative with the materials under study. The low pH of the silage did not impose a major restriction of food intake in either the sheep or the cow (Farhan & Thomas, 1977) despite evidence of the importance of pH with other ensiled materials (see Wilkins, 1974). Similarly, in sheep, food intake was unaffected by intravenous supplements of methionine (Kelly & Thomas, 1975) although methionine has given marked increases in silage intake under other conditions (Barry *et al.*, 1973). In contrast in both sheep (Thomas *et al.*, 1976*b*) and cows (Farhan & Thomas, 1977) there was evidence that the intake of the formic acid silages was controlled by the physical restriction of the capacity of the rumen. When silage was given in a minced form as opposed to a chopped form, intake increased markedly, presumably because of an enhanced rate of passage of the silage particles from the rumen. On the other hand, when rumen volume was reduced, by placing water-filled balloons in the rumen, silage intake was depressed.

Utilization of energy

Calorimetric evaluations of perennial ryegrass silages were made in three experiments, two with silages prepared from autumn regrowth cut at an early or later stage of maturity and one with silage made in the spring (see Thomas *et al.*, 1976*c*). The utilization of each silage was studied with castrated sheep at two levels of feeding, one providing the energy required for maintenance and the other about 1·5 times the energy required for maintenance. For the spring silage and for the later autumn-cut silage the effect of supplementation with barley at the lower level of feeding was investigated also. The silages had gross energy contents varying from 18·73 MJ/kg dry matter to 19·15 MJ/kg dry matter and the metabolizable energy contents expressed as percentages of the gross energy were 57·7, 66·0 and 56·9% for the spring, early autumn and later autumn silages respectively. Corresponding values for the metabolizable energy expressed as percentages of the digestible energy were 84·9, 85·4 and 83·7%.

The efficiency of utilization of metabolizable energy for maintenance was 69% for the spring silage and 72 and 74% for the autumn silages, values in close agreement with those predicted from tables which relate the metabolizable energy content of a food to the efficiency of utilization of metabolizable energy for maintenance (Agricultural Research Council, 1965). Similar agreement was also obtained for the efficiency of utilization of energy for fattening with the autumn silages, the efficiency being 61% for the first cut and 55% for the later cut material. In contrast, the efficiency for the spring silage was exceptionally low at 24%.

This low efficiency of energy utilization was linked in some way with the ingestion of high levels of silage since, when the lower level of silage was supplemented with barley, the efficiency of utilization of the barley, calculated by difference, was 60%, a value agreeing well with that obtained for barley in combination with other foods. The three silages differed in chemical composition in several respects but especially in their concentration of lactic acid. The spring silage contained 11·6% lactic acid whereas the autumn silages each contained about 4·0%. Moreover, in the autumn silages, about 30% of the total lactic acid was in the D(−) form whereas in the spring silage this isomer accounted for 42% of the total. D(−) Lactic acid is thought to be a causal factor in the development of lactic acid acidosis during grain engorgement (Telle & Preston, 1971) and since it is metabolized only slowly in the tissues (Mackenzie, 1967) it may in some way impair energy utilization.

Milk protein content

In feeding trials with cows given silage diets, milk protein contents have frequently been low, even in animals receiving diets that apparently provided adequate levels of energy. Whilst there was no unequivocal evidence that the silage was responsible there was suspicion that this was partly the case. Milk protein content is known to be sensitive to the amount and form of energy in the diet (see Thomas, 1975) and is frequently, although not invariably, stimulated by an increase in energy intake (Rook & Line, 1961; Wright *et al.*, 1974; Thomas & Kelly, 1976). This effect has been considered to operate through the influence of diet on the molar proportion of propionic acid in the rumen – intraruminal infusions of propionic acid increase milk protein content (Rook & Balch, 1961). However, intraruminal infusions of acetic acid depress milk protein content (see Thomas & Rook, 1977) and recent experiments at the Institute have indicated that the dietary stimulation of the milk protein depends not so much on an increased proportion of propionic acid in the rumen as on an increased ratio of propionic acid to acetic acid. Milk protein content can be raised by appropriate changes in the composition of the diet even when the total energy intake is moderately deficient (J. S. Chalmers, P. C. Thomas & M. E. Kelly, unpublished work). However, the replacement with flaked maize of up to 45% of the barley in a silage and barley (1 : 1) diet, a change normally expected to increase the ratio of propionic acid to acetic acid in the rumen, had no effect on rumen fermentation or on milk protein content. More drastic changes in diet involving the reduction in the proportion of silage to 25% increased the propionic acid : acetic acid ratio and the level of milk protein but even under these conditions the increases in both were small (J. S. Chalmers, P. C. Thomas & M. E. Kelly, unpublished work).

Utilization of silage nitrogen

There is considerable circumstantial evidence in the literature that the nitrogen in silages is poorly utilized (see Wilkins, 1974) but there is little direct information on the subject. With the formic acid silages sheep cannulated in the rumen, duodenum and ileum were used to determine the net absorption of nitrogen from the rumen and lower gut. Supplementation of ryegrass silage with barley increased the duodenal flow of nitrogen and enhanced its absorption in the small intestine (P. C. Thomas, N. C. Kelly & D. G. Chamberlain, unpublished work). On the other hand, when red clover silage was supplemented with mixtures containing barley, or barley and groundnut, or barley and urea, the composition of the supplement had no effect on the duodenal nitrogen flow (Thomas *et al.*, 1976*a*). Taken together, the available results indicate that when the dietary ratio of nitrogen to digestible organic matter is 3·1 to 3·7 there is little net absorption of nitrogen in the rumen. With higher ratios, ammonia is absorbed and wastefully excreted (Thomas *et al.*, 1976*c*). A large part of the duodenal nitrogen is microbial in origin but a substantial portion of the protein in the silage passes through the rumen unfermented and this also contributes to the duodenal flow. This unfermented protein is, however, digested in the small intestine since on average 69% of the nitrogen flowing into the duodenum is absorbed before the terminal ileum (Thomas *et al.*, 1976*c*). Information on the uptake of individual amino acids in the small intestine is not yet available and nitrogen digestion has yet to be examined in the dairy cow, where the levels of feeding are especially high. These topics are important since there is evidence from feeding trials that diets based on silage provide the cow with a less than optimal supply of amino acids and consequently impair milk production (see review by Castle & Reid, in this volume p. 107).

REFERENCES

Agricultural Research Council (1965) *The Nutrient Requirements of Farm Livestock, No. 2, Ruminants*, Agricultural Research Council, London

Armstrong, D. G. & Blaxter, K. L. (1957) *Br. J. Nutr.* **11**, 413–425

Barry, T. N., Fennessy, P. F. & Duncan, S. J. (1973) *N.Z. J. Agric. Res.* **16**, 64–68

Blaxter, K. L. (1962) *The Energy Metabolism of Ruminants*, Hutchinson, London

Blaxter, K. L. & Wainman, F. W. (1964) *J. Agric. Sci.* **63**, 113–128

Bull, L. S., Reid, J. T. & Johnson, D. E. (1970) *J. Nutr.* **100**, 262–276

Chalmers, J. S., Thomas, P. C. & Kelly, M. E. (1977) *Proc. Nutr. Soc.*, in press

Chamberlain, D. G. & Thomas, P. C. (1974) *Proc. Nutr. Soc.* **33**, 59A–60A

Chamberlain, D. G., Thomas, P. C. & Wilson, A. G. (1976) *J. Sci. Food Agric.* **27**, 231–238

Eadie, J. M., Hyldgaard-Jensen, J., Mann, S. O., Reid, R. S. & Whitelaw, F. G. (1970) *Br. J. Nutr.* **24**, 157–177

Farhan, S. M. A. & Thomas, P. C. (1977) *J. Br. Grassl. Soc.*, in press

Harrison, D. G., Beever, D. E. & Thomson, D. J. (1974) *Proc. Nutr. Soc.* **33**, 43A–45A

Hodgson, J. C. & Thomas, P. C. (1972) *Proc. Nutr. Soc.* **31**, 57A–58A

Hodgson, J. C. & Thomas, P. C. (1975) *Br. J. Nutr.* **33**, 447–456

Hodgson, J. C., Thomas, P. C. & Wilson, A. G. (1976) *J. Agric. Sci.* **87**, 297–302

Holter, J. B., Heald, C. W. & Colovos, N. F. (1970) *J. Dairy Sci.* **53**, 1241–1247
Ishaque, M., Thomas, P. C. & Rook, J. A. F. (1971) *Nat. New Biol.* **231**, 253–256
Johnson, D. E. (1972) *J. Nutr.* **102**, 1093–1100.
Kelly, N. C. & Thomas, P. C. (1975) *J. Br. Grassl. Soc.* **30**, 237–239
Lawrence, P. R. & Thomas, P. C. (1972) *Proc. Nutr. Soc.* **31**, 49A
Lawrence, P. R. & Thomas, P. C. (1973) *Anim. Prod.* **17**, 209–212
Mackenzie, D. D. S. (1967) *J. Dairy Sci.* **50**, 1772–1786
Robb, J. & Reid, J. T. (1972) *Br. J. Nutr.* **28**, 249–259
Rook, J. A. F. & Balch, C. C. (1961) *Br. J. Nutr.* **15**, 361–369
Rook, J. A. F. & Line, C. (1961) *Br. J. Nutr.* **15**, 109–119
Telle, P. P. & Preston, R. L. (1971) *J. Anim. Sci.* **33**, 698–705
Thomas, P. C. (1973) *Proc. Nutr. Soc.* **32**, 85–91
Thomas, P. C. (1975) *World Rev. Anim. Prod.* **11** (4), 33–44
Thomas, P. C. (1977) *Proc. 2nd EAAP Symp. Protein Nutrition*, pp. 47-50
Thomas, P. C. & Kelly, M. E. (1976) *J. Dairy Res.* **43**, 1–7
Thomas, P. C. & Rook, J. A. F. (1974) *Proc. 6th EAAP Symp. Energy Metabolism of Farm Animals*, pp. 429–434, Universität Hohenheim Dokumentationsstelle, Stuttgart
Thomas, P. C. & Rook, J. A. F. (1977) *Proc. 11th Univ. Nottingham Conf. Feed Manufacturers*, in press
Thomas, P. C., Chamberlain, D. G. & Alwash, A. H. (1976*a*) *J. Br. Grassl. Soc.* **31**, 123–128
Thomas, P. C., Kelly, N. C. & Wait, M. K. (1976*b*) *J. Br. Grassl. Soc.* **31**, 19–22
Thomas, P. C., Kelly, N. C., Chamberlain, D. G. & MacDonald, L. (1976*c*) *Proc. 7th Symp. Energy Metabolism of Farm Animals*, pp. 245–248, de Bussac, Clermont-Ferrand
Thomson, D. J., Beever, D. E., Mundell, D. C., Elderfield, M. L. & Harrison, D. G. (1975) *Proc. Nutr. Soc.* **34**, 111A–112A
Tyrell, H. F., Reynolds, P. J. & Moe, P. W. (1976) *Proc. 7th Symp. Energy Metabolism of Farm Animals*, pp. 57–60, de Bussac, Clermont-Ferrand
Wilkins, R. J. (1974) *Proc. 8th Univ. Nottingham Conf. Feed Manufacturers*, p. 167, Butterworth, London
Wright, J. A., Rook, J. A. F. & Wood, P. D. P. (1974) *J. Dairy Res.* **41**, 155–164

THE METABOLISM OF AROMATIC COMPOUNDS IN RUMINANTS

A. K. MARTIN

THE most abundant aromatic compound excreted in the urine of ruminants is usually benzoylglycine (hippuric acid). It was first found in cow's urine by Roelle in 1784 (Smith & Williams, 1970). Herbivores excrete large quantities of this acid, of phenaceturic acid (Vasiliu *et al.*, 1938) and of phenols such as *p*-cresol (Campbell & Hey, 1944) in their urine. In addition, small amounts of two other aromatic acids, 3-phenylpropionic and cinnamic acids, and of other phenols (Martin, 1975) and phenolic acids (Suemitsu *et al.*, 1968) have been found in ruminant urine.

At the Hannah Research Institute, interest in the metabolism of aromatic compounds arose from a number of incidental observations made during energy metabolism studies with sheep and cattle. Ruminal metabolism of protein was shown to cause increased urinary output of aromatic acids (Blaxter & Martin, 1962), although at that time the acids were not identified. This observation reminded us of the large quantities of hippuric acid found in ruminant urine and suggested that it might be possible to use the conjugation of benzoic acid with glycine to measure the energy cost to the animal of the synthesis of the high-energy phosphate bonds of adenosine triphosphate (ATP). The energy requirement of two high-energy phosphate bonds of ATP per mole of hippuric acid synthesized was known (Kellerman, 1958) and it was calculated that if sheep could synthesize 100 g/day of hippuric acid, the resultant heat increment could be determined accurately. The necessary rate of hippuric acid synthesis was not achieved (Martin, 1966) but in the control periods of this experiment the observation of urinary outputs of up to 18 g hippuric acid per kg dried grass intake aroused interest in the origin of the benzoic acid moiety of the hippuric acid. Further studies showed that the quantity of urinary aromatic acids excreted by cattle and sheep varied widely with the type of food eaten; Blaxter *et al.* (1966) summarized the results of experiments in which the daily losses of urine energy, nitrogen (N) and carbon were determined when forty-three different foods were eaten. The energy to N ratio in urine varied from 29 to 193 kJ/g N and this variation was shown to be due largely to the considerable differences in the quantities of hippuric acid (302·7 kJ/g N) excreted when different foods were eaten.

Microbial fermentation of food in the intestine has been held responsible for the large urinary output of aromatic compounds by herbivores (Hawk *et al.*, 1947) but a survey of the output of individual aromatic acids by different species proved difficult. This was due either to the lack of specificity of the analytical methods used or to a failure to take account of the various forms in which compounds such as benzoic acid can be excreted. Output of the glycine conjugate could have been determined but the possibilities of the free acid or glucuronide conjugates being excreted were ignored. However, in a survey conducted by Martin (1969*a*), a comparison was made of the total urinary output of aromatic acids expressed as benzoic acid equivalent (BAE) in ruminant and non-ruminant animals: the former excreted between 96 and 1475 mg BAE/kg body weight$^{0\cdot75}$ per day and the latter between 14 and 115 mg BAE/kg body weight$^{0\cdot75}$ per day.

Although it was established that herbivores, and in particular ruminants, excrete considerable quantities of aromatic compounds, little was known of their identity and less about possible precursors in the diet. Numerous suggestions for the identity of aromatic acid precursors have been made; these include the alcohol-soluble precursors of lignin (Brigl & Pfähler, 1929), lignin (Pazur & DeLong, 1948), fibre (Schiemann *et al.*, 1965) and aromatic amino acids (Scott *et al.*, 1964). We are therefore studying the nature and amounts of aromatic compounds excreted by ruminants and the identity of their precursors. Only when these are known will it be possible to assess the significance of this aspect of ruminant metabolism.

DETERMINATION OF AROMATIC ACIDS

In the inter-species survey of aromatic acid output (Martin, 1969*a*), the analytical methods used could be divided into four categories. In order of increasing specificity these were:

1. The titratable acidity of unhydrolysed samples; extracts made with polar solvents

Ethyl acetate (Hampton, 1948) or diethyl ether (e.g. Folin & Flanders, 1912; Vasilev, 1960) have been used to extract acids and phenols from acidified samples. Both free and amino acid conjugated aromatic and phenolic acids and some aliphatic acids are included in this type of assay. Glucuronic acid conjugates are not extracted.

2. Determination of the N content of group 1 extracts

Griffith (1926) removed extracted urea and the N content of the residue gave an estimate of the amino acid conjugated aromatic and phenolic acids extracted.

3. The titratable acidity of hydrolysed samples; extracts made with non-polar solvents

Only aromatic acids are extracted from acidified hydrolysates by these solvents; as the samples are hydrolysed, both conjugated and unconjugated acids are determined. Kingsbury & Swanson (1921) used chloroform as a solvent; Bray *et al.* (1946) used light petroleum and subsequently toluene (Bray *et al.*, 1947).

4. Specific analyses

Gaffney *et al.* (1954) developed a colorimetric assay for glycine conjugates of aromatic acids separated by paper chromatography.

Other specific assay systems for particular aromatic acids were developed but only in a few instances did these methods permit the total output of individual aromatic acids to be determined. Some workers (e.g. Chandler & Lewis, 1932) did attempt to identify the aromatic acids excreted but the methods used were laborious and subject to interference.

DETERMINATION OF PHENOLIC COMPOUNDS

The wide variety of closely related phenols and phenolic acids excreted by ruminants (Suemitsu *et al.*, 1968) and the difficulties involved in their quantitative analysis mean that few reports of their output are found in the older literature. Most reports depend on methods involving partition of solvent extracted phenolic compounds into neutral (phenol) and acidic (phenolic acid) fractions by washing the extracts with sodium bicarbonate solutions (Barness *et al.*, 1963). Unfortunately, phenols do not partition in this way: about 20% of monohydroxyphenols and all of the more polar trihydroxyphenols are extracted with the acids into the bicarbonate phase (A. K. Martin, unpublished observations). Total phenols in each fraction are then determined colorimetrically by reaction with a non-specific reagent such as the Folin-Ciocalteau reagent. The phenol content of each was calculated from standard curves of an appropriate reference phenol with this reagent. The choice of standard is critical as the responses of the Folin-Ciocalteau reagent to different phenols may vary by a factor of three (A. K. Martin, unpublished observation). The application of such an analytical scheme did show that between 5 and 28% of the titratable ether-soluble acids excreted in the urine of sheep were phenolic (*p*-hydroxyphenylacetic acid reference) and that between 0·2 and 0·8 g/day of phenols (*p*-cresol reference) were excreted. Nehring & Zelck (1965) using a similar non-specific method with cattle urine, found that between 37 and 74% of the ether-soluble acids excreted was hippuric acid and that the output of phenolic acids in some instances exceeded that of hippuric acid.

DEVELOPMENT OF SPECIFIC ANALYTICAL TECHNIQUES

Aromatic acids

Samples were hydrolysed in dilute alkali (Martin, 1969*b*) and either extracted continuously with diethyl ether for 16 hours at pH 1 for aromatic, phenolic and aliphatic acid determination or the solvent was replaced by light petroleum or toluene for aromatic acid assay. With the last solvent, monohydroxy-phenols or *o*-hydroxyaromatic acids may interfere with the assay in some samples. The former can be removed partially by bicarbonate extraction.

Benzoic acid assay

This acid absorbs u.v. light strongly at 230 nm in acidified aqueous solution and use was made of this to determine the benzoic acid content of light petroleum extracts of urine (Martin, 1969*b*; 1973). Other aromatic acids could not be estimated by this procedure.

Partition of aromatic acids

Gas chromatographic techniques have been developed for the separation of free aromatic acids (Martin, 1970*a*) but more efficient separations may be obtained with their methyl esters on polar columns (Martin, 1973).

Phenols

It has not proved possible to obtain adequate resolution of phenols by thin-layer chromatography (t.l.c.). Micro-crystalline cellulose layers developed in solvent systems based on those described by Pastuska & Petrowitz (1962) have, however, been useful in confirming the identity of phenols separated by gas-liquid chromatography (g.l.c.). T.l.c. of diazotized *p*-nitroaniline derivatives has also proved useful in separating phenol isomers difficult to resolve by g.l.c. (Smith & Sullivan, 1964).

G.l.c. of phenolic compounds

More than forty phenolic compounds have been found in urine samples (Armstrong *et al.*, 1956). To improve resolution of phenolic compounds it is necessary to separate the phenolic acid fraction from phenols. The most successful method was to hydrolyse samples with a β-glucuronidase plus sulphatase enzyme preparation (Bakke & Scheline, 1969) and then to extract phenols only with diethyl ether at pH 7. Bakke & Scheline (1969) maintained that to obtain complete recovery of phenols, extractions should be made at pH 1 but we have found that, providing the pH did not exceed 7·0, complete recoveries of phenols were obtained. Extracted phenols were acetylated and separated by g.l.c. on polar and non-polar columns.

Preparative g.l.c.

The identity of some compounds separated by g.l.c. has been confirmed by isolating the appropriate peaks after injecting large samples in to a chromatograph equipped with a manual preparative kit (Christie, 1974).

THE EFFECT OF DIET ON THE URINARY OUTPUT OF AROMATIC ACIDS

Some examples of the influence of diet on the total urinary output of aromatic acids and phenols by sheep are summarized in Tables 1 and 2 respectively. These experiments have shown that the output of aromatic acids (Martin, 1970*a*) and of some phenols (Martin *et al.*, 1975) is related directly to the amount of a particular food consumed and so, for comparative purposes, urinary outputs are expressed as output/kg food intake.

Aromatic acid output

Between 95 and 99% of the total titratable acids of the aromatic acid fraction are accounted for by the individual components of the fraction partitioned by g.l.c. or u.v. spectrophotometry (Table 1). Sometimes small amounts of cyclohexanecarboxylic acid equivalent to less than 1% of the total aromatic acid output have been found confirming the observations of others that this acid may be excreted by ruminants (Suemitsu *et al.*, 1971).

Benzoic acid excretion was greatest with young leafy forages and declined as their maturity increased but smaller amounts were excreted when protein concentrates were fed and even smaller amounts when cereal concentrates were given. Heather appeared to be a rich source of the precursors of the benzoic acid excreted in the urine of sheep. Phenylacetic acid excretion was greatest when young leafy forages or protein concentrates were eaten.

The quantities of 3-phenylpropionic and cinnamic acids excreted were characteristic of the animal rather than of the diet. Some reports (Williams & Sweeley, 1964) state that cinnamic acid is not found in mammalian urine but Lederer & Polonsky (1948) have isolated it as a minor component of equine urine. In this laboratory it has been isolated by preparative g.l.c. and its identity confirmed by u.v. spectrophotometry.

Phenol output

The data in Table 2 show that the principal phenol excreted by sheep is *p*-cresol and this agrees with similar observations by Braden *et al.* (1967) with sheep and Suemitsu *et al.* (1970) with dairy cows. Of the other phenols detected in ruminant urine, quinol and orcinol have been found in little more than trace amounts except when diets containing heather were eaten.

TABLE 1

Daily urinary output of aromatic acids by sheep, expressed as g benzoic acid equivalent per kg food intake

	Titratable acids of hydrolysed urine extracts		Acids determined by specific analyses			
	Soluble in diethyl ether	Soluble in light petroleum	Benzoic	Phenylacetic	3-Phenyl-propionic	Cinnamic
*Roughages**						
Young leafy grass	21·9	16·3	14·4	1·46	—	—
Dried grass – good	18·6	—	10·3	—	—	—
Dried grass – poor	16·3	10·1	7·4	2·00	0·03	0·12
Grass-stem emergence	—	10·3	6·9	2·97	—	—
Hay	9·9	6·2	5·7	0·20	—	—
Straw	8·5	5·7	5·4	0·25	—	—
Heather†						
Heather only	32·3	—	13·7	0·19	0·01	0·27
Heather – grass (13 : 87 w/w)	15·6	—	7·8	0·69	0·01	0·07
Concentrates‡						
(i) Protein concentrates						
Linseed meal + hay (2 : 3 w/w)	15·1	10·3	7·8	2·90	—	—
Beans + hay (1 : 1 w/w)	10·1	6·2	3·6	2·50	—	—
(ii) Carbohydrate concentrates						
Sugar beet pulp + hay (4 : 1 w/w)	—	—	4·2	0·50	0·02	0·17
Flaked maize + hay (2 : 1 w/w)	7·5	—	3·8	0·21	0·00	0·09
Barley + hay (1 : 1 w/w)	6·8	3·3	2·7	0·50	—	—

Dashes indicate that analyses were not made.
* Martin (1970*a*) and unpublished work.
† Martin *et al.* (1975).
‡ Martin (1973) and unpublished work.

TABLE 2

*Daily urinary output of phenols by sheep, expressed as mg phenols per kg food intake**

	Phenol	*m*-, *p*-Cresol	*m*-, *p*-Ethyl phenol	Catechol	Resorcinol	Quinol	4-Methyl catechol	Orcinol
Dried grass – good	38	1436	26	226	0	0	51	0
Dried grass – poor	23	1603	0	101	0	0	34	0
Hay + sugar beet pulp (4 : 1 w/w)	29	551	22	120	7	0	18	0
Sugar beet pulp + hay (1)	73	593	3	169	33	Trace	33	14
Sugar beet pulp + hay (2)	20	887	2	66	5	Trace	10	0
Heather	33	1211	0	210	0	159	297	459
Grass + heather (87 : 13 w/w)	3	670	0	160	0	28	83	24

* Martin *et al.* (1975).

AROMATIC ACIDS IN RUMEN LIQUOR

In contrast with urine, the major aromatic acid in rumen liquor is not benzoic acid but 3-phenylpropionic acid (Scott *et al.*, 1964; Cmelik & Matthews, 1965; Patton & Kesler, 1967). To extend these observations, the diurnal variation in the distribution of aromatic acids in rumen liquor has been studied (A. K. Martin, unpublished work). The animals used were fed twice daily: 3-phenylpropionic acid concentration rose to a maximum 4 hours after feeding and then declined; that of benzoic acid did not comprise more than 2% of the total aromatic acids at any sampling time whereas the concentration of phenylacetic acid was a maximum just before feeding and declined to a minimum between 4 and 6 hours after feeding. Cyclohexanecarboxylic acid was also found in rumen liquor in amounts equivalent to 14% of the total aromatic acids present: its diurnal concentration changes were similar to those of 3-phenylpropionic acid.

PRECURSORS OF AROMATIC ACIDS AND PHENOLS

The effect of fasting

On withholding food the urinary benzoic acid output of sheep declined rapidly and by the third day of fasting had reached outputs of between 6 and 11 mg BAE/kg body weight$^{0.75}$ per day (Martin, 1969*b*). These outputs are similar to those reported for man and for rats when fasted (Armstrong *et al.*, 1955). In contrast, the output of phenylacetic acid increased at first on starvation and then decreased to a stable but high level of output for fasting periods of up to 10 days' duration (Martin, 1973). This stable level was two to four times greater than the phenylacetic acid output of normal adult man when fed (Seakins, 1971) or fasted (Stein *et al.*, 1954). Amongst phenols only *p*-cresol could be detected in significant quantity in the urine of fasted sheep (A. K. Martin, unpublished work).

Phenylacetic acid precursors

Comparison of the urinary aromatic acid outputs of sheep given six different cuts of perennial ryegrass showed that there was an exponential relationship between phenylacetic acid excretion and digestible N intake (Martin, 1973). The shape of the curve was considered to be due to the effect of different diets on the efficiency with which rumen micro-organisms could use phenylacetic acid produced in the rumen for their own phenylalanine synthesis (Allison, 1965). Scott *et al.* (1964) have postulated that phenylalanine and tyrosine undergo Stickland reactions in the rumen: in this reaction pairs of amino acids are deaminated, one loses a carbon atom from its side chain and acts as a hydrogen donor to the other which does not lose carbon. These workers found that phenylalanine acted as a hydrogen donor yielding phenylacetic acid; tyrosine acted partly as a hydrogen donor yielding *p*-hydroxyphenylacetic acid and partly as a hydrogen acceptor yielding 3- (*p*-hydroxyphenyl-) propionic acid. Ruminal metabolism of these phenolic acids would yield *p*-cresol and 3-phenylpropionic acid respectively with the latter being oxidized in the liver to produce benzoic acid. The quantitative significance of these reactions was not studied *in vivo* by Scott *et al.* (1964). Experiments were conducted, therefore, in which a readily fermentable protein was infused into either the rumen or the abomasum of sheep (Martin, 1973). These studies showed that when casein was infused into the rumen, 95% of the constituent phenylalanine was excreted as phenylacetic acid and that 36% of the constituent tyrosine was excreted as *p*-cresol (A. K. Martin, unpublished work). No increment in urinary benzoic acid was found on the ruminal infusion of protein whereas on the abomasal infusion of protein, urinary output of aromatic compounds did not change. It was concluded that the production of phenylacetic acid from protein was confined to the rumen and that in these experiments tyrosine acted only as a hydrogen donor in the Stickland reaction. The maximum phenylalanine content of infused protein was 7·5 g/day. Woolf (1951) suggested that phenylacetic acid may be a minor end-product of phenylalanine catabolism in man: the lack of any phenylacetic acid increment on abomasal infusion of casein to sheep suggests that endogenous phenylalanine catabolism is of negligible importance as a precursor of urinary phenylacetic acid in sheep.

Free phenylacetic acid does not occur in plant foods but phenolic phenylacetic acids are found in forages and may be produced also in the rumen on microbial metabolism of flavan-3-ols (Cheng *et al.*, 1971). There is, however, no evidence that these acids can be dehydroxylated by rumen micro-organisms to yield phenylacetic acid. On present evidence it may be concluded that most urinary phenylacetic acid excreted by ruminants is derived from microbial metabolism of phenylalanine and that the quantity excreted represents an equilibrium between the rates of production and re-utilization of phenylacetic acid by rumen micro-organisms.

Benzoic acid precursors

Several possible benzoic acid precursors are present in plants. After ruminal or abomasal administration of some of these compounds, changes in the aromatic acid and phenol composition of urine and rumen liquor have been studied (Martin, 1975). Some results are summarized in Fig. 1.

Reports of large outputs of urinary hippuric acid by man are usually associated with consumption of foods rich in quinic (Mosonyi *et al.*, 1948) or chlorogenic (3-0 (3, 4-dihydroxycinnamoyl-) quinic) acids (Oomen,

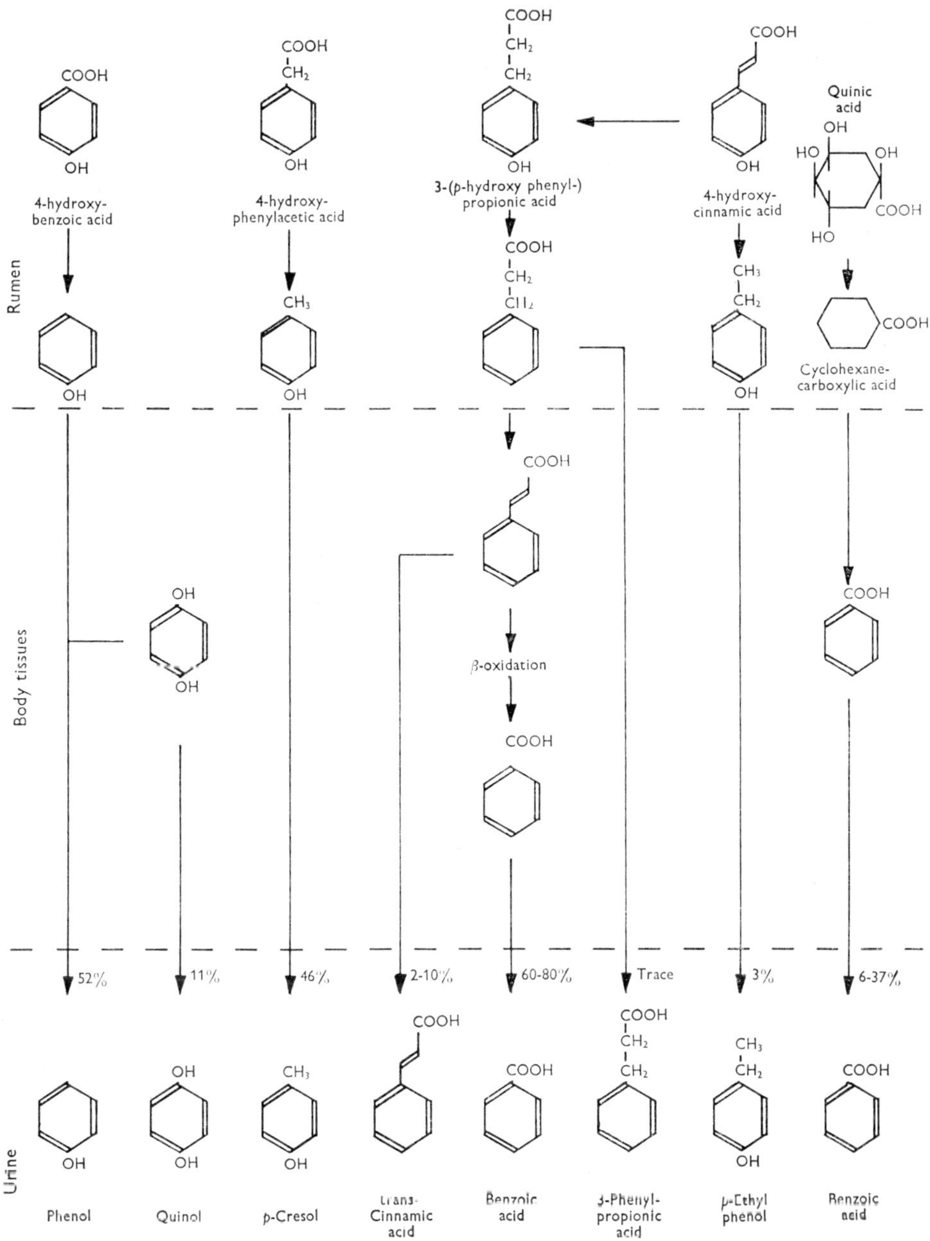

Fig. 1. *The metabolism of some precursors of phenols and aromatic acids illustrating the role of the rumen microflora* (*top section*) *and body tissues* (*middle section*) *on metabolites finally excreted in the urine* (*bottom section*). The percentages of ruminally infused acids recovered as urinary metabolites are given.

1967). Asatoor (1965) has shown that the microflora of the intestine are necessary for this 'aromatization' of quinic (1,3,4,5-tetrahydroxy-cyclohexanecarboxylic acid) and shikimic acids (3,4,5-trihydroxycyclohex-l-enecarboxylic acid). It might be expected then that ruminally infused quinic acid would be extensively metabolized to benzoic acid and excreted as such in the urine. Contrary to expectations there was no increase in the normally very small quantities of benzoic acid in rumen liquor but there were significant increases in concentrations of cyclohexanecarboxylic acid; smaller increases were found also when any cinnamic acid derivative was infused but not in experiments when other aromatic or phenolic acids were administered. In addition only rather small increments in urinary benzoic acid were obtained (See Fig. 1). No increments were observed when cinnamic acid derivatives were infused into the abomasum. Cyclohexanecarboxylic acid is aromatized to benzoic acid in the liver of many species (Beer *et al.*, 1951) and when this acid was infused ruminally or abomasally to sheep, urinary increments of benzoic acid equivalent to 40% of the infused acid were obtained. These experiments suggest that cyclohexanecarboxylic acid is produced in the rumen and then aromatized after absorption: the role of intestinal micro-organisms in the metabolism of shikimic acid is similar (Brewster *et al.*, 1976). Complete aromatization of all the alicyclic acids in perennial ryegrass given to sheep could account for only a part of the observed output of benzoic acid in the urine (Martin, 1970*a*, *b*). As only partial aromatization of quinic acid has been observed after ruminal infusion, other precursors must be present in quantity in ruminant foods.

The phenolic cinnamic acids which are distributed widely in plant tissues (Harborne & Simmonds, 1964) appeared to be likely precursors of the 3-phenylpropionic acid found in rumen liquor. However, although the metabolism of these compounds by the intestinal microflora of a number of species has been studied, these investigations have been confined to phenolic metabolites (e.g. Scheline, 1968*a*) and the extent of aromatic acid production has not been determined. The intestinal microflora can decarboxylate phenolic derivatives of benzoic, phenylacetic or cinnamic acids if they contain a 4-hydroxyl group; similar derivatives of 3-phenylpropionic acid are not decarboxylated (Scheline, 1968*a*, *b*). These results were confirmed (Martin, 1975) but decarboxylation of phenolic cinnamic acids was small (about 3%) presumably because the side-chain was hydrogenated rapidly to yield a 3-phenylpropionic acid derivative (see Fig. 1). Benzoic or phenylacetic acid derivatives containing 2- or 3-hydroxy or 2,5-dihydroxy substituents were not metabolized by rumen micro-organisms in contrast with analogous phenolic cinnamic acids, all of which were extensively hydrogenated and dehydroxylated yielding elevated concentrations of 3-phenylpropionic acid in the rumen. As expected, both cinnamic and 3-phenylpropionic acid were oxidized extensively to benzoic acid by the body tissues after both ruminal or abomasal infusion (70 to 100%). Microbial metabolism was an essential stage for metabolism of their phenolic derivatives to benzoic acid; for most of these acids this meant exposure to metabolism in the rumen. Some phenolic cinnamic acids, however, were partly (13 to 47%) excreted in the urine as benzoic acid following abomasal infusion; 3,4-dihydroxy- and 3-methoxy-4-hydroxy cinnamic acids and chlorogenic acid were excreted in the bile and thus exposed to microbial metabolism in the large intestine. Associated with the observed urinary benzoic acid increments, small but significant increments in cinnamic acid were found; these were equivalent to between 2 and 10% of the acid infused. The cinnamic acid found normally in ruminant urine is probably an intermediate in the β-oxidation of 3-phenylpropionic acid (Ranganathan & Ramasarma, 1971) in the liver. The significant differences between animals noted earlier suggest that individual variations in the efficiency of oxidation of 3-phenylpropionic acid may be responsible for these differences in cinnamic acid output.

Phenol precursors

Sheep rumen liquor extensively decarboxylated 4-hydroxy-containing benzoic and phenylacetic acids (Fig. 1). The phenols reported in Table 2 as normal urinary constituents are probably all derived by decarboxylation of the appropriate phenolic acid in the rumen. Thus, 4-hydroxybenzoic acid was decarboxylated (63%) on ruminal infusion, 17% of the phenol produced being hydroxylated to quinol in the body tissues. The major urinary phenol, *p*-cresol, was produced in the rumen on decarboxylation of 4-hydroxyphenylacetic acid (Fig. 1) and catechol was formed in the rumen on demethoxylation and decarboxylation of 3-methoxy-4-hydroxy-benzoic acid (64%) or on decarboxylation of 3,4-dihydroxybenzoic acid (79%). Gallic acid was metabolized in the rumen (48%) to yield pyrogallol (77%), which was further dehydroxylated to yield resorcinol.

Simple phenols are not common constituents of plants. In the *Ericaceae*, however, orcinol (Harborne & Williams, 1971) and quinol glycosides (Thieme & Winkler, 1971) are found. These phenols are not metabolized by ruminants and are excreted quantitatively; urinary orcinol output may thus be used as an indicator of the quantity of heather eaten if its orcinol content is known (Martin *et al.*, 1975 and Fig. 2).

SIGNIFICANCE OF AROMATIC ACID AND PHENOL PRODUCTION IN RUMINANTS

In young grass the concentration of soluble phenolic acids can account for the greater part of the observed benzoic acid output of sheep (Martin,

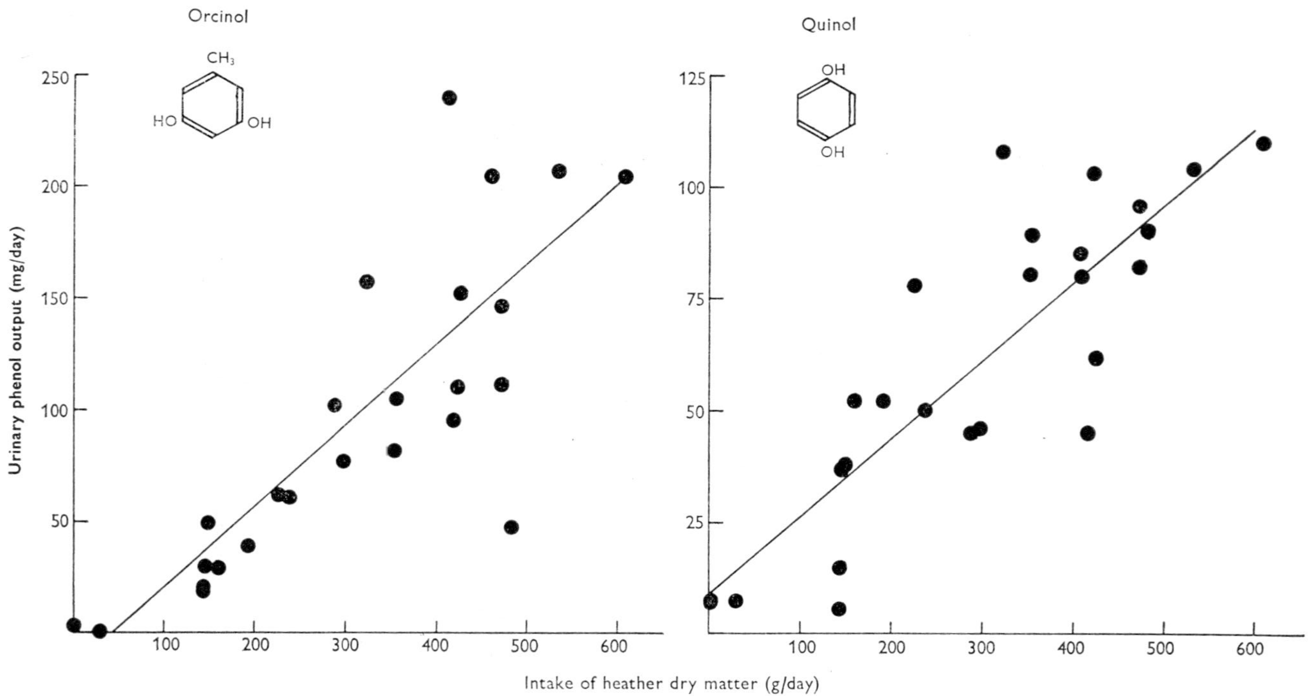

Fig. 2. *Relationships between heather intake and urinary excretion of phenols.* Regression equations relating mg phenol output (Y) to g heather dry matter intake (X) were: orcinol, $Y = 0{\cdot}365X - 15$; quinol, $Y = 0{\cdot}191X + 4$. Both were significant at $P < 0{\cdot}001$.

1970*a*). With increasing maturity of the grass, the concentration of soluble phenolic compounds decreased more rapidly than did the urinary output of benzoic acid. Bound forms of phenolic cinnamic acids are particularly common in the *Gramineae*. Their release during digestion probably accounts for most of the remaining aromatic acid production in ruminants. Bound phenolic cinnamic acids are found in lignin (Higuchi *et al.*, 1967), in plant cutins and suberins (Riley & Kolattukudy, 1975), in non-lignified plant cell walls (Harris & Hartley, 1976), in proteins (van Sumere *et al.*, 1973) and may be involved in grass hemicellulose cross-linkages (Morrison, 1974).

Aromatic acids have been regarded as end products of metabolism in mammals but some of those produced in the rumen may not be. Williams & Evans (1975) have outlined an anaerobic reductive pathway of ring-fission involving cyclohexanecarboxylic acid as an intermediate. The finding of this acid when ruminal infusions of quinic or cinnamic acid derivatives were made suggests that the incomplete recoveries of urinary metabolites (Fig. 1) obtained with those acids may be due to their use as an energy source by rumen micro-organisms.

The experiments reviewed suggest that urinary benzoic acid excreted by ruminants is derived from dietary precursors of no nutritive value which in non-ruminants would either not be digested or be excreted in the urine as a variety of phenolic compounds.

In contrast, phenylacetic acid and *p*-cresol are produced in quantity from useful nutrients, namely the aromatic amino acids phenylalanine and tyrosine. Other studies have shown that ruminal metabolism of tryptophan (Yokoyama *et al.*, 1975) can yield the highly toxic 3-methylindole (skatole). It may be one cause of a respiratory disorder of cattle known as fog fever, which is sometimes found when cattle are transferred abruptly from poor to good quality pasture. Skatole is produced by decarboxylation of indolyl-3-acetic acid (Yokoyama & Carlson, 1974) which may be produced from tryptophan by a reaction analogous to those already described for tyrosine and phenylalanine. Reasons why aromatic amino acids may occasionally be metabolized in the rumen to yield large quantities of potentially toxic metabolites are little understood. Two observations in the experiments described may be relevant. Firstly, phenylacetic acid concentration in the rumen was greatest immediately before feeding and declined thereafter and secondly, both phenylacetic acid and *p*-cresol production in fasted sheep remained high. These findings indicate that limitations of the supply of available energy may inhibit the re-utilization of non-toxic intermediates such as *p*-hydroxyphenylacetic acid and indolyl-3-acetic acid for microbial protein synthesis (Allison *et al.*, 1974). In consequence, the production of toxic metabolites from aromatic amino acids may be increased. Clearly, factors controlling the metabolism of aromatic amino acids in the rumen are insufficiently understood.

REFERENCES

Allison, M. J. (1965) *Biochem. Biophys. Res. Commun.* **18**, 30–35

Allison, M. J., Robinson, I. M. & Baetz, A. L. (1974) *J. Bacteriol.* **117**, 175–180

Armstrong, M. D., Chao, F.-C., Parker, V. J. & Wall, P. E. (1955) *Proc. Soc. Exp. Biol. Med.* **90**, 675–679

Armstrong, M. D., Shaw, K. N. F. & Wall, P. E. (1956) *J. Biol. Chem.* **218**, 293–303

Asatoor, A. M. (1965) *Biochim. Biophys. Acta* **100**, 290–292

Bakke, O. M. & Scheline, R. R. (1969) *Anal. Biochem.* **27**, 451–462

Barness, L. A., Mellman, W. J., Tedesco, T., Young, D. G. & Nocho, R. (1963) *Clin. Chem.* **9**, 600–607

Beer, C. T., Dickens, F. & Pearson, J. (1951) *Biochem. J.* **48**, 222–237

Blaxter, K. L. & Martin, A. K. (1962) *Br. J. Nutr.* **16**, 397–407

Blaxter, K. L., Clapperton, J. L. & Martin, A. K. (1966) *Br. J. Nutr.* **20**, 449–460

Braden, A. W. H., Hart, N. K. & Lamberton, J. A. (1967) *Aust. J. Agr. Res.* **18**, 335–348

Bray, H. G., Neale, F. C. & Thorpe, W. V. (1946) *Biochem. J.* **40**, 134–139

Bray, H. G., Ryman, B. E. & Thorpe, W. V. (1947) *Biochem. J.* **41**, 212–218

Brewster, D., Jones, R. S., & Parke, D. V. (1976) *Biochem. Soc. Trans.* **4**, 518–521

Brigl, P. & Pfähler, A. (1929) *Tierernaehrung* **1**, 30–36

Campbell, N. R. & Hey, D. H. (1944) *Nature* (*Lond.*) **153**, 745

Chandler, J. P. & Lewis, H. B. (1932) *J. Biol. Chem.* **96**, 619–636

Cheng, K.-J., Krishnamurty, H. G., Jones, G. A. & Simpson, F. J. (1971) *Can. J. Microbiol.* **17**, 129–131

Christie, W. W. (1974) *Scan* (*Pye Unicam Ltd*), **4**, 21

Cmelik, S. H. W. & Matthews, R. J. (1965) *Rhod. Zambia Malawi J. Agric. Res.* **3**, 95–99

Folin, O. & Flanders, F. F. (1912) *J. Biol. Chem.* **11**, 257-264

Gaffney, G. W., Shreier, K., DiFerrante, N. & Altman, K. I. (1954) *J. Biol. Chem.* **206**, 695–698

Griffith, W. H. (1926) *J. Biol. Chem.* **69**, 197–208

Hampton, J. W. F. (1948) *S. Afr. J. Med. Sci.* **13**, 45

Harborne, J. B. & Simmonds, N. W. (1964) in *Biochemistry of Phenolic Compounds* (Harborne, J. B., ed.), p. 77, Academic Press, London

Harborne, J. B. & Williams, C. A. (1971) *Phytochemistry* (*Oxf.*) **10**, 2727–2744

Harris, P. J. & Hartley, R. D. (1976) *Nature* (*Lond.*) **259**, 508–510

Hawk, P. B., Oser, B. L. & Summerson, W. H. (1947) *Practical Physiological Chemistry*, 12th edn, pp. 855 and 967, Churchill, London

Higuchi, T., Ito, Y. & Kawamura, I. (1967) *Phytochemistry* (*Oxf.*) **6**, 875–881

Kellerman, G. M. (1958) *J. Biol. Chem.* **231**, 427–443

Kingsbury, F. B. & Swanson, W. W. (1921) *J. Biol. Chem.* **48**, 13–20

Lederer, E. & Polonsky, J. (1948) *Biochim. Biophys. Acta* **2**, 431–438

Martin, A. K. (1966) *J. Sci Food Agric.* **17**, 496–500

Martin, A. K. (1969*a*) *Br. J. Nutr.* **23**, 389–399

Martin, A. K. (1969*b*) *Br. J. Nutr.* **23**, 715–725

Martin, A. K. (1970*a*) *Br. J. Nutr.* **24**, 943–959

Martin, A. K. (1970*b*) *J. Sci. Food Agric.* **21**, 496–501

Martin, A. K. (1973) *Br. J. Nutr.* **30**, 251–267

Martin, A. K. (1975) *Proc. Nutr. Soc.* **34**, 69A–70A

Martin, A. K., Milne, J. A. & Moberley, P. (1975) *Proc. Nutr. Soc.* **34**, 70A–71A

Morrison, I. M. (1974) *Biochem. J.* **139**, 197–204

Mosonyi, L., Oblatt, E. & Surján, M. G. (1948) *Hung. Acta Med.* **1**, 1–8

Nehring, K. & Zelck, U. (1965) *Arch. Tierernaehr.* **15**, 25–43

Oomen, H. A. P. C. (1967) *Trop. Geogr. Med.* **19**, 31–47

Pastuska, G. & Petrowitz, H.-J. (1962) *Chemikerzeitung* **36**, 811
Patton, S. & Kesler, E. M. (1967) *J. Dairy Sci.* **50**, 1505–1508
Pazur, J. H. & DeLong, W. H. (1948) *Sci. Agric.* **28**, 39–46
Ranganathan, S. & Ramasarma, T. (1971) *Biochem. J.* **122**, 487–493
Riley, R. G. & Kolattukudy, P. E. (1975) *Plant Physiol.* (*Bethesda*) **56**, 650–654
Scheline, R. R. (1968*a*) *Acta Pharmacol. Toxicol.* **26**, 189–205
Scheline, R. R. (1968*b*) *J. Pharm. Sci.* **57**, 2021–2037
Schiemann, R., Zelck, U. & Nehring, K. (1965) *Arch. Tierernaehr.* **15**, 81–91
Scott, T. W., Ward, P. F. V. & Dawson, R. M. C. (1964) *Biochem. J.* **90**, 12–24
Seakins, J. W. T. (1971) *Clinica Chim. Acta* **19**, 121–131
Smith, G. A. L. & Sullivan, P. J. (1964) *Analyst* (*Lond.*) **89**, 312–318
Smith, R. L. & Williams, R. T. (1970) in *Metabolic Conjugation and Metabolic Hydrolysis* (Fishman, W. H., ed.), p. 8, Academic Press, New York
Stein, W. H., Paladini, A. C., Hirs, C. H. W. & Moore, S. (1954) *J. Am. Chem. Soc.* **76**, 2848–2849
Suemitsu, R., Fujita, S., Mishima, Y. & Yoshimura, M. (1968) *Bull. Chem. Soc. Jpn* **41**, 1381–1385
Suemitsu, R., Fujita, S., Yoshimura, M., Gen, H., Yuasa, A. & Ushijima, J. (1970) *Agric. Biol. Chem.* **34**, 957–960
Suemitsu, R., Fujita, S. & Kamata, T. (1971) *Agric. Biol. Chem.* **35**, 1950–1954
Thieme, H. & Winkler, H.-J. (1971) *Pharmazie* **26**, 235–243
van Sumere, C. F., de Pooter, H., Ali, H. & Degrauw-van Bussel, M. (1973) *Phytochemistry* (*Oxf.*) **12**, 407–411
Vasiliu, H., Huber, Z., Timoşencu, A., Zaimov, C. & Coteleu, V. (1938) *Bull. Fac. Ştiinte Agric. Chişinau, Commun. Lab. Chim. Agric.* **2**, 27–48
Vasilev, G. (1960) *C. R. Acad. Bulg. Sci.* **13**, 685–688
Williams, C. M. & Sweeley, C. C. (1964) in *Biomedical Aspects of Gas Chromatography* (Symanski, H. A., ed.), p. 238, Plenum Press, New York
Williams, R. J. & Evans, W. C. (1975) *Biochem. J.* **148**, 1–10
Woolf, L. I. (1951) *Biochem. J.* **49**, ix–x
Yokoyama, M. T. & Carlson, J. R. (1974) *Appl. Microbiol.* **27**, 540–548
Yokoyama, M. T., Carlson, J. R. & Dickinson, E. O. (1975) *Am. J. Vet. Res.* **36**, 1349–1352

THE METABOLISM OF POLYUNSATURATED FATTY ACIDS BY THE NEONATAL RUMINANT

R. C. NOBLE and J. H. MOORE

ESSENTIAL FATTY ACID DEFICIENCY

BURR & BURR (1929) described a new deficiency disease in rats given a lipid-free diet; this disease did not occur in rats given the lipid-free diet supplemented with lard. The following year Burr & Burr (1930) discovered that certain polyunsaturated fatty acids, notably linoleic acid, 18 : 2 ($n-6$)*, contained in the dietary lard were responsible for preventing the appearance of the deficiency symptoms. Accordingly, Burr & Burr (1930) described 18 : 2 ($n-6$) as an essential fatty acid (EFA). More recent investigations have shown that polyunsaturated fatty acids other than 18 : 2 ($n-6$) possess EFA activity (Holman, 1968; Guarnieri & Johnson, 1970; Houtsmuller, 1973) and have confirmed that these acids cannot be synthesized *de novo* in animal tissues (Bernhard & Schoenheimer, 1940; Bernhard *et al.*, 1942; Mead *et al.*, 1953). The strict exclusion of EFA from the diet of rats results in a reduced growth rate associated with a decreased pigmentation of the skin and the appearance of a scaly dermatitis; other physiological changes include an impairment of reproduction, an increased metabolic rate and an increased permeability of cellular membranes (Holman, 1970). Similar symptoms have been reported in mice, guinea-pigs, hamsters, rabbits, dogs, swine and monkeys given diets deficient in EFA (Holman, 1970). In man, a dermatitis characteristic of EFA deficiency has been recorded by Hansen *et al.*,1958, in infants given a skim milk diet of low 18 : 2 ($n-6$) content, and by Holman (1973) in infant and adult patients maintained for prolonged periods by intravenous alimentation with fat-free preparations. Although it has not been possible to induce symptoms of EFA deficiency in cows, Lambert *et al.* (1954) have reported growth retardation and the development of diarrhoea and a scaly dermatitis in young dairy calves given a fat-free diet.

INTERCONVERSION OF POLYUNSATURATED FATTY ACIDS IN ANIMAL TISSUES

The polyunsaturated fatty acids of normal mammalian tissues may be differentiated into two separate groups, 18 : 2 ($n-6$) and its biosynthetic

* See IUPAC-IUB Commission on Biochemical Nomenclature (1967) *Eur. J. Biochem.* **2**, 127.

derivatives, and 18 : 3 $(n-3)$ and its biosynthetic derivatives. Neither 18 : 2 $(n-6)$ nor 18 : 3 $(n-3)$ can be synthesized *de novo* in animal tissues, but each of these polyunsaturated fatty acids may undergo chain elongation and further desaturation to give two separate series of fatty acids as follows:

Linoleic series

18 : 2 $(n-6)$→18 : 3 $(n-6)$→20 : 3 $(n-6)$→20 : 4 $(n-6)$→22 : 4 $(n-6)$

Linolenic series

18 : 3 $(n-3)$→18 : 4 $(n-3)$→20 : 4 $(n-3)$→20 : 5 $(n-3)$→22 : 5 $(n-3)$

The enzyme systems involved in these metabolic interconversions can be detected in the microsomal fractions of most animal tissues (Nugteren, 1962, 1965; Stoffel, 1963; Stoffel & Ach, 1964). There is considerable evidence that, for the animal species studied so far, fatty acids of the $n-6$ series possess maximum EFA potency whereas fatty acids of the $n-3$ series possess only low or medium EFA potency (Guarnieri & Johnson, 1970; Holman, 1970; Houtsmuller, 1973). Nevertheless there have been claims that fatty acids of the $n-3$ series have specific physiological functions such as in the prevention of multiple sclerosis (Bernsohn & Stephanides, 1967). The chain elongation and desaturase systems in the microsomal fractions of animal tissues have high substrate specificities for 18 : 2 $(n-6)$ and 18 : 3 $(n-3)$ but only very low substrate specificities for monoenoic fatty acids such as oleic and palmitoleic acids (Mead, 1968). In EFA deficiency, when animal tissues contain only very small concentrations of 18 : 2 $(n-6)$ or 18 : 3 $(n-3)$, the microsomal chain elongation and desaturase systems will then accept oleic acid, 18 : 1 $(n-9)$, as substrate and thus give rise to a third series of polyunsaturated fatty acids as follows:

Oleic series

18 : 1 $(n-9)$→18 : 2 $(n-9)$→20 : 2 $(n-9)$→20 : 3 $(n-9)$→22 : 3 $(n-9)$

Mead & Slaton (1956) discovered that the trienoic acid, 20 : 3 $(n-9)$, characteristically accumulated in the tissues of animals deficient in EFA. When normal rats were given a diet deficient in EFA, Mohrhauer & Holman (1963) noted that there were gradual decreases in the concentrations of 18 : 2 $(n-6)$ and 20 : 4 $(n-6)$ in the tissues and a gradual increase in the concentration of 20 : 3 $(n-9)$ with the onset of dermatitis. External symptoms of EFA deficiency began to appear once the 20 : 3 $(n-9)$/20 : 4 $(n-6)$ ratio in the liver was in excess of 0·4; such values for this ratio were observed in the tissues only when 18 : 2 $(n-6)$ constituted 1 to 2% or less of the total dietary energy. The 20 : 3 $(n-9)$/20 : 4 $(n-6)$ ratio in the tissues has now been used successfully to assess the EFA status of many non-ruminant animals (Holman, 1968, 1973). It may be concluded therefore that for non-ruminant animals, the minimum dietary requirement for EFA is met when 18 : 2 ($n-6$) constitutes 1 to 2% or more of the total dietary energy.

UNUSUAL COMPOSITION OF POLYUNSATURATED FATTY ACIDS IN THE TISSUES OF THE YOUNG RUMINANT

Interest in the possibility that these relationships did not apply to the young ruminant was stimulated in our laboratory by an incidental observation that emerged in 1966 from a study of the effect of cold exposure on the lipid metabolism of the neonatal lamb. The plasma lipids of the newborn lamb were found to contain very low concentrations of 18 : 2 ($n-6$) and 20 : 4 ($n-6$) but relatively high concentrations of a C_{20} trienoic acid, the identity of which was established as 20 : 3 ($n-9$); the 20 : 3 ($n-9$)/20 : 4 ($n-6$) ratio in the plasma lipids was well in excess of 0·4. Similar results had appeared in a brief report published 2 years earlier by Leat (1964) who noted also that during the first 48 hours after birth there was a pronounced increase in the 18 : 2 ($n-6$) content of the plasma fatty acids of the suckling lamb. Leat (1964) suggested that this increase was due to the polyunsaturated fatty acids provided by the colostrum. At the time, this suggestion seemed unacceptable since preliminary work in our laboratory had shown that 18 : 2 ($n-6$) constituted 0·5% or less of the total energy in the colostrum and milk of the ewe. These initial observations formed the basis of a series of experiments that have been conducted at the Hannah Research Institute over the past 8 years or so; the results of these experiments together with those obtained by other workers have established that during foetal development and the early postnatal period there are marked differences in the requirement and metabolism of EFA by the ruminant and non-ruminant animal.

There is now ample evidence that the tissues of the newborn lamb, calf and kid contain only small concentrations of EFA (Noble, 1973; Moore & Noble, 1975). For instance, Noble *et al.* (1972*b*) showed that the total carcass of a newborn lamb weighing 5·2 kg contained only 0·3 g 18 : 2 ($n-6$) and 1·3 g 20 : 4 ($n-6$). Much higher concentrations of EFA have been found, for example, in the tissues of the newborn rat (Sklan *et al.*, 1972). The results of Leat (1966) indicated that the 20 : 3 ($n-9$)/20 : 4 ($n-6$) ratios in the total plasma lipids of the newborn kid and lamb were 2·8 and 1·7 respectively; 20 : 3 ($n-9$)/20 : 4 ($n-6$) ratios of 2·4 and 3·4 were obtained for the total lipids in liver and heart tissues of the foetal lamb. Noble *et al.* (1971*b*) reported a 20 : 3 ($n-9$)/20 : 4 ($n-6$) ratio of 1·0 in the total plasma lipids of the newborn lamb; ratios of 1·27 and 0·58 were found in the plasma cholesteryl esters and phospholipids respectively.

In the tissue lipids of the newborn lamb 20 : 3 $(n-9)$/20 : 4 $(n-6)$ ratios well in excess of 0·4 have been observed by Body *et al.* (1970), Noble *et al.* (1971*a*, *c*, *d*), and Noble (1973), In the plasma lipids of the newborn calf Leat (1966) and Sklan *et al.* (1972) were unable to detect the presence of 20 : 3 $(n-9)$ but we have detected small concentrations of this trienoic acid (about 1·0% of the total plasma fatty acids) and have found 20 : 3 $(n-9)$/20 : 4 $(n-6)$ ratios of 0·66 (Moore & Noble, 1975).

Thus, on the criteria formulated from tissue fatty acid compositions for non-ruminant animals, the newborn ruminant and in particular, the newborn lamb, may be classified as being deficient in EFA. The reason for this anomaly seems to be that the unesterified fatty acid fraction is the only maternal plasma lipid fraction that is able to traverse the placenta (Van Duyne & Havel, 1960; Van Duyne *et al.*, 1960, 1962; Hershfield & Nemeth, 1968; Portman *et al.*, 1968). Although the plasma lipids of the maternal ruminant contain high concentrations of EFA, these are confined almost exclusively to the cholesteryl ester and phospholipid fractions, neither of which are available to the developing foetus (Noble *et al.*, 1971*e*). In contrast, the partition of polyunsaturated fatty acids between the various plasma lipid classes is much less evident in non-ruminant animals (Moore & Williams, 1963). The unesterified fatty acid fraction in sow plasma, for example, contains about 15% 18 : 2 $(n-6)$; appreciable concentrations of EFA are thus found in the tissues of the newborn piglet (Leat, 1966). With regard to the distribution of C_{18} and C_{20} polyunsaturated fatty acids of the $(n-6)$ series, there exists a characteristic difference between the tissues of the maternal ewe and those of the newborn lamb. For example, in the plasma phospholipids the 20 : 4 $(n-6)$/18 : 2 $(n-6)$ ratios found for the maternal ewe and newborn lamb were 0·39 and 4·9 respectively; the corresponding ratios for the plasma cholesteryl esters were 0·085 and 0·47 respectively (Noble *et al.*, 1971*b*, *e*). It seems unlikely that this difference could result from the preferential transport of 20 : 4 $(n-6)$ across the placenta; under normal conditions, 20 : 4 $(n-6)$ is hardly detectable in the plasma unesterified fatty acid fraction in the maternal ewe (Noble *et al.*, 1971*e*). It seemed possible, therefore, that the enzyme system involved in the synthesis of 20 : 4 $(n-6)$ from 18 : 2 $(n-6)$ was more active in the tissues of the foetal and neonatal lamb than it was in the tissues of the ewe. However, experiments *in vitro* with tissue homogenates and ^{14}C-labelled 18 : 2 $(n-6)$ have shown that the synthesis of 20 : 4 $(n-6)$ from 18 : 2 $(n-6)$ in the livers of foetal and neonatal lambs is much less active than it is in the liver of the ewe (R. C. Noble, J. H. Shand & J. H. Moore, unpublished work). On the other hand, our recent results (Shand *et al.*, 1977) suggest that placental tissue might be an active site for the synthesis of 20 : 4 $(n-6)$ from 18 : 2 $(n-6)$.

DIET AND THE RETENTION OF POLYUNSATURATED FATTY ACIDS BY THE YOUNG RUMINANT

In lambs reared on ewes' milk, there were small but progressive increases in the concentrations of 18 : 2 ($n-6$) in the plasma triglycerides and unesterified fatty acids during the first 8 days after birth (Noble *et al.*, 1971*b*). Increases in the 18 : 2 ($n-6$) contents of the liver triglycerides and unesterified fatty acids were more pronounced (Noble *et al.*, 1971*c*). Marked changes were observed in the plasma phospholipids of the lambs, the 18 : 2 ($n-6$) contents of which increased from 1 to about 10% during the first 24 hours of life, attained a maximum level of 17·6% on day 4 and then tended to decrease somewhat to 15·0% on day 8 (Noble *et al.*, 1971*b*). The 18 : 2 ($n-6$) concentration in the plasma cholesteryl esters increased from 2·7 to only 6·7% during the first 24 hours of life, but then increased rapidly to 22·2% on day 4 and subsequently more slowly to 26·2% on day 8 (Noble *et al.*, 1971*b*). These results are consistent with those obtained by Leat (1970). The increases in the 18 : 2 ($n-6$) contents of the liver phospholipids and cholesteryl esters in the suckling lambs were less pronounced than in the corresponding plasma lipids (Noble *et al.*, 1970*a*, 1971*c*, *d*). During the first 8 days of life, no major changes occurred in the concentrations of 20 : 4 ($n-6$) in the plasma lipids of the neonatal lamb, but in the liver phospholipids the concentration of 20 : 4 ($n-6$) increased from 5·7 to 10·5%. In the newborn lamb a 20 : 4 ($n-6$)/18 : 2 ($n-6$) ratio of 3·2 was found in the total liver phospholipids but after 2, 4 and 8 days on a diet of ewes' milk, this ratio decreased to 1·9, 1·3 and 1·1 respectively; this decrease in the 20 : 4 ($n-6$)/18 : 2 ($n-6$) ratio was correlated with a decrease in the phosphatidyl ethanolamine/phosphatidyl choline ratio in the liver phospholipids (Noble *et al.*, 1970*a*, 1971*d*). Of the two major metabolic pathways that are available for the biosynthesis of phosphatidyl choline, the one involving the methylation of phosphatidyl ethanolamine results in the formation of phosphatidyl choline molecules that are relatively rich in C_{20} and C_{22} polyunsaturated fatty acids, whereas the one involving cytidine diphosphate choline as an intermediate results in the formation of phosphatidyl choline containing high concentrations of C_{18} polyunsaturated fatty acids (Balint *et al.*, 1967; Rytter *et al.*, 1968). In studies with pre- and postnatal rats and rabbits, low activities of cytidine diphosphate choline transferase and phosphatidyl methyl transferase have been found in the liver before birth; although the acitivities of both enzymes increased after birth, the increase in the activity of cytidine diphosphate choline transferase was more pronounced than that of phosphatidyl methyl transferase (Baldwin & Cornatzer, 1968; Weinhold, 1969). An analogous increase in the activity of cytidine diphosphate choline transferase, together with an increase in the availability of 1,2 diglycerides containing 18 : 2 ($n-6$) would explain the changes in the phosphatidyl ethanolamine/phosphatidyl choline ratio, and in the fatty acid compositions of these two phospholipids observed in the liver of the lamb during the first 8 days after birth (Noble *et al.*, 1970*a*, 1971*d*), During the first 48 hours of life, the concentration of 20 : 3 ($n-9$) decreased from 5·2 to 0·25% in the plasma phospholipids of the lambs, and from 1·4% to trace amounts in the plasma cholesteryl esters (Noble *et al.*, 1971*b*). In the liver lipids of the lambs, the concentration of 20 : 3 ($n-9$) decreased less rapidly than it did in the plasma lipids (Noble *et al.*, 1971*c*). From the results obtained by Noble *et al.* (1971*b*, *c*) it may be calculated that during the first 2 days of life, the 20 : 3 ($n-9$)/20 : 4 ($n-6$) ratio decreased from 1·02 to 0·15 in the total plasma fatty acids and from 0·88 to 0·22 in the total liver fatty acids. A detailed analysis of the ewes' milk consumed by the lambs during these experiments showed that even on the day of parturition, 18 : 2 ($n-6$) constituted only between 0·5 and 0·4% of the total dietary energy (Noble *et al.*, 1970*b*). Thus rapid and pronounced improvements in EFA status are achieved by the young lamb when consuming a diet that would presumably result in the appearance of external symptoms of EFA deficiency when given to the young rat, for example. Noble & Moore (1971) and Noble *et al.* (1972*b*) conducted further experiments in which groups of lambs were killed immediately after birth whereas others were hand-reared on a diet of cows' milk and were killed at 10, 20 and 30 days after birth; the total dietary intakes of 18 : 2 ($n-6$) were determined as were the contents of 18 : 2 ($n-6$), 20 : 4 ($n-6$) and 20 : 3 ($n-9$) in the whole carcasses of the lambs. Although 18 : 2 ($n-6$) constituted only about 1% of the total energy in the cows' milk, the 20 : 3 ($n-9$)/20 : 4 ($n-6$) ratio in the carcasses of the lambs decreased from 1·6 at birth to 0·38, 0·21 and 0·15 on the 10th, 20th and 30th days after birth respectively. In spite of the inevitable errors that are inherent in comparative slaughter experiments, it was found that during the first 20 days the total dietary intake of 18 : 2 ($n-6$) was similar to the increment in the EFA, 18 : 2 ($n-6$) plus 20 : 4 ($n-6$), content of the carcass. During the period between days 20 and 30, however, the total dietary intake of 18 : 2 ($n-6$) was about five times greater than the increment of the EFA content of the carcass. It seemed reasonable to conclude therefore that there was a highly efficient absorption and retention of dietary 18 : 2 ($n-6$) by the lamb during the first 20 days of life when the 20 : 3 ($n-9$)/20 : 4 ($n-6$) ratios in excess of 0·4 obtained for the carcass lipids indicated that there was a deficit of EFA in the tissues. Once this deficit had been made good, as indicated by a reduction in the 20 : 3 ($n-9$)/20 : 4 ($n-6$) ratio in the carcass lipids to 0·38 on the 20th day after birth, the subsequent retention of dietary 18 : 2 ($n-6$) by the lamb was much less efficient. Similar but less extreme changes have been reported by Moore & Noble (1975) and Noble *et al.* (1975*a*) in the

20 : 3 $(n-9)$/20 : 4 $(n-6)$ ratio in the total plasma lipids of calves reared on cows' milk, the 18 : 2 $(n-6)$ content of which amounted to 1% of the total energy; the 20 : 3 $(n-9)$/20 : 4 $(n-6)$ ratio decreased from 0·55 at birth to 0·41 and 0·24 after 24 and 48 hours respectively. In contrast, the 20 : 3 $(n-9)$/20 : 4 $(n-6)$ ratio in the serum of human infants given a diet of cows' milk rose from 0·38 on the day of birth to about 1·0 on the 40th day after birth. Over a similar period there was a marked decrease in the 20 : 3 $(n-9)$/20 : 4 $(n-6)$ ratio in the serum of infants given human milk, the 18 : 2 $(n-6)$ content of which amounted to 3·5% of the total energy (Holman *et al.*, 1965; Pikaar & Fernandes, 1966; Olegård & Svennerholm, 1971). In a most elegant study of comparative nutrition, Sklan *et al.* (1972) fed newborn calves and rats on various diets including those in which 18 : 2 $(n-6)$ provided 0·01, 0·32 or 1·00% of the total energy. When any of these three diets was given to calves the concentrations of 18 : 2 $(n-6)$ and 20 : 4 $(n-6)$ in the fatty acids of the erythrocytes and of the plasma cholesteryl esters and phospholipids increased rapidly during the first few days of the experiment; increases were least in the calves given the diet with the smallest 18 : 2 $(n-6)$ content. Very different results were obtained when exactly the same diets were given to rats. When the diet in which 18 : 2 $(n-6)$ provided 0·32% of the total energy was given to rats for a period of 50 days there were marked decreases in the 18 : 2 $(n-6)$ and 20 : 4 $(n-6)$ concentrations in the plasma cholesteryl esters and phospholipids, and marked increases in the 20 : 3 $(n-9)$/20 : 4 $(n-6)$ ratios in these blood lipid fractions. That the young ruminant animal is sensitive to diets of extremely low EFA content was shown by Noble *et al.* (1971*b*, *c*) who fed lambs from birth to 8 days on a diet in which 18 : 2 $(n-6)$ provided only 0·001% of the total energy. During this period the 20 : 3 $(n-9)$/20 : 4 $(n-6)$ ratios in the plasma phospholipids and cholesteryl esters increased from 0·88 to 2·86 and from 1·26 to 5·2 respectively; in the liver lipids the 20 : 3 $(n-9)$/20 : 4 $(n-6)$ ratio increased from 1·6 to 2·6 in the cholesteryl esters, from 0·84 to 1·46 in the phospholipids, from 0·98 to 1·36 in the unesterified fatty acids and from 0·76 to 1·23 in the triglycerides. It would have been interesting to ascertain whether the characteristic external symptoms of EFA deficiency would have appeared in lambs that were maintained for prolonged periods on diets of such low EFA content. Unfortunately, it has been our experience that when lambs are given diets that provide no more than 0·001% of the total energy as 18 : 2 $(n-6)$ they become highly susceptible to infectious disease, and death from such causes occurs usually after 2 to 3 weeks.

There is clear evidence then that the level of 18: 2 $(n-6)$ in the diet required to satisfy the minimum requirement for essential fatty acids in the young lamb and calf is far less than that required by non-ruminant animals. This difference may be due to the fact that EFA appear to be utilized more efficiently in the tissues of the ruminant animal. It has been established that certain EFA act as precursors in the biosynthesis of prostaglandins (Ramwell *et al.*, 1968) but the complete biochemical function of EFA is yet to be elucidated. However, there is evidence that polyunsaturated fatty acids of the $n-6$ series are utilized for the synthesis of those phospholipids and cholesteryl esters that are essential structural components of normal cellular and subcellular membranes (Aaes-Jørgensen, 1961; van Deenen, 1966). In ruminant animals given normal diets, the highly selective incorporation of polyunsaturated fatty acids into tissue cholesteryl esters and phospholipids (Moore, 1974) ensures that virtually all the EFA in the tissues are in a form that enables the animal to obtain maximum benefit from their specialized function in membrane structure and function. On the other hand, the preferential incorporation of polyunsaturated fatty acids into the phospholipids and cholesteryl esters of the tissues in non-ruminant animals is only partial, and appreciable concentrations of EFA occur in the tissue triglycerides and unesterified fatty acids (Moore & Williams, 1963; Swell & Treadwell, 1963) which serve mainly as a source of energy. It is likely therefore that only a proportion of the EFA in the tissues of non-ruminant animals is in a form that can exert maximum biological activity. The highly efficient utilization of EFA in the tissues of ruminant animals is a process that presumably has evolved to counteract the fact that, due to the hydrogenating activity of rumen microorganisms (Viviani, 1970; Dawson & Kemp, 1970), only small amounts of dietary EFA reach the small intestine. Paradoxically this highly efficient utilization of EFA in the tissues of the maternal ruminant does appear to present the foetus and neonate with certain problems since the foetus can only obtain EFA from the unesterified fatty acids of the maternal plasma, and the suckling neonate can only obtain EFA indirectly from the triglycerides of the maternal plasma; under normal conditions low-density lipoprotein triglycerides constitute the only maternal plasma lipid fraction that donates fatty acids for milk fat synthesis in the mammary gland.

POLYUNSATURATED FATTY ACIDS IN THE SKIN SURFACE LIPIDS

The triglycerides isolated from the various tissues of ruminant animals normally contain only 2 to 3% 18 : 2 ($n-6$); a notable exception to this characteristic feature of adult ruminant tissue lipids is provided by the skin surface triglycerides which are of epidermal and sebaceous origin and have been found by Noble *et al.* (1974, 1975*a*) to contain up to 25% 18 : 2 ($n-6$). It can only be assumed that this unusually high concentration of 18 : 2 ($n-6$) in the skin surface triglycerides of the adult ruminant animal is not without physiological significance. Although the precise function of the skin surface lipids is unknown, it is generally accepted that they afford some protection against the environment, in particular against invasion of the skin surface by microbial or other exogenous agents. Hydrolysis of these skin surface triglycerides would give rise to free 18 : 2 ($n-6$) which would undoubtedly influence the survival of bacteria on the skin surface (Nieman, 1954). Moreover, EFA are known to play an important part in regulating the loss of moisture from the skin (Holman, 1968). In view of the poor EFA status of the neonatal ruminant it was of interest therefore to investigate the variation in the 18 : 2 ($n-6$) content of the skin surface triglycerides during the first few weeks after birth. Noble *et al.* (1975*a*) found that in the calf the skin surface triglycerides contained only 2 to 3% 18 : 2 ($n-6$) during the first 3 weeks after birth; between the 3rd and 4th week there was an abrupt increase in the 18 : 2 ($n-6$) content of the skin surface triglycerides and this continued until a level of about 20% was attained by the 6th week after birth. Comparison of these changes in the fatty acid composition of the skin surface triglycerides with those observed in the fatty acid compositions of the plasma lipid fractions in the calves during the same period suggested that the 18 : 2 ($n-6$) in the skin surface triglycerides could have been derived only from the plasma phospholipids or cholesteryl esters. Determination of the structure of the skin surface triglycerides showed that 18 : 2 ($n-6$) was almost exclusively esterified in position 2; this structure would be consistent with the skin surface triglycerides being synthesized from a 1,2 diglyceride that had been derived from the degradation of plasma phosphatidyl choline.

HEAT EXPOSURE AND THE METABOLISM OF POLYUNSATURATED FATTY ACIDS

Interest in the effect of heat stress on the metabolism of polyunsaturated fatty acids by the ruminant animal was stimulated by J. C. O'Kelly, a visiting research worker from the CSIRO Cattle Research Laboratory in Queensland, Australia. From experiments in Australia, O'Kelly (1968) had observed that the 18 : 2 ($n-6$) content of the plasma cholesteryl esters of heat-stressed cattle was considerably less than that in the plasma cholesteryl esters of control animals. This observation was confirmed and extended in experiments conducted at the Hannah Research Institute (Noble *et al.*, 1972*a*, 1973). In steers exposed to a high temperature (35°C) the 18 : 2 ($n-6$) content of the plasma cholesteryl esters was markedly less than that in the plasma cholesteryl esters of steers maintained at 22°C, but exposure to the high temperature had little or no effect on the 18 : 2 ($n-6$) content of the plasma phosphatidyl choline. A high proportion of the cholesteryl linoleate in bovine plasma is synthesized by the activity of lecithin-cholesterol-acyl transferase in the plasma (Noble *et al.*, 1972*a*), and from experiments *in vitro* it was found that the synthesis of cholesteryl linoleate by this mechanism was greatly reduced in plasma obtained from steers exposed to a temperature of 35°C (Noble *et al.*, 1973). The results of these experiments led to a study of the effects of heat exposure on the metabolism of polyunsaturated fatty acids in the young lamb. In lambs maintained at a temperature of 22°C and given a diet of cows' milk the 18 : 2 ($n-6$) content of the plasma cholesteryl esters increased from 2·8 to 41% during the first 7 days of life but in lambs maintained for the same period at 30°C the 18 : 2 ($n-6$) content of the plasma cholesteryl esters increased to only 26% (Noble & Moore, 1974). Again, there was evidence that the exposure to the high temperature decreased the synthesis of cholesteryl linoleate by the lecithin-cholesterol-acyl transferase system in the plasma of the lamb (Noble *et al.*, 1975*b*). The effect of exposure to a high temperature on the 18 : 2 ($n-6$) content of the plasma lipids was greatly reduced when the lambs were given a dietary supplement of safflower oil (Noble & Moore, 1974). In the high temperatures naturally encountered in Queensland, deaths due to hyperthermia occur frequently in lambs during the first 8 days or so after birth (Moule, 1954; J. C. O'Kelly, personal communication). It seemed of interest therefore to investigate the effect of a dietary supplement of 18 : 2 ($n-6$) on the ability of lambs to withstand heat stress. In a preliminary experiment, lambs were given a diet of cows' milk or cows' milk supplemented with 18 : 2 ($n-6$) and exposed to a temperature of 35°C (R. C. Noble, J. A. McLean & J. H. Moore, unpublished observations). The lambs given the diet of cows' milk exhibited signs of heat stress but those given the supplement of 18 : 2 ($n-6$) showed no signs of heat stress whatsoever. Moreover, the heat loss through the skin of the lambs given the dietary supplement of 18 : 2 ($n-6$) was considerably greater than that observed for the lambs given the diet of cows' milk alone.

IMPROVEMENT OF THE ESSENTIAL FATTY ACID STATUS OF THE YOUNG RUMINANT

Many of the physiological changes that result from EFA deficiency may be attributed to a breakdown in membrane structure and function; these changes include an increased fragility and permeability of cellular membranes and a lowered resistance to infectious disease (Holman, 1968). Moreover, it is known also that polyunsaturated fatty acids such as 18 : 2 ($n-6$) and 20 : 4 ($n-6$) possess antibacterial activity (Nieman, 1954), and there is evidence that the presence of these fatty acids in the small intestine may provide a mechanism whereby the numbers of potential pathogenic micro-organisms in the alimentary tract are controlled (Fuller & Moore, 1967). Therefore, an improvement in the EFA status of the young calf or lamb during the first few critical days after birth may be of practical significance. Such improvements can result from the simple incorporation of 18 : 2 ($n-6$) into the post-parturient diet of the lamb when considerable increases in the EFA concentrations in the plasma and tissue lipids may be readily achieved within a short period (Noble & Moore, 1974). However, an improved EFA status during foetal development would seem to present the greatest advantage. The biohydrogenation of EFA in the rumen may be avoided when the diet contains polyunsaturated oils 'protected' with a coating of protein treated with formaldehyde (Scott *et al.*, 1971). The inclusion of protected polyunsaturated supplements in the diet of adult sheep increases the concentrations of 18 : 2 ($n-6$) in all plasma lipid fractions including the unesterified fatty acids and low-density lipoprotein triglycerides (Noble *et al.*, 1977*c*). Consequently, when a dietary supplement of protected polyunsaturated oil was given to ewes during the last 2 months of pregnancy, the elevated levels of unesterified 18 : 2 ($n-6$) in the maternal plasma resulted in relatively high concentrations of 18 : 2 ($n-6$) and 20 : 4 ($n-6$), and a 20 : 3 ($n-9$)/20 : 4 ($n-6$) ratio of less than 0·4 in the plasma and tissues of the newborn lamb (Noble *et al.*, 1977*b*). Furthermore, the elevated levels of 18 : 2 ($n-6$) in the low density lipoprotein triglycerides of the maternal plasma resulted in increased concentrations of 18 : 2 ($n-6$) in the colostrum and milk secreted by the ewes; thus larger amounts of EFA were available to the suckling lambs (Noble *et al.*, 1977*c*).

Throughout the series of experiments on the metabolism of polyunsaturated fatty acids by the young lamb we have observed consistently that in blood samples taken immediately after birth there was a marked tendency for the erythrocytes to disrupt during normal handling in the laboratory. The fact that this tendency gradually disappeared during the first week after birth coupled with the observation by MacMillan & Sinclair (1958) that EFA deficiency in rats resulted in an increased fragility of the red blood cells suggested that the erythrocyte fragility in the newborn lamb might be related to poor EFA status. This suggestion prompted a direct comparison of erythrocyte fragilities in newborn lambs obtained from ewes given a dietary supplement of protected polyunsaturated oil with those in newborn lambs obtained from ewes given a conventional diet. There was clear evidence that the improved EFA status of the lambs born to the ewes given the supplement of protected polyunsaturated oil was associated with a highly significant reduction in erythrocyte fragility (Noble *et al.*, 1977*a*). As far as we are aware this is the first indication that the high 20 : 3 ($n-9$)/20 : 4 ($n-6$) ratio normally observed in the tissues of the newborn lamb might be related to a structural or functional abnormality.

From the evidence presented above, it could be argued that practical benefits might accrue from an improvement in the EFA status of the young ruminant during the critical period immediately after birth (Noble &

Moore, 1975). However, we have been unable to detect any significant improvement in the general health and growth rate of suckling lambs given a dietary supplement of 18 : 2 ($n-6$) during the first week of life. On the other hand, the results of a preliminary experiment conducted in New Zealand (K. Jaguch, personal communication) have shown that during the first 28 days after birth the growth rate of lambs given a dietary supplement of 18 : 2 ($n-6$) was significantly greater than that of control animals; the response to dietary 18 : 2 ($n-6$) appeared to be greater in those lambs that were weaker at birth. Whether or not there is a practical response to a dietary supplement of 18 : 2 ($n-6$) could well depend on the degree of environmental adversity encountered by the lamb during the first few days of life. A possible approach to future work might involve an investigation into the effect of a dietary supplement of 18 : 2 ($n-6$) in lambs deliberately exposed to a challenge by pathogenic micro-organisms.

REFERENCES

Aaes-Jørgensen, E. (1961) *Physiol. Rev.* **41**, 1–51

Baldwin, J. & Cornatzer, W. E. (1968) *Lipids* **3**, 361–367

Balint, J. A., Beeler, D. A., Treble, D. H. & Spitzer, H. L. (1967) *J. Lipid Res.* **8**, 486–493

Bernhard, K. & Schoenheimer, R. (1940) *J. Biol. Chem.* **133**, 707–712

Bernhard, K., Steinhauser, H. & Bullet, F. (1942) *Helv. Chim. Acta* **25**, 1313–1318

Bernsohn, L. & Stephanides, L. M. (1967) *Nature (Lond.)* **215**, 821–823

Body, D. R., Shorland, F. B. & Czochanska, Z. (1970) *J. Sci. Food Agric.* **21**, 220–225

Burr, G. O. & Burr, M. M. (1929) *J. Biol. Chem.* **82**, 345–368

Burr, G. O. & Burr, M. M. (1930) *J. Biol. Chem.* **86**, 587–622

Dawson, R. M. C. & Kemp, P. (1970) in *Physiology of Digestion and Metabolism in the Ruminant* (Phillipson, A. T., ed.), p. 504, Oriel Press, Newcastle upon Tyne

Fuller, R. & Moore, J. H. (1967) *J. Gen. Microbiol.* **46**, 23–41

Guarnieri, M. & Johnson, R. M. (1970) *Adv. Lipid Res.* **8**, 115–174

Hansen, A. E., Haggard, M. E., Boelsche, A. N., Adam, D. J. D. & Wiese, H. F. (1958) *J. Nutr.* **66**, 565–576

Hershfield, M. S. & Nemeth, A. M. (1968) *J. Lipid Res.* **9**, 460–468

Holman, R. T. (1968) *Prog. Chem. Fats Other Lipids* **9**, 275–348

Holman, R. T. (1970) *Prog. Chem. Fats Other Lipids* **9**, 607–682

Holman, R. T. (1973) in *Dietary Lipids and Postnatal Development* (Galli, C., Jacini, G. & Pecile, A., eds), p. 127, Raven Press, New York

Holman, R. T., Hayes, H. W., Rinne, A. & Söderhjelm, L. (1965) *Acta Paediatr. Scand.* **54**, 573–577

Houtsmuller, U. M. T. (1973) in *Dietary Lipids and Postnatal Development* (Galli, C., Jacini, G. & Pecile, A., eds), p. 145, Raven Press, New York

Lambert, M. R., Jacobson, N. L., Allen, R. S. & Zaletel, J. H. (1954) *J. Nutr.* **52**, 259–272

Leat, W. M. F. (1964) *Biochem. J.* **93**, 22P

Leat, W. M. F. (1966) *Biochem. J.* **98**, 598–603

Leat, W. M. F. (1970) in *Physiology of Digestion and Metabolism in the Ruminant* (Phillipson, A. T., ed.), p. 211, Oriel Press, Newcastle upon Tyne

MacMillan, A. L. & Sinclair, H. M. (1958) in *Essential Fatty Acids* (Sinclair H. M., ed.), p. 208, Butterworth, London
Mead, J. F. (1968) *Prog. Chem. Fats Other Lipids* **9**, 159–192
Mead, J. F. & Slaton, W. H. (1956) *J. Biol. Chem.* **219**, 705–709
Mead, J. F., Steinberg, G. & Howton, D. R. (1953) *J. Biol. Chem.* **205**, 683–689
Mohrhauer, H. & Holman, R. T. (1963) *J. Lipid Res.* **4**, 151–159
Moore, J. H. (1974) in *Industrial Aspects of Biochemistry* (Spencer, B., ed.), pp. 835–863, North-Holland, Amsterdam
Moore, J. H. & Noble, R. C. (1975) in *Digestion and Metabolism in the Ruminant* (McDonald, I. W. & Warner, A. C. I., eds), p. 465, University of New England Publishing Unit, Armidale, Australia
Moore, J. H. & Williams, D. L. (1963) *Can. J. Biochem. Physiol.* **41**, 1821–1835
Moule, G. R. (1954) *Aust. Vet. J.* **30**, 153–171
Nieman, C. (1954) *Bacteriol. Rev.* **18**, 147–163
Noble, R. C. (1973) *World Rev. Anim. Prod.* **9** (1), 18–27
Noble, R. C. & Moore, J. H. (1971) *Proc. Nutr. Soc.* **30**, 61A–62A
Noble, R. C. & Moore, J. H. (1974) *Res. Vet. Sci.* **17**, 204–209
Noble, R. C. & Moore, J. H. (1975) *IRCS Med. Sci.* **3**, 1–4
Noble, R. C., Steele, W. & Moore, J. H. (1970*a*) *Biochim. Biophys. Acta* **218**, 359–362
Noble, R. C., Steele, W. & Moore, J. H. (1970*b*) *J. Dairy Res.* **37**, 297–301
Noble, R. C., Christie, W. W. & Moore, J. H. (1971*a*) *J. Sci. Food Agric.* **22**, 616–619
Noble, R. C., Steele, W. & Moore, J. H. (1971*b*) *Lipids* **6**, 26–34
Noble, R. C., Steele, W. & Moore, J. H. (1971*c*) *Br. J. Nutr.* **26**, 97–105
Noble, R. C., Steele, W. & Moore, J. H. (1971*d*) *Lipids* **6**, 926–929
Noble, R. C., Steele, W. & Moore, J. H. (1971*e*) *Res. Vet. Sci.* **12**, 47–53
Noble, R. C., O'Kelly, J. C. & Moore, J. H. (1972*a*) *Biochim. Biophys. Acta* **270**, 519–528
Noble, R. C., Steele, W. & Moore, J. H. (1972*b*) *Br. J. Nutr.* **27**, 503–508
Noble, R. C., O'Kelly, J. C. & Moore, J. H. (1973) *Lipids* **8**, 216–223
Noble, R. C., Crouchman, M. L. & Moore, J. H. (1974) *Res. Vet. Sci.* **17**, 372–376
Noble, R. C., Crouchman, M. L., Jenkinson, D. McE. & Moore, J. H. (1975*a*) *Lipids* **10**, 128–133
Noble, R. C., Crouchman, M. L. & Moore, J. H. (1975*b*) *Biol. Neonate* **26**, 117–121
Noble, R. C., Shand, J. H., Drummond, J. T. & Moore, J. H. (1977*a*) *IRCS Med. Sci.* **5**, 196
Noble, R. C., Shand, J. H. & Moore, J. H. (1977*b*) *IRCS Med. Sci.* **5**, 41
Noble, R. C., Vernon, R. G., Christie, W. W., Moore, J. H. & Evans, A. J. (1977*c*) *Lipids* **12**, 423–433
Nugteren, D. H. (1962) *Biochim. Biophys. Acta* **60**, 656–657
Nugteren, D. H. (1965) *Biochim. Biophys. Acta* **106**, 280–290
O'Kelly, J. C. (1968) *Aust. J. Biol. Sci.* **21**, 1013–1024
Olegård, R. & Svennerholm, L. (1971) *Acta Paediatr. Scand.* **60**, 505–511
Pikaar, N. A. & Fernandes, J. (1966) *Am. J. Clin. Nutr.* **19**, 194–204
Portman, O. W., Behrman, R. E. & Soltys, P. (1968) *Am. J. Physiol.* **216**, 143–147
Ramwell, P. W., Shaw, J. E., Clarke, G. B., Grostic, M. F., Kaiser, D. G. & Pike, J. E. (1968) *Prog. Chem. Fats Other Lipids* **9**, 231–273
Rytter, D., Miller, J. E. & Cornatzer, W. E. (1968) *Biochim. Biophys. Acta* **152**, 418–421
Scott, T. W., Cook, L. J. & Mills, S. C. (1971) *J. Am. Oil Chem. Soc.* **48**, 358–364
Shand, J. H., Noble, R. C., Bell, A. W., Thompson, G. E. & Moore, J. H. (1977) *IRCS Med. Sci.* **5**, 453
Sklan, D., Volcani, R. & Budowski, P. (1972) *Br. J. Nutr.* **27**, 365–374
Stoffel, W. (1963) *Hoppe-Seyler's Z. Physiol. Chem.* **333**, 71–88

Stoffel, W. & Ach, K. L. (1964) *Hoppe-Seyler's Z. Physiol. Chem.* **337**, 123–132
Swell, L. & Treadwell, C. R. (1963) in *Atherosclerosis and its Origin* (Sandler, M. & Bourne, G. H., eds), p. 301, Academic Press, New York
van Deenen, L. L. M. (1966) *Prog. Chem. Fats Other Lipids* **8**, 1–127
Van Duyne, C. M. & Havel, R. J. (1960) *Clin. Res.* **8**, 111
Van Duyne, C. M., Parker, H. R., Havel, R. J. & Holm, L. W. (1960) *Am. J. Physiol.* **199**, 987–990
Van Duyne, C. M., Havel, R. J. & Felts, J. M. (1962) *Am. J. Obstet. Gynecol.* **84**, 1069–1074
Viviani, R. (1970) *Adv. Lipid Res.* **8**, 267–346
Weinhold, P. A. (1969) *Biochim. Biophys. Acta* **187**, 85–93

THE METABOLISM OF ADIPOSE TISSUE IN THE RUMINANT

R. G. VERNON

INTRODUCTION

ADIPOSE tissue in ruminants, as in other animal species, is the major store of energy of the body. Lipid is accumulated in adipose tissue by esterification of fatty acids derived from the plasma unesterified fatty acid pool or from plasma lipoproteins following the hydrolysis of their constituent triglycerides by the action of the enzyme lipoprotein lipase. Fatty acids may also be synthesized *de novo* in adipose tissues. Fatty acids and glycerol may be released from adipose tissue triglycerides by the process of lipolysis.

The pathways of fatty acid metabolism in ruminant adipose tissue appear to be the same as those found for other animal species with the exception of the precursors that are used for fatty acid synthesis *de novo* (Bauman, 1976). In non-ruminants both glucose and acetate are used for fatty acid synthesis; but in the ruminant, acetate is the only important precursor (Ballard *et al.*, 1969). This was shown initially by comparing the relative rates of incorporation of ^{14}C into fatty acids from various ^{14}C-labelled precursors (Ballard *et al.*, 1969), and has recently been confirmed by comparing the incorporation of ^{3}H from $^{3}H_2O$ into fatty acids, which is regarded as a measure of the total rate of fatty acid synthesis (Jungas, 1968; Salmon *et al.*, 1974), with the rate of incorporation of ^{14}C from [1-^{14}C] acetate (Vernon, 1976).

Glucose does stimulate the rate of fatty acid synthesis in ruminant tissues, but this appears to be due to its ability to stimulate the formation of NADPH which is required for fatty acid synthesis (Hanson & Ballard, 1967; Ingle *et al.*, 1972). Also, glucose is the principal precursor of α-glycerophosphate which is required for fatty acid esterification (Hood *et al.*, 1972); esterification of fatty acids would prevent their accumulation within the cell which could result in the feed-back inhibition of fatty acid synthesis (Vernon, 1977*a*).

Several hormones (insulin, glucagon, adrenaline and noradrenaline) have been shown to influence ruminant adipose tissue metabolism (Bauman, 1976). The effects of these hormones, for example, the stimulation of lipogenesis by insulin or of lipolysis by glucagon or adrenaline, are qualitatively similar to those observed in non-ruminants but quantitatively they are much smaller (Bauman, 1976). However, the effect on ruminant adipose tissue metabolism of a variety of hormones, such as growth hormone, prolactin, thyroid hormones and glucocortcoids, which influence adipose tissue lipid metabolism in non-ruminants (Winegrad *et al.*, 1960; Volpe & Vagelos, 1976) does not appear to have been investigated to date. Again, it is well known that both the quantity and nature of the diet can markedly influence adipose tissue metabolism in non-ruminants (Volpe & Vagelos, 1976), but there have been few comparable studies in ruminants. Nevertheless, it appears that high-carbohydrate diets can result in increased rates of lipogenesis in sheep adipose tissue, whereas fasting or high-fat diets result in reduced rates of lipogenesis (Baldwin *et al.*, 1976). Thus in these respects the response of ruminant adipose tissue to diet is again qualitatively similar to that of non-ruminants.

In general, however, the regulation of adipose tissue metabolism in ruminants is poorly understood, especially when compared with our knowledge of the mechanisms which regulate lipid metabolism in non-ruminants such as the rat.

In subsequent sections, some aspects of ruminant adipose tissue metabolism under investigation in the Institute are described.

EFFECT OF DIETARY POLYUNSATURATED FATTY ACIDS ON OVINE ADIPOSE TISSUE METABOLISM

The supposed relationship between dietary saturated fats and cardiovascular degeneration in man led to an interest in the possibility of increasing the levels of polyunsaturated fatty acids in ruminant meat and milk (Scott *et al.*, 1970; Cook *et al.*, 1972; Pan *et al.*, 1972; Faichney *et al.*, 1973; Dinius *et al.*, 1974). Little was known of the effects of increased amounts of dietary polyunsaturated fatty acids on the metabolism of ruminant tissues. Studies with rodents, however, had shown that diets supplemented with linoleic acid resulted in a significantly greater inhibition of the rate of lipogenesis in liver slices from adult (Bartley & Abraham, 1972) and neonatal (Smith & Abraham, 1970) rats and mice than did diets containing equivalent amounts of saturated fatty acids. Lipogenesis in adipocytes from rats that were deficient in essential fatty acids and that were then given polyunsaturated fatty acids was also lower than that of adipocytes from such rats given saturated or mono-unsaturated fatty acids (Du & Kruger, 1972). In contrast, with rats receiving adequate levels of essential fatty acids, increased levels of dietary saturated fatty acids resulted in a greater inhibition of the rate of adipose tissue lipogenesis than did equivalent amounts of dietary polyunsaturated fatty acids (Waterman *et al.*, 1975). As yet the reason for the differential effect of dietary saturated and polyunsaturated fatty acids on lipogenesis is unknown.

It seemed pertinent, therefore, to investigate the effects of increased levels of dietary polyunsaturated fatty acids on the rate of lipogenesis in tissues from ruminant animals. As rates of hepatic lipogenesis are very low in ruminants (Hanson & Ballard, 1967; Ingle *et al.*, 1972), the studies were confined to adipose tissue lipogenesis.

Initial studies were carried out with neonatal lambs to avoid the problem of biohydrogenation of polyunsaturated fatty acids by rumen microflora (Reiser, 1951; Shorland *et al.*, 1957; Ward *et al.*, 1964). Preliminary studies with lambs showed that the rate of adipose tissue lipogenesis was barely detectable at birth, but after a lag-phase of 2 to 3 days it increased over ten-fold by 10 days of age (Vernon, 1975). Oral administration of 5 ml safflower oil (70% linoleic acid) each day from birth for 11 days reduced significantly the rate of lipogenesis by more than 50%, but it had no effect on the rate of acetate oxidation (Vernon, 1975). The amount of safflower oil administered represented less than 10% of the total fat intake of the lambs and had no effect on either the rate of growth or the total food intake of the lambs. This study suggested that dietary linoleic acid had a specific effect on lamb adipose tissue lipogenesis.

The availability of dietary lipid supplements (donated by Alta Lipids (UK) Ltd) that were protected against hydrolysis and hydrogenation in the rumen (Scott *et al.*, 1971) permitted an investigation of the effects of dietary polyunsaturated fatty acids on adipose tissue metabolism in sheep with a functioning rumen (Vernon, 1976). Basically, four groups of five 6-month old Cheviot sheep were fed for 6 weeks either on a control diet of hay plus concentrates, or on diets in which a proportion of concentrates was replaced by either 'protected' tallow, 'protected' polyunsaturated fatty acids (about 70% sunflower seed oil and 30% soybean oil), or maize oil. The four groups of sheep did not differ in their rates of growth. Feeding with protected tallow or maize oil, but not with protected polyunsaturated fatty acids, resulted in reduced rates of fatty acid synthesis in both perirenal and subcutaneous adipose tissue slices. In contrast, feeding with the various dietary supplements had no effect on the rate of glyceride glycerol synthesis. The activities of various enzymes associated with fatty acid biosynthesis were measured in homogenates of adipose tissue from the sheep. There was, however, no obvious relationship between the effects of the various fat-supplemented diets on the rate of fatty acid biosynthesis and on the activities of the enzymes examined. But the enzymes were assayed under optimal conditions, whereas their activities in adipose tissue may have been modulated by a variety of factors. For example, acetyl CoA carboxylase, which is thought to have a critical role in regulating the rate of fatty acid synthesis, occurs in both active and inactive forms (Denton, 1974), whereas the assay procedure used measured total activity. The correlation between the rate of fatty acid biosynthesis and total acetyl CoA carboxylase activity found in fasted and re-fed sheep (Ingle *et al.*, 1973) was not observed in this study.

Feeding with the three fat-supplemented diets had little effect on the concentration of plasma lipids of the sheep (Noble *et al.*, 1977). However, feeding with the protected polyunsaturated fatty acid diet did increase significantly the proportion of linoleic acid in several lipid classes in plasma (Noble *et al.*, 1977) and other tissues (Christie *et al.*, 1975*a*, *b*), indicating that the protection against biohydrogenation was effective and that increased quantities of linoleic acid were being absorbed by the animal.

The response of fatty acid biosynthesis in adipose tissue from adult sheep to increased levels of dietary polyunsaturated fatty acids was markedly different from that of neonatal lambs. The reason for this difference is not yet clear. Considering the differential response to dietary polyunsaturated fatty acids of adipose tissue lipogenesis in normal rats (Waterman *et al.*, 1975) and in rats deficient in essential fatty acids (Du & Kruger, 1972) it would appear that the essential fatty acid status of the animal may influence its response to dietary polyunsaturated fatty acids, for the neonatal lamb is relatively deficient in essential fatty acids (Noble *et al.*, 1972).

The reasons for the differential effects of dietary saturated and polyunsaturated fatty acids on lipogenesis are unknown. Several possibilities have been suggested, such as alteration of membrane lipid properties, modification of prostaglandin metabolism, and the formation of a lipid-containing repressor for genes for lipogenic enzymes (see Musch *et al.*, 1974). Another suggestion, that polyunsaturated fatty acids alter appetite and exert their effect by changing food intake (Gozukara *et al.*, 1972) has now been excluded both for rats (Musch *et al.*, 1974) and sheep (Vernon, 1975, 1976).

The varied effects of dietary fatty acids on lipogenesis, however, could have been due simply to the differential inhibition of the activity of one or more of the enzymes of fatty acid synthesis by the exogenous fatty acids, or more probably by their acyl CoA esters.

A comparison of the effects of different fatty acids or their acyl CoA esters on the activities of the enzymes of fatty acid biosynthesis is complicated since, being detergents, fatty acids and their acyl CoA esters can inhibit enzymic activities by non-physiological processes (Srere, 1965; Taketa & Pogell, 1966; Dorsey & Porter, 1968; Pande & Mead, 1968; Parvin & Dakshinamurti, 1970). Furthermore, both fatty acids and their acyl CoA esters are probably bound to proteins or membranes within the cell so that their effective concentration is uncertain (Ockner *et al.*, 1972; Mishkin *et al.*, 1972; Mishkin & Turcotte, 1974; Sumper, 1974). Thus it seemed preferable to compare the effects of different fatty acids on lipogenesis using intact cells. Also, it had been shown that addition of palmitic acid to the incubation medium reduced the rate of lipogenesis in rat adipocytes (Saggerson & Tomassi, 1971; Sooranna & Saggerson, 1975). So, when adipose tissue slices from either rats or sheep were incubated in Krebs-Ringer bicarbonate solutions supplemented with glucose (5 mM), acetate (2 mM) and albumin (4%, w/v), it was found that addition of either palmitic or linoleic acids reduced the rate of lipogenesis (Vernon, 1977*a*). The range of concentration of added fatty acids was 0·1 to 1 mM, which, with the endogenous fatty acids from the albumin, gave a final concentration range of 0·17 to 1·17 mM. This is comparable to the physiological range of unesterified fatty acids for both rat plasma (0·3 to 1·7 mM (Scow & Chernick, 1970)) and sheep plasma (0·2 to 1·9 mM (Noble *et al.*, 1971; Jarrett *et al.*, 1974)).

The characteristics of the reduction of the rate of lipogenesis by addition of palmitic and linoleic acid (for example the speed with which the inhibition occurred and its ease of reversal) were consonant with the fatty acids exerting their effects by inhibiting the activity of one or more of the enzymes of lipogenesis. The observation suggested also that the two fatty acids exerted their effects by the same mechanism.

A comparison of the reductions in the rates of lipogenesis resulting from the addition of palmitic, linoleic, stearic, oleic and linolenic acids and mixtures of fatty acids, the composition of which was based on that of the plasma unesterified fatty acid fraction of sheep and rat plasma, showed that stearic acid was the most effective inhibitor of lipogenesis in both rat and sheep adipose tissue; palmitic, oleic and the mixtures of fatty acids appeared to be similarly effective in their ability to reduce the rate of lipogenesis, whereas linoleic and linolenic acids were the least effective.

However, although stearic acid was a more potent inhibitor of fatty acid biosynthesis *in vitro* than the other fatty acids listed, this would not appear to account for the effects of different dietary fats on fatty acid biosynthesis in ovine adipose tissue *in vivo*. Dietary tallow resulted in a significantly greater decrease in the rate of fatty acid biosynthesis in sheep adipose tissue than did a mixture of sunflower and soybean oil (see above), but both diets increased the plasma unesterified stearic acid level to the same extent relative to that of control animals receiving a low-fat diet (Noble *et al.*, 1977). The only fatty acid whose plasma concentration was changed differently by the two dietary fats was linoleic acid, but the differences were small and although linoleic acid was a less effective inhibitor of ovine lipogensis than palmitic and oleic acids, the difference appears to be too slight to offer a reasonable explanation for the differential effects of dietary saturated and polyunsaturated fatty acids.

With regard to the effect of different dietary fatty acids on the rate of lipogenesis in rat adipose tissue, the situation is less clear. Unfortunately, in the studies described by Du & Kruger (1972) and by Waterman *et al.* (1975), the composition of the plasma unesterified fatty acids was not reported. Stearic acid, however, is a relatively minor component (about

12% by weight) of the unesterified fatty acid fraction of rat plasma (Rose *et al.*, 1964) and at this concentration its effect on fatty acid biosynthesis would be swamped by the effects of the other fatty acids. Furthermore, although linoleic acid was a less effective inhibitor of lipogenesis than were palmitic and oleic acids, the difference was small. In the experiments described above, however, adipose tissue was exposed to fatty acids for only short periods and the effects of the fatty acids were probably due to inhibition of enzyme acitivity. So, the possibility remains that over longer periods the various fatty acids may have differential effects on the rates of enzyme synthesis or degradation. To investigate this possibility it was necessary to develop a system for maintaining rat and sheep adipose tissue in organ culture for periods of several days. This is described in a subsequent section.

DEVELOPMENT OF THE PERIRENAL FAT PADS IN THE NEONATAL LAMB

As described above, administration of safflower oil to neonatal lambs impaired the development of lipogenesis in the perirenal fat pad. But it was not clear from these studies whether safflower oil influenced only lipid metabolism or whether it retarded the development of the fat pad in general. The development of the perirenal fat pad in the lamb was of further interest, as studies at the Institute and elsewhere had shown that the tissue had the morphological characteristics of brown adipose tissue at birth, but, by about 10 days of age the tissue had attained the appearance of white adipose tissue (Thompson & Jenkinson, 1969; Gemmell *et al.*, 1972). This change was associated with a loss of noradrenaline-stimulated oxygen uptake, presumably due to a loss of the capacity of non-shivering thermogenesis in the whole animal (Alexander & Williams, 1968; Thompson & Jenkinson, 1969). Also, the change from brown to white fat could be delayed by cold exposure (Gemmell *et al.*, 1972). These observations suggested that it would be worth while to examine the effects of dietary safflower oil on adipose tissue development in the neonatal lamb.

Lambs were given either cow's milk or cow's milk supplemented with 5 ml of safflower oil per day from birth until they were killed at 17 days of age. Other animals were killed at birth and at 9 days *post partum*. Changes in the morphology, the lipid, protein, RNA and DNA content, the adipocyte size and number, and the rate of protein, RNA and DNA synthesis of the fat pads were determined (Vernon, 1977*b*).

During the first 9 days after birth the perirenal adipose tissue grew proportionately at the same rate as the whole animal; 85% of the increase in adipose tissue weight was due to lipid deposition. There was no change in either the DNA content of, or the number of, adipocytes in the fat pads, the increased lipid content being accommodated by a two-fold increase in the volume of the adipocytes. During this period the transformation from 'brown' to 'white' fat occurred. The lack of change in the number of adipocytes and the tissue DNA content indicated that the change from brown to white fat was due to a differentiation of brown into white adipocytes rather than a replacement of the brown adipocytes by a new population of cells. This finding supports a similar conclusion (Gemmell *et al.*, 1972) based on morphological evidence (principally the absence of degenerating cells). Whether or not the development of white adipose tissue in other species involves a brown adipocyte-like phase is still controversial (see Smith & Horwitz, 1969; Derry *et al.*, 1972; Hahn & Novak, 1975), but for the lamb at least there appears to be such a transformation.

Between days 9 and 17 the mass of the perirenal fat pads increased proportionately more rapidly than the total body weight. This was again mainly due to deposition of lipid. Once more, there was no change in the number of adipocytes as they increased further in size, but there was a 65% increase in the amount of DNA in the tissue. Surprisingly, this increase in the DNA content of the fat pads appeared to be due to an increase in the amount of DNA per adipocyte. A similar observation was made with hepatocytes from neonatal rats (Bibbiani *et al.*, 1973). There is no explanation for these observations at present but they certainly suggest that changes in the amount of DNA in a tissue from neonatal animals may not reflect changes in cell numbers, as is usually assumed.

The number of adipocytes of the perirenal fat pad found in this study (10^9 per lamb) is similar to that found in perirenal fat pads of yearling cross-bred western castrated lambs (2×10^9 per lamb (Haugebak *et al.*, 1974)); the difference in adipocyte number could well be due to breed. Thus it seems likely that the number of adipocytes of the perirenal fat pad in sheep is determined before birth and so should not be open to postnatal manipulation. It is clear, however, that the adipocyte number of other fat pads does change after birth (Haugebak *et al.*, 1974; Allen, 1976).

The development of the rat epididymal adipose tissue has been divided into three phases: proliferation, differentiation and maturation (Greenwood & Hirsch, 1974). It appears that, for the lamb perirenal fat pad, proliferation is complete by birth. This is followed in the immediate postnatal period by a phase of differentiation as the brown adipocytes are transformed into white adipocytes. When this is complete, the adipocytes enter a mature phase when further changes in the fat pad mass are due to changes in cell size. These phases appear to be conveniently synchronized in the lamb and it seems that the lamb would provide an excellent model for a study of adipose tissue development.

Administration of safflower oil to the lambs had no effect on the rate of lipid deposition between birth and day 9 when the adipocytes were differentiating. But dietary safflower oil decreased markedly the rate of lipid deposition between days 9 and 17; this was associated with a reduction in adipocyte volume but not number. The reduced rate of lipid deposition was not due to a decrease in food intake or growth rate of the lambs receiving the safflower oil. Furthermore, the decrease in the rate of lipogenesis following the administration of safflower oil, observed previously, although presumably associated with the decreased rate of lipid deposition, cannot account for it. This is because the rate of fatty acid biosynthesis in control lambs, 1·7 μmol acetate incorporated into fatty acid/hour per g tissue, at 11 days of age, would account for only 2 to 3% of the fatty acids being deposited in the perirenal adipose tissue (about 2·5 g fat deposited/day per lamb at 11 days of age corresponding to approximately 10 μmol fatty acid esterified/hour per g tissue). Most of the fatty acids esterified in the tissue would presumably have come from plasma lipids, the lambs consuming about 70 g fat/day at 11 days of age. Thus the safflower oil must have affected other processes as well as adipose tissue fatty acid biosynthesis.

And so, just how dietary safflower oil exerts its effects remains unknown. Clearly it would be worth investigating the effects on other pathways of lipid metabolism, in particular the activity of lipoprotein lipase, as a reduction in the activity of this enzyme would effectively decrease the rate of fatty acid uptake by the tissue.

Administration of dietary safflower oil had no effect on the number of adipocytes, or on the DNA, RNA and protein content of the tissue. Also, it had no detectable effect on the transformation of brown to white fat as judged by microscopic examination of the tissue at magnifications up to 1000-fold. Thus, although dietary safflower oil has an effect on the lipid metabolism of perirenal adipose tissue from neonatal lambs, it does not appear to affect the overall development of the tissue.

MAINTENANCE OF ADIPOSE TISSUE IN ORGAN CULTURE

During the preceding studies it became increasingly apparent that it would be extremely useful to be able to maintain adipose tissue in organ culture even for periods of just a few days. Such a technique could permit an investigation of the effects of different fatty acids on the synthesis and degradation of lipogenic enzymes. It would also provide an approach for further examining the effects of dietary safflower oil on adipose tissue development in the neonatal lamb, and for exploring the other factors, in particular hormones, which may influence the development of the tissue and the transformation from brown to white fat.

A method for maintaining adipose tissue in organ culture would obviously have many other applications and should be of especial value for studying ruminant metabolism when the number of animals available may be limited for economic reasons.

Trowell (1959) demonstrated that white adipose tissue from rats could be maintained for 9 days in organ culture without any morphological evidence of degeneration. Subsequent studies have confirmed that rodent adipose tissue will survive in organ culture for several days using either morphological (Slavin & Elias, 1969, 1970) or metabolic criteria (Herrera *et al.*, 1965; Frohlich *et al.*, 1972; Hallowes *et al.*, 1973; Meyuhas *et al.*, 1976) for measuring tissue viability. Various culture media were used, usually supplemented with either serum or a protein such as albumin or gelatin. Survival of the tissue varied; Frohlich *et al.* (1972) using several criteria, including the rate of incorporation of [^{3}H]thymidine into DNA and [U-^{14}C]glucose into lipids, found that adipose tissue remained metabolically active for at least 6 days in culture in a medium supplemented with serum. In contrast, Hallowes *et al.* (1973) measured the rate of [1-^{14}C] acetate incorporation into fatty acids and found that there was no detectable activity after 48 hours in culture in a chemically defined medium. However, inclusion of serum or even albumin in the culture medium has the disadvantage that the medium is no longer chemically defined.

The initial objective of the subsequent study was to determine the optimal condition for maintaining sheep and rat adipose tissue in organ culture in a chemically defined medium. The principal criterion of viability used was the oxidation of glucose to CO_2 as this provided a rapid, quantitative measure of survival.

Preliminary studies with rat adipose tissue showed that of a variety of tissue culture media tested, Medium 199, buffered with 25 mM Hepes and supplemented with insulin (10 μg/ml) appeared to be the most effective in maintaining the tissue. The tissue culture media tested included Waymouth's medium and Trowell's medium as used by others in earlier studies. Addition to the culture medium of serum or albumin as described previously in fact resulted in lower rates of glucose oxidation by adipose tissue slices after 3 days in organ culture. However, there was always a gradual decrease in the rate of glucose oxidation of about 40% over the 3 days in organ culture, even in Medium 199. This decrease was not due to depletion of substrates, as changing the culture medium daily had no beneficial effect and increasing the glucose concentration of the medium actually resulted in lower rates of glucose oxidation after the 3 days in culture. Reducing the temperature of the organ culture from 37 to 33°C resulted in a better survival of the tissue and inclusion of insulin in the culture medium also markedly stimulated its survival as judged by the rate of glucose oxidation.

Further studies have shown that the decrease in the rate of glucose oxidation during maintenance of the tissue in culture is probably not due to degeneration of the tissue. Thus over the first 24 hours in culture the rate of fatty acid synthesis for example increased by about 50%; this was followed by a subsequent decline but certainly not to the very low levels reported by Hallowes *et al.* (1973). In contrast, the rate of glycerol synthesis showed little change during the initial 24 hours in culture, but then increased by about 60% over the following 2 days.

Preliminary studies have been performed with sheep adipose tissue, in which the rate of fatty acid synthesis was measured after various times in organ culture. Sheep adipose tissue appears to survive well in organ culture in Medium 199 plus insulin. Basically the rate of lipogenesis increased by about 50% over the first 48 hours and then declined. Again a temperature of 33°C seemed preferable to one of 37°C. When the tissue was cultured in the absence of insulin, the rate of lipogenesis again increased by about 50% but this time it occurred during the first 24 hours in culture; during the next 2 days the rate decreased to about 40% of the rate found in fresh tissue. The reason for the changes in the rate of lipogenesis during organ culture in the presence and, in particular, the absence of insulin remain to be elucidated, but the system seems to be suitable for investigating the control of adipose tissue metabolism in ruminants.

Concurrent studies with adipose tissue from newborn lambs have shown that when the tissue is maintained in organ culture in Medium 199 supplemented with insulin, lipogenesis develops over 5 days in a manner similar to that observed previously *in vivo* (Vernon, 1975). With this system, however, the rate of lipogenesis developed more rapidly when maintained at 37°C than at 33°C. This suggests that certain enzymes were being synthesized *de novo*. The development of lipogenesis appears to depend on insulin. Also, addition of 5 μg of ovine growth hormone (NIH-GH-S11, 0·56 IU/mg, kindly donated by R. W. Bates) per ml prevented the increase in the rate of lipogenesis despite the presence of insulin. Surprisingly, there do not appear to be any previous reports of a direct effect of growth hormone on ruminant adipose tissue, despite the fact that administration of growth hormone elicits an increase in plasma fatty acid concentration in ruminants (Williams *et al.*, 1963) and also that growth hormone can stimulate lipolysis in rodent adipose tissue (Fain & Saperstein, 1970).

So, although the investigations with adipose tissue maintained in organ culture are still at an early stage, the potential of the system is clearly illustrated by the studies with tissue from neonatal lambs.

CONCLUSIONS AND FUTURE STUDIES

The studies described above clearly demonstrate that ruminant, or at least, ovine, adipose tissue metabolism is sensitive to the nature of the dietary fat. They also reveal a complication in the apparent age effect or possibly the essential fatty acid status of the animal.

To date no satisfactory explanation can be offered for the different effects of dietary saturated and polyunsaturated fats. However, one possibility, i.e. the differential inhibition of the activity of one or more of the enzymes of fatty acid biosynthesis by dietary fatty acids, now appears to be excluded. The method for maintaining adipose tissue from both rats and sheep in organ culture would permit a further investigation of the idea that the various fatty acids exert differential direct effects on adipose tissue metabolism. It would also seem worthwhile to investigate the effect of dietary fats on lipoprotein lipase activity of adipose tissue, and on the properties of plasma lipoproteins, both in adult and neonatal sheep.

The significance to the animal as a whole of these metabolic adaptations to different dietary fats is still unclear, although it appears that administration of safflower oil confers some benefits to neonatal lambs. However, as long as there is an interest in altering the fatty acid composition of meat and dairy products, it would seem worth while trying to understand the metabolic consequences of such manipulations.

For the future, studies on adipose tissue metabolism will be concerned with its regulation in the lactating animal. During early lactation in most species, and especially in the dairy cow, adipose tissue lipids are used by the animal both for energy and milk production (Spray, 1950; Beaton *et al.*, 1954; Robinson, 1973; Bauman, 1976). During later stages of lactation there is accumulation of adipose tissue as the animal partitions its nutrients between milk production and body growth (Broster *et al.*, 1969; Bauman, 1976). The mechanisms which regulate the utilization of lipids during early lactation and which regulate the partition of nutrients during later stages of lactation are not known, but it is probable that hormones have a major role in these mechanisms. Thus, studies are being initiated to determine which hormones can affect adipose tissue metabolism in ruminants and to examine whether the response to these hormones changes during lactation.

This work on adipose tissue metabolism is but part of a study of the regulation of ruminant metabolism during lactation; other aspects include investigation of the factors regulating the metabolism of the mammary gland and liver. Further, apart from the role of hormones in regulating the metabolism of these tissues, the role of the central nervous system is being studied also. It is hoped that this integrated approach will provide a much greater understanding of the mechanisms that determine the fate of nutrients during lactation.

REFERENCES

Alexander, G. & Williams, D. (1968) *J. Physiol.* (*Lond.*) **198**, 251–276

Allen, C. E. (1976) *Fed. Proc.* **35**, 2302–2307

Baldwin, R. L., Yang, Y. T., Crist, K. & Grichting, G. (1976) *Fed. Proc.* **35**, 2314–2318

Ballard, F. J., Hanson, R. W. & Kronfeld, D. S. (1969) *Fed. Proc.* **28**, 218–231
Bartley, J. C. & Abraham, S. (1972) *Biochim. Biophys. Acta* **280**, 258–266
Bauman, D. E. (1976) *Fed. Proc.* **35**, 2308–2313
Beaton, G. H., Beare, J., Ryu, M. H. & McHenry, E. W. (1954) *J. Nutr.* **54**, 291–304
Bibbiani, C., Tongiani, R. & Viola-Magni, M. P. (1973) *Histochemie* **35**, 189–198
Broster, W. H., Broster, V. J. & Smith, T. (1969) *J. Agric. Sci.* **72**, 229–245
Christie, W. W., Hunter, M. L., Moore, J. H., Noble, R. C. & Vernon, R. G. (1975*a*) *Lipids* **10**, 649–651
Christie, W. W., Moore, J. H., Noble, R. C. & Vernon, R. G. (1975*b*) *Lipids* **10**, 645–648
Cook, L. J., Scott, T. W. & Pan, Y. S. (1972) *J. Dairy Res.* **39**, 211–218
Denton, R. M. (1974) *Biochem. Soc. Trans.* **2**, 1208–1213
Derry, D. M., Morrow, E., Sadre, N. & Flattery, K. V. (1972) *Dev. Biol.* **27**, 204–216
Dinius, D. A., Oltjen, R. R. & Satter, L. D. (1974) *J. Anim. Sci.* **38**, 887–892
Dorsey, J. A. & Porter, J. W. (1968) *J. Biol. Chem.* **243**, 3512–3516
Du, J. T. & Kruger, F. A. (1972) *J. Nutr.* **102**, 1033–1038
Faichney, G. J., Scott, T. W. & Cook, L. J. (1973) *Aust. J. Biol. Sci.* **26**, 1179–1188
Fain, J. N. & Saperstein, R. (1970) in *Adipose Tissue: Regulation and Metabolic Function* (Jeanrenaud, B. & Hepp, D., eds), pp. 20–27, Academic Press, London
Frohlich, J., Vost, A. & Hollenberg, C. H. (1972) *Biochim. Biophys. Acta* **280**, 579–587
Gemmell, R. T., Bell, A. W. & Alexander, G. (1972) *Am. J. Anat.* **133**, 143–164
Gozukara, E. M., Frohlich, M. & Holten, D. (1972) *Biochim. Biophys. Acta* **286**, 155–163
Greenwood, M. R. C. & Hirsch, J. (1974) *J. Lipid Res.* **15**, 474–483
Hahn, P. & Novak, M. (1975) *J. Lipid Res.* **16**, 79–91
Hallowes, R. C., Wang, D. Y., Lewis, D. J., Strong, C. R. & Dils, R. (1973) *J. Endocrinol.* **57**, 265–276
Hanson, R. W. & Ballard, F. J. (1967) *Biochem. J.* **105**, 529–536
Haugebak, C. D., Hedrick, H. B. & Asplund, J. M. (1974) *J. Anim. Sci.* **39**, 1016–1025
Herrera, M. G., Phillips, G. R. & Renold, A. E. (1965) *Biochim. Biophys. Acta* **106**, 221–233
Hood, R. L., Thompson, E. H. & Allen, C. E. (1972) *Int. J. Biochem.* **3**, 598–606
Ingle, D. L., Bauman, D. E. & Garrigus, U. S. (1972) *J. Nutr.* **102**, 609–616
Ingle, D. L., Bauman, D. E., Mellenberger, R. W. & Johnson, D. E. (1973) *J. Nutr.* **103**, 1479–1488
Jarrett, I. G., Filsell, O. H. & Ballard, F. J. (1974) *Horm. Metab. Res. Suppl. Ser.* **4**, 111–116
Jungas, R. L. (1968) *Biochemistry* **7**, 3708–3717
Meyuhas, O., Reshef, L., Ballard, F. J. & Hanson, R. W. (1976) *Biochem. J.* **158**, 9–16
Mishkin, S. & Turcotte, R. (1974) *Biochem. Biophys. Res. Commun.* **57**, 918–924
Mishkin, S., Stein, L., Gratmaitan, Z. & Arias, I. M. (1972) *Biochem. Biophys. Res. Commun.* **47**, 997–1003
Musch, K., Ojakian, M. A. & Williams, M. A. (1974) *Biochim. Biophys. Acta* **337**, 343–348
Noble, R. C., Steele, W. & Moore, J. H. (1971) *Res. Vet. Sci.* **12**, 47–53
Noble, R. C., Steele, W. & Moore, J. H. (1972) *Br. J. Nutr.* **27**, 503–508
Noble, R. C., Vernon, R. G., Christie, W. W., Moore, J. H. & Evans, A. T. (1977) *Lipids*, **12**, 423–433
Ockner, R. K., Manning, J. A., Poppenhausen, R. B. & Ho, W. K. L. (1972) *Science* (*Wash. D.C.*) **177**, 56–58
Pan, Y. S., Cook, L. J. & Scott, T. W. (1972) *J. Dairy Res.* **39**, 203–210
Pande, S. V. & Mead, J. F. (1968) *J. Biol. Chem.* **243**, 6180–6185
Parvin, R. & Dakshinamurti, K. (1970) *J. Biol. Chem.* **245**, 5773–5778
Reiser, R. (1951) *Fed. Proc.* **10**, 236 (Abstr.)

Robinson, J. J. (1973) *Vet. Rec.* **92**, 602–606
Rose, H., Vaughan, M. & Steinberg, D. (1964) *Am. J. Physiol.* **206**, 345–350
Saggerson, E. D. & Tomassi, G. (1971) *Eur. J. Biochem.* **23**, 109–117
Salmon, D. M. W., Bowen, N. L. & Hems, D. A. (1974) *Biochem. J.* **142**, 611–618
Scott, T. W., Cook, L. J., Ferguson, K. A., McDonald, I. W., Buchanan, R. A. & Loftus Hills, G. (1970) *Aust. J. Sci.* **32**, 291–293
Scott, T. W., Cook, L. J. & Mills, S. C. (1971) *J. Am. Oil Chem. Soc.* **48**, 358–364
Scow, R. O. & Chernick, S. S. (1970) in *Comprehensive Biochemistry*, (Florkin, M. & Stotz, E. H., eds), vol. 18, pp. 19–49, Elsevier, Amsterdam
Shorland, F. B., Weenink, R. O., Johns, A. T. & McDonald, I. R. C. (1957) *Biochem. J.* **67**, 328
Slavin, B. G. & Elias, J. J. (1969) *Anat. Rec.* **164**, 141–150
Slavin, B. G. & Elias, J. J. (1970) *Anat. Rec.* **167**, 213–223
Smith, R. E. & Horwitz, B. A. (1969) *Physiol. Rev.* **49**, 330–425
Smith, S. & Abraham, S. (1970) *Arch. Biochem. Biophys.* **136**, 112–121
Sooranna, S. R. & Saggerson, E. D. (1975) *Biochem. J.* **150**, 441–451
Spray, C. M. (1950) *Br. J. Nutr.* **4**, 354–360
Srere, P. A. (1965) *Biochim. Biophys. Acta* **106**, 445–455
Sumper, M. (1974) *Eur. J. Biochem.* **49**, 469–475
Taketa, K. & Pogell, B. M. (1966) *J. Biol. Chem.* **241**, 720–726
Thompson, G. E. & Jenkinson, D. McE. (1969) *Can. J. Physiol. Pharmacol.* **47**, 249–253
Trowell, O. A. (1959) *Exp. Cell Res.* **16**, 118–147
Vernon, R. G. (1975) *Lipids* **10**, 284–289
Vernon, R. G. (1976) *Lipids* **11**, 662–669
Vernon, R. G. (1977*a*) *Int. J. Biochem.* **8**, 517–523
Vernon, R. G. (1977*b*) *Biol. Neonate* **32**, 15–23
Volpe, J. J. & Vagelos, P. R. (1976) *Physiol. Rev.* **56**, 339–417
Ward, P. F. V., Scott, T. W. & Dawson, R. M. C. (1964) *Biochem. J.* **92**, 60–68
Waterman, R. A., Romsos, D. R., Tsai, A. C., Miller, E. R. & Leveille, G. A. (1975) *Proc. Soc. Exp. Biol. Med.* **150**, 347–351
Williams, W. F., Lee, S. D., Head, H. H. & Lynch, J. (1963) *J. Dairy Sci.* **46**, 1405–1408
Winegrad, A. I., Shaw, W. N., Lukens, F. D. W. & Stadie, W. C. (1960) *Am. J. Clin. Nutr.* **8**, 651–665

METABOLISM IN THE COLD

G. E. THOMPSON

WINTER, in Britain, has marked effects on the nutrition and metabolism of ruminant animals. Throughout winter, plant growth virtually stops and so the animals are given conserved food. Also, there are brief periods of intense cold that stimulate directly the animal's metabolism. The latter effect has been studied at the Hannah Institute during the last 20 years, first in the Nutrition Department and then in the Physiology Department. The experiments carried out in the Nutrition Department (see the review by Blaxter in this volume, p. 50) were designed to develop equations for predicting the amount by which the animal's metabolism, and therefore food requirement, increases as environmental temperature falls, and how much this is altered by natural occurrences such as wind, rain, shearing the animal or varying its food intake. The experiments carried out in the Physiology Department, and reviewed here, were designed to improve our understanding of the mechanism by which metabolism is increased in a cold environment, to identify which tissues in the body have an increased metabolic rate and the substrates that they use, and to describe how cold exposure affects some of the productive processes in the body.

THE CONTROL OF METABOLIC RATE IN A COLD ENVIRONMENT

The animal's first physiological responses, when the environmental temperature falls, reduce the rate at which it loses heat to the environment. These responses become inadequate if the environmental temperature continues to fall below a 'critical' value, when the animal loses more heat to the environment and the ingested food and water than is released in the tissues as a by-product of resting metabolism. Endogenous nutrients are then mobilized and metabolized, with no productive benefit in the animal, until the extra heat from their metabolism keeps body temperature within the normal limits. During acute cold exposure these adjustments are probably brought about by the nervous system.

Cold-sensory receptors have not been identified properly in any ruminant but, judging from the timing and the size of an animal's response to local cooling of various parts of its body, it can be inferred that such receptors operate in the peripheral tissues of sheep (Webster, 1966), in the hypothalamus of the steer (Calvert *et al.*, 1972) and the spinal cord of sheep (Clough *et al.*, 1973), but not in the tissues of the abdomen of the steer (D. T. Calvert, unpublished work). The routes taken by impulses leaving any cold receptors have also not been studied, but there must be controlling nervous pathways in the brain with synapses close to the cerebral ventricles, probably in the hypothalamus. If the synaptic neurotransmitter noradrenaline is injected into the cerebral ventricles of cattle or sheep it has no effect unless the animal is exposed to a cold environment when the injection stops the animal shivering, reduces its metabolic rate and body temperature (Findlay & Thompson, 1968) and also stops the increases in blood pressure (Darling & Thompson, 1973) and in the concentration of unesterified fatty acids in blood plasma (Clough & Thompson 1972) that normally occur in a cold environment. Experiments, carried out at the Institute of Animal Physiology, Babraham, showed that injecting the synaptic transmitter acetylcholine, with eserine to stop its immediate destruction, into the cerebral ventricles of sheep produces shivering when the animal is not exposed to cold (Bligh *et al.*, 1971). At the Hannah Research Institute it was found that these injections also increase blood pressure (Darling & Thompson, 1973) and increase the concentration of unesterified fatty acids in plasma, an effect that is mediated by splanchnic and sympathetic nerves (Clough & Thompson, 1972). Taken together, these experiments with intraventricular injections suggest that noradrenaline and acetylcholine may be central inhibitory and facilitatory transmitters, respectively, in the control of metabolic rate and substrate mobilization in the cold.

The efferent nerves that control shivering have not been identified in the ruminant. Those that control the increase in the level of plasma unesterified fatty acids in the cold are probably sympathetic nerves; when steers are acutely exposed to mild cold the increased concentration of unesterified fatty acids in their blood plasma is prevented by the sympathetic-nerve-blocking drug bethanidine (Thompson & Clough, 1972). Sympathetic nerves have been identified in the adipose tissue of lambs. They probably contain noradrenaline, but when the noradrenaline in this tissue was being isolated it was found that its precursor, dopamine, is much more prevalent than noradrenaline, and the dopamine is probably located in heparin-containing mast cells which occur among the adipose cells (Thompson *et al.*, 1971). Dopamine and heparin-containing mast cells are found also in the adipose tissue of cattle, and intravenous injections of dopamine, like noradrenaline, increase the concentration of unesterified fatty acids in the blood plasma (Clough *et al.*, 1971) probably by mobilizing adipose-tissue triglycerides. It has not been established, however, that these mast cells are controlled by the sympathetic nervous system or release their catecholamine during cold exposure.

THE TISSUES THAT PRODUCE EXTRA HEAT IN THE COLD AND THE SUBSTRATES THAT THEY USE

Experiments, carried out about the same time in Australia (Alexander & Williams, 1968) and in the Hannah Institute (Jenkinson & Thompson, 1968) have shown the presence of brown adipose tissue in newborn lambs during the first 2 weeks of life, and the capacity for 'non-shivering thermogenesis' when lambs of this age are injected with noradrenaline. Brown adipose tissue occurs also in newborn goats (Thompson & Jenkinson, 1970); early experiments on newborn calves failed to show it in this species (Jenkinson *et al.*, 1968) but further investigations, in Australia, showed that it is present in the first few days of life (Alexander *et al.*, 1975); this was confirmed in our laboratory (Thompson & Bell, 1976). At this early age the tissue produces extra heat if noradrenaline is injected into the animal (Thompson & Bell, 1976) and probably oxidizes extra stored fatty acids. When the animal is older the tissue loses this capacity.

Another tissue that produces extra heat during cold exposure, in the newborn and the older ruminant, is shivering muscle. When 10-month-old steers were exposed for about 2 hours to a mildly cold environment, which increased their total metabolic rate by 85%, blood flow to the hind limb increased two-and-a-half fold, 91% of which was due to increased blood flow in muscle, and the oxygen consumption of the hind limb increased more than four-fold (Bell *et al.*, 1976). The uptake of extra substrates, to support this extra metabolism in the cold, has been studied by comparing the substrate concentrations in blood entering and leaving the leg, and multiplying the difference by the rate at which blood flows through the leg. Circulating unesterified fatty acids are probably major substrates for the extra metabolism of the shivering leg (Bell & Thompson, 1974) but their arterio-venous concentration difference includes unesterified fatty acid uptake by the leg and unesterified fatty acid release from adipose tissue in the leg. These two processes are now being studied separately by labelling the fatty acids which enter the leg, with radioactive isotopes. The uptake of glucose by the leg increases substantially during acute cold exposure, and not much of it is released as lactate or pyruvate (Bell *et al.*, 1975*b*); the uptake of volatile fatty acids by the leg also increases in the cold (Bell *et al.*, 1974). These metabolic changes in the leg have been followed during 4 days of continuous cold exposure (Bell *et al.*, 1975*a*). As well as showing the likely effects of wintering young weaned cattle outdoors these studies have also provided new information on the normal fluctuations of substrate uptake by the leg between feeds.

The liver has, in the past, been considered a possible source of extra heat in the body during cold exposure, but when sheep were acutely exposed to a cold environment and their whole-body oxygen consumption increased by an average of 99% the liver's oxygen consumption increased by 36% and this increase accounted for only 5% of the body's increase (Thompson *et al.*, 1975).

White adipose tissue is probably the source of the unesterified fatty acids which normally appear in high concentrations in the circulation during cold exposure, but it is difficult to isolate adipose tissue and study its metabolism in the conscious animal; blood flow to adipose tissue is faster in steers if they are exposed to cold (Bell *et al.*, 1976). The concentration of glucose in the blood plasma also increases when steers (Bell *et al.*, 1975*b*) or sheep are exposed to cold and, in sheep, the liver increases its output of glucose (G. E. Thompson, W. Manson, P. L. Clarke & A. W. Bell, unpublished work). Some of this glucose may derive from glycogen stored in the liver but the liver's uptake of some known precursors of glucose increases also during acute exposure, e.g. glycerol (Thompson *et al.*, 1975) and propionate. This extra propionate is released from the viscera, presumably the rumen, when food intake has not been altered in the cold (G. E. Thompson, J. M. Bassett & A. W. Bell, unpublished work) and the reason for this is not clear. Some amino acids are also known, in certain circumstances, to be gluconeogenic in the ruminant liver but the severities and durations of cold exposure that have been studied up to now have not stimulated any large-scale movement of amino acids out of the leg of the steer (Bell *et al.*, 1975*b*) or into the liver of the sheep (G. E. Thompson, W. Manson, P. L. Clarke & A. W. Bell, unpublished work).

THE EFFECTS OF COLD EXPOSURE ON THE METABOLISM OF THE UDDER

Experiments carried out in the United States of America (Cobble & Herman, 1951) and Canada (MacDonald & Bell, 1958) on lactating cows have shown that milk yield is reduced and milk composition changed during chronic cold exposure. Similar effects have been observed, at the Hannah Institute, with lactating goats exposed for 2 days to a mildly cold environment that raised total metabolic rate by only 28% but reduced significantly the yield of milk, and of milk lactose, crude protein and fat. The proportion of the individual fatty acids in the milk fat changed also, those with a chain length C_8 to C_{14} becoming less prevalent and the C_{18} acids more prevalent (Clarke *et al.*, 1976). It is not known why milk yield is reduced by such a mild cold stress; in a subsequent experiment it has been found that udder blood flow and oxygen consumption remain unchanged, so the reduction may be due to a neuroendocrine change in the body which has a deleterious effect on the gland. In colder conditions, where total metabolic rate is increased by an average of 50% and milk yield is further reduced, udder blood flow and the udder's fraction of the cardiac output are both reduced. The udder's efficiency, i.e. milk secretion per unit blood flow, is reduced and its oxygen consumption reduced also, though not significantly (E. M. Thomson and G. E. Thompson, unpublished work). In this situation the supply of substrates to the udder is probably impeded, with a local effect of cold on the gland.

This effect of cold exposure, reducing the metabolism of a peripherally situated organ, contrasts with its effect on total body metabolism. The importance of these effects on the udder, for commercial milk production and for the survival of cold-exposed newborn animals in the field, is not known.

REFERENCES

Alexander, G. & Williams, D. (1968) *J. Physiol.* (*Lond.*) **198**, 251–276

Alexander, G., Bennett, J. W. & Gemmell, R. T. (1975) *J. Physiol.* (*Lond.*) **244**, 223–234

Bell, A. W. & Thompson, G. E. (1974) *Res. Vet. Sci.* **17**, 265–267

Bell, A. W., Gardner, J. W. & Thompson, G. E. (1974) *Br. J. Nutr.* **32**, 471–477

Bell, A. W., Clarke, P. L. & Thompson, G. E. (1975*a*) *Q. J. Exp. Physiol. Cogn. Med. Sci.* **60**, 267–284

Bell, A. W., Gardner, J. W., Manson, W. & Thompson, G. E. (1975*b*) *Br. J. Nutr.* **33**, 207–217

Bell, A. W., Hilditch, T. E., Horton, P. W. & Thompson, G. E. (1976) *J. Physiol.* (*Lond.*) **257**, 229–243

Bligh, J., Cottle, W. H. & Maskrey, M. (1971) *J. Physiol.* (*Lond.*) **212**, 377–392

Calvert, D. T., Clough, D. P., Findlay, J. D. & Thompson, G. E. (1972) *Life Sci.* **11**, 223–229

Clarke, P. L., Thompson, G. E. & Thomson, E. M. (1976) *J. Physiol.* (*Lond.*) **263**, 176P

Clough, D. P. & Thompson, G. E. (1972) *Neuroendocrinology* **9**, 365–374

Clough, D. P., Kenny, J. D. R., Scobie, A. & Thompson, G. E. (1971) *Res. Vet. Sci.* **12**, 228–233

Clough, D. P., Darling, K. F., Findlay, J. D. & Thompson, G. E. (1973) *Pfluegers Arch. Ges. Physiol.* **342**, 137–144

Cobble, J. W. & Herman, H. A. (1951) *Mo. Agric. Exp. Stn. Res. Bull.* no. 485

Darling, K. F. & Thompson, G. E. (1973) *J. Physiol.* (*Lond.*) **232**, 93P–94P

Findlay, J. D. & Thompson, G. E. (1968) *J. Physiol.* (*Lond.*) **194**, 809–816

Jenkinson, D. McE. & Thompson, G. E. (1968) *J. Physiol.* (*Lond.*) **198**, 88P

Jenkinson, D. McE., Noble, R. C. & Thompson, G. E. (1968) *J. Physiol.* (*Lond.*) **195**, 639–646

MacDonald, M. A. & Bell, J. M. (1958) *Can. J. Anim. Sci.* **38**, 160–170

Thompson, G. E. & Bell, A. W. (1976) *Biol. Neonate* **28**, 375–381

Thompson, G. E. & Clough, D. P. (1972) *Q. J. Exp. Physiol. Cogn. Med. Sci.* **57**, 192–198

Thompson, G. E. & Jenkinson, D. McE. (1970) *Res. Vet. Sci.* **11**, 102

Thompson, G. E., Clough, D. P., Scobie, A. & Kenny, J. D. R. (1971) *Biol. Neonate* **17**, 394–398

Thompson, G. E., Gardner, J. W. & Bell, A. W. (1975) *Q. J. Exp. Physiol. Cogn. Med. Sci.* **60**, 107–121

Webster, A. J. F. (1966) *Res. Vet. Sci.* **7**, 454–465

SKIN BIOLOGY

D. McEwan Jenkinson

INTRODUCTION

At the Hannah Research Institute, interest in the biology of the skin arose in the late 1940s as an integral part of the developing programme on the reactions of cattle to their environment (see the review by Bligh in this volume p. 73). It stemmed from a comprehensive survey by J. D. Findlay (1950) of the existing literature on the effects of temperature, humidity, air movement and solar radiation on the behaviour of farm animals. This review illustrated the potential importance to bovine thermoregulation of coat type and colour, peripheral vascular changes and cutaneous evaporation. The conclusions at this stage were based largely on evidence from field experiments and comparative data from man; information, even on the basic structure of the skin of domestic animals, was scanty. It was clear that data on the structure and function of cattle skin were required before the effectiveness of its role in thermoregulation could be assessed.

THE FIRST DECADE (1948 TO 1958)

The initial research effort concentrated on investigation of the anatomy of cattle skin, the nature of its blood supply and the development of facilities and techniques that would enable a study of the physiological reactions of cattle to be monitored under controlled environmental conditions.

Skin histology

The first paper on the 'capillary distribution in cow skin' was published by Findlay and S. H. Yang (1948) and this was followed by the initial description of the 'hair follicle unit' (Findlay & Yang, 1950) which they showed consisted of four closely associated structures, a hair follicle, its arrector muscle, a bilobular sebaceous gland and a tubular sweat gland. Findlay and Yang made extensive measurements of the density of the hair coat, which in view of the fact that each hair follicle had an associated sweat gland, also gave estimates of sweat gland density. They determined the area of secreting surface of the sweat glands in different body regions and demonstrated that they were composed of two layers of cells, an outer myoepithelium and an inner glandular epithelium. They concluded from their histological evidence that the inner layer of epithelium secreted in an apocrine manner, i.e. by the loss of the cell apices which aggregate to form the secretion in the glandular lumen.

Practical applications of the work were also being considered. The new techniques and basic knowledge already obtained were applied immediately to testing the validity of an earlier claim by Burcev (1937) that the number of sweat glands in the helix of the cow's ear was significantly correlated with milk yield. The investigation by Findlay, Yang and A. M. Goodall (later Mrs Nisbet) (Findlay *et al.*, 1950) showed that the number of sweat glands varied greatly from one part of the ear to another and decreased with age. Their study thus illustrated that sweat gland density could not be used in animal selection as a reliable index of milk yielding capacity. Over the next few years they gave a detailed description of the vascular supply to the hair follicle unit and showed that the sweat glands of cattle have a poor blood supply. Direct shunts between the arterioles and venules (arterio-venous anastomoses) were described in peripheral body regions such as the ear and the glabrous teat. The myoepithelial cells of the sweat gland were studied in greater detail and the composition of the sweat produced by cattle was investigated histochemically. Yang (1952) assessed the degree of cutaneous pigmentation in twenty-three body regions of Ayrshire and Zebu cattle and found the skin of Zebu to be more heavily pigmented. He demonstrated that the dorsal and lateral surfaces were more pigmented than were the ventral and leg regions in both cattle types.

By the end of this decade, the department was rapidly acquiring an international reputation for its work on the skin of cattle as well as for its studies in environmental physiology and was then in a position to undertake collaborative work on practical problems in the tropics. Nisbet, in collaboration with Sir Charles Lovatt Evans and K. A. Ross (Lovatt Evans *et al.*, 1957) investigated the histology of the sweat glands of normal and anhidrotic horses. They discovered that 10% of the sweat gland units in dry-coated horses were blocked by keratinous plugs which they considered to be a consequence of the condition rather than its cause. The histological studies during this period provided a solid foundation for the physiological experiments which ensued.

Skin physiology

By 1954 Findlay, then joined by W. R. Beakley, had completed the construction of a climatic chamber and was in a position to study the effects of ambient temperature and humidity on the physiological behaviour of cattle under carefully controlled conditions. Among the early studies (Beakley & Findlay, 1955) were investigations of the effects of climate on the skin surface temperature of Ayrshire calves. These were made using a specially developed temperature measuring system and provided an equation that described the behaviour of mean skin trunk temperature with ambient temperature and humidity. The general applicability of the equation was subsequently tested on the ear and the significance of the previously described arterio-venous anastomoses in this region was discussed. Their work on scrotal skin temperature highlighted the practical implications of high environmental temperature on reproductive function. These studies and later experiments by G. C. Whittow (1962) illustrated that the extremities of the ox have a thermoregulatory function but showed also that skin temperature did not itself hold promise of providing a useful index of heat tolerance for cattle. They raised the interesting question of the effect of the animal's coat on heat loss by cutaneous evaporation in warm conditions. This question was partly answered by W. Bianca (1959) who demonstrated that clipping the coat increased heat tolerance, probably due to improved evaporative loss from the skin. Attention was drawn, therefore, to the role of cutaneous evaporation in thermoregulation. The question 'do cattle sweat?' had just recently been answered. The early belief that sweat glands in cattle were probably vestigial was based partly on the failure to observe moisture on cattle skin during heat exposure save on the muzzle. Ferguson & Dowling (1955), using a stereo-microscope and indicator papers, provided direct evidence for the fact that cattle sweat glands respond to heat and to injection of the hormone adrenaline. It was then evident that cattle sweat glands were functional organs, but the question 'does cutaneous evaporation, including sweating, contribute significantly to thermoregulation in cattle?' still remained. Even by 1958 the early view that the so-called apocrine glands were unlikely to function effectively in heat regulation was still held although it was being challenged at the Hannah Research Institute and in Australia.

THE SECOND DECADE (1958 to 1968)

The development by J. A. McLean, during the early part of this decade, of a ventilated capsule technique for quantitation of the evaporation from a known area of skin, enabled McLean (1963) to demonstrate that the rate of sweating varied in different parts of the body and that the rise in moisture vaporization rate during the early part of exposure to high temperature occurred in a series of steps which were attributed to bursts in sweat gland activity. He partitioned the heat loss from cattle under different ambient conditions and illustrated that cutaneous and respiratory evaporation were of similar magnitude at low air temperatures but cutaneous evaporation was the major means of heat loss under conditions of heat stress. He showed that cutaneous evaporation was controlled quantitatively by physiological means near to a level that made up the difference between heat production and other forms of heat loss. These results, backed by histological evidence in 1964 from Findlay and D. McE. Jenkinson which showed that the sweat glands decreased in volume on exposure of calves to heat, led to the conclusion that the sweat glands of cattle are functional organs of major importance to heat loss and hence to thermoregulation. The moisture droplets, which could be seen easily on the muzzle in warm environments, had previously been shown by A. M. Mackie & Nisbet (1959) to be derived from multilobular compound tubulo-acinar glands that produced a watery mucin-containing fluid. They provided the first description of the structure of the muzzle, its blood supply including the presence of arterio-venous anastomoses, its innervation and aspects of its histochemistry.

McLean showed that changes in blood flow exerted little influence on non-evaporative heat loss at temperatures above 20°C and that non-evaporative heat loss was determined primarily by insulation of the coat and the physical properties of the environment. These findings and his illustration of the importance of sweating in cattle to thermoregulation led to more detailed studies of sweat gland control, sweat gland function and the potential use of sweat gland and hair coat indices for the selection of cattle for hot climates.

Sweat gland control

The evidence from Australia, confirmed by Findlay & Jenkinson (1964), that the sweat glands of cattle respond to adrenaline, the finding by D. L. Ingram *et al.* (1963) that cutaneous evaporative loss is increased by heating the hypothalamus of the brain, and comparative evidence from man, all suggested that the sympathetic nervous system might have an important role in sudomotor control. This possibility was tested by Findlay & D. Robertshaw (1965) who, using the techniques of sympathectomy and splanchnicotomy, not only established that heat-induced sweating was indeed controlled by the sympathetic nervous system but also showed that hormones of the adrenal medulla were not involved in the mechanism.

This result prompted histological study of the sympathetic nerves to the glands and of the enzymes concerned with the removal or destruction of adrenaline and acetylcholine, the most likely activating transmitters. This work by Jenkinson, B. P. Sengupta and P. S. Blackburn (Jenkinson *et al.*, 1966) led to the surprising conclusion that cattle sweat glands are not innervated and revealed the presence of a peripheral component in the sudomotor control mechanism. This component was believed to be non-neural although, paradoxically, enzymes concerned with the metabolism of catecholamines and of acetylcholine were detected on the glands. The histological experiments on skin from normal and sympathectomized cattle also revealed that the innervation of the arrector pili muscle and of the vascular supply was largely, if not entirely, sympathetic in nature and that the pallisade nerve ending around the hair follicle had a motor component and was not purely a sensory organ as previously supposed.

The histological evidence thus indicated that the sweat glands of cattle differed from those of primates and of the footpads of the dog and cat which had been shown previously to have a closely associated nerve supply. The sweat glands of the horse, the only other species which had been studied, were also reputed to be innervated and it thus seemed possible that the mode of control of sweating in ruminants differed from that of non-ruminants. This hypothesis was tested partly by histological study of the skin innervation of other ruminant and non-ruminant species. Jenkinson & Blackburn (1967*a*, *b*; 1968), in developing the comparative approach, discovered that the sweat glands of the sheep, goat, cat, dog and pig had no nerve supply and illustrated that a peripheral component in the sudomotor mechanism was also characteristic of some non-ruminants; the glands of the horse and primates, and the cat and dog footpads, with their closely associated nerve supply, appeared to be the exception rather than the rule. The comparative work raised several important questions which were the subject of investigations during the following decade.

Sweat gland function

The early histological findings suggested that cattle sweat glands functioned in an apocrine manner but conflicting evidence on their morphology led Findlay and Jenkinson to investigate their mode of secretion. They discovered that the glands did not empty during activity and always had a flattened epithelium. The absence of sequential stages of necrobiotic discharge led them to the conclusion that cattle sweat glands were not apocrine and subsequently Jenkinson (1967) challenged the existence of such a mode of secretion in the sweat glands of any species. Bianca demonstrated that the onset of sweating was delayed for up to 1½ hours and the water content of the skin reduced by about 5% during dehydration. Over-hydration on the other hand had no measurable effect on skin water content. Bianca & J. R. S. Hales (1970) showed that skin moisture loss was higher in newborn calves which tended to be more heat tolerant than 1-year-old calves due to their greater sweating capacity per unit of metabolic size. The observation by McLean of a peculiar stepwise recruitment at the onset of sweating in cattle was followed up in Cambridge by a former member of staff of the Institute, J. Bligh, who found that in sheep there were only intermittent discharges of sweat, which gradually declined in magnitude. Continuous secretion did not occur during heat exposure and the glands rapidly fatigued. Bligh (1967) proposed that this pattern was the result of myoepithelial expulsion of preformed secretion, as opposed to the pattern in cattle, which could be explained by periodic contractions of the myoepithelium against a background of increasing sweat production by the secretory epithelium. This theory provided the basis for further studies of sweat gland function at the Hannah Research Institute and elsewhere during the ensuing decade.

Skin indices in cattle selection

The demonstration of the importance of cutaneous evaporation and coat type to bovine thermoregulation led Jenkinson and a collaborator in Australia, T. Nay, to examine the possibility of using skin measurements as indices for the selection of cattle suitable for hot climates. There was at that time a need for a simple index of heat tolerance that could be used in breeding and selection in the tropics. The sampling phase of a world survey of the skin of cattle and related bovidae, the results of which appeared in the following decade, was undertaken during this period. The skin types of temperate breeds of cattle were compared with those of tropical breeds.

Nay and Jenkinson (1964) also re-examined the possibility of using a sweat gland measurement as an index of milk-producing capacity in cattle. Although sweat gland density had been shown earlier to be an unreliable index, other parameters such as sweat gland shape and volume had not been tested. The close phylogenetical relationship between sweat and mammary glands and their similar histological appearances suggested the possibility that circulating hormones would have a similar effect on both glands. They found no significant relationship between sweat gland measurements and milk yield, but discovered that cows with a shallow hair follicle depth could be expected to give more milk. However, although this was a statistically significant relationship it had only a limited value in practice.

The techniques developed during this work were applied to the solution of other practical problems. Jenkinson, with Blackburn and R. Proudfoot (Jenkinson *et al.*, 1967), located the source of the offensive rutting odour in the male goat and their study led to subsequent attempts to depress the odour by glandular extirpation in the young animal.

THE THIRD DECADE (1968 TO 1978)

At the beginning of this period, the research programme on skin biology still encompassed the three main lines developed during the previous decade.

Sweat gland control

The previous histological evidence on the innervation of the sweat glands stimulated comparative physiological studies of the role of the sympathetic nervous system in sweat gland control. Robertshaw (1968) illustrated that the glands of the sheep and goat, like those of the cow, were controlled by the sympathetic nervous system and thus also appeared to have a peripheral component in their control systems. In 1969, while on secondment in Africa, and, in collaboration with C. R. Taylor of Harvard University (Robertshaw & Taylor, 1969), he demonstrated that heat-induced sweating in the equine was also controlled by the sympathetic nervous system although circulating adrenal hormones were involved in the control of sweating due to exercise. These and later studies of the monkey and kangaroo in the United States of America by Robertshaw and collaborators clearly established the importance of the sympathetic nervous system and of adrenergic mechanisms of sweat control in most species. This work threw new light on the control mechanism which, as a result of early studies on man, had been considered to be a cholinergic system. The evidence for adrenergic control and of enzymes which metabolized catecholamines on the glands led to investigation of the potential role of adrenaline in the peripheral control mechanism. An attempt was made to locate catecholamines on the gland using histochemical techniques involving fluorescence under ultraviolet light. G. E. Thompson, Robertshaw and Findlay (Thompson *et al.*, 1969) described the nonadrenergic innervation of the arrector pili muscle and the presence of fluorescent cells distributed randomly throughout the skin. Unfortunately, owing to a natural autofluorescence of the sweat glands they were unable to determine whether catecholamines were present on the glands. The fluorescent cells within the skin were shown by Jenkinson *et al.* (1970) to be mast cells which were unlikely to have a specific sudomotor function. R. M. Mabon & Jenkinson (1971) found catecholamine metabolites in cattle skin washings but were unable to associate these with a peripheral sweating mechanism.

The comparative approach to the problem led gradually to the thesis presented by Jenkinson (1973) that even in man and the horse the glands were not innervated. He postulated that the nerve supply was to the capillaries and suggested that the mechanism of control was basically the same in all species, the more profuse sweating in man and the horse resulting from the closer proximity of the vascular supply to the glands. This thesis is being tested in detailed electron microscopical studies of the innervation of the sweat glands of a number of species including man and the horse by Jenkinson in collaboration with I. Montgomery and H. Y. Elder of the University of Glasgow.

Sweat gland function

By 1969, Bligh and his colleagues, and Robertshaw and Taylor had developed the comparative approach and had discovered a wide range of patterns of sweat discharge consistent with different combinations of continuous and intermittent discharge in a variety of species. These patterns could all be explained in terms of Bligh's theory on the relative importance of the myo- and secretory epithelia to the sweating process. Jenkinson & Robertshaw (1971) investigated the cause of the sweat gland fatigue in the goat and concluded that the decrement in sweat output was due to a decline in the amount of sweat produced by each gland. They showed that the volume of sweat produced was greater than the corresponding decrease in sweat gland volume and concluded that the fatigue was due to the rate of expulsion exceeding the rate of sweat production. In the same year Jenkinson (1971) compared the myoepithelial cell diameters in different species and postulated that the frequency of myoepithelial contraction was of greater importance to sweat delivery than the relative strength of each contraction. Bligh's theory is now being examined by more detailed studies of the mode of secretion of, and substrate transport to the glands of different species using a combination of electron microscopy, histochemistry and autoradiography. The findings will also provide comparative evidence on mammary gland activity.

Skin indices in cattle selection

The initial results of the survey of cattle skin types were published by Jenkinson & Nay (1968). The effects of factors such as age, sex and season on sweat gland and hair follicle measurements were first examined, and sweat gland shape and hair follicle depth were shown to be reliable measures of skin type in adult cattle. Using these parameters they described between 1972 and 1975 the skin structures of cattle and related bovidae throughout the world. They identified a skin type characteristic of cattle more tolerant to heat, and heritability studies in 1975 established the value of this skin type as an index for the selection of cattle for warm climates. The question of whether animals with this type of skin also exhibit a satisfactory resistance to tropical diseases still must be answered before the index can be applied widely in practice.

During this period Jenkinson, with colleagues at the West of Scotland Agricultural College, also examined the effects of freeze-branding on the skin of young beef animals (Day *et al.*, 1971).

THE SKIN SURFACE AND BODY DEFENCE

By 1973 the observation made during earlier studies that the sweat glands of some species and in specialized body regions were not of major importance to thermoregulation, and a change of emphasis in the general research programme of the department had led to the investigation of one of the other important functions of the skin, its role in body defence against foreign organisms.

The skin is more than a simple barrier to infection. The healthy skin surface supports an ecosystem which includes bacteria, fungi and various parasites. An understanding of the role of the skin surface in body defence thus requires a knowledge of its structure, chemistry and inhabitants, and their interactions with each other and with the external environment. The first stage of the investigation was, therefore, devoted to determining the nature of the skin surface secretions, the structure of the epidermis and to identifying the normal surface microflora.

Skin surface secretions

The effectiveness of the barrier provided by the epidermis depends partly on its integrity and partly on the chemical nature of the surface secretions. The potential importance of chemical defence on the skin surface prompted studies to identify some of its constituents and determine their source within the skin. Jenkinson & Mabon (1973) established that skin pH was fairly stable except in very hot humid environments when it became more alkaline. They detected sodium, potassium, magnesium, calcium and chloride on the skin surface and concluded that they were derived largely from sweat. They demonstrated that in cattle, in contrast to man, the sodium/potassium ratio fell during sweating. Jenkinson *et al.* (1974*a*) examined the loss of nitrogenous substances from the skin of cattle and found that most of the nitrogen lost was derived from protein. This work was extended to other species and protein was shown to be a general constituent of sweat. More extensive study of the skin surface proteins of cattle by D. H. Lloyd, Mabon and Jenkinson (Lloyd *et al.*, 1977) using immunological methods enabled a tentative evaluation of all the proteins and positive identification of albumin and immunoglobulins G_1, G_2 and A to be made. The presence of glycoprotein on the skin surface was confirmed also. The source of the immunoglobulins and the effects on their output of challenge by pathogens are now being studied. Investigation of the source of these substances and of histamine, which is also present on the skin surface, will provide evidence on the mode of function and control of sweat output.

Sebaceous gland function

Corticosteroids and lipid detected on the skin surface were derived apparently from sebum, and studies of sebum output and composition were initiated. M. E. Smith & Jenkinson (1975) found that sebum output was similar in cattle of different ages but tended to be higher in castrated males than in females and was lower in winter than in summer. They found that neither sebum output nor its fatty acid composition was influenced appreciably by short-term changes in temperature. Prolonged heat exposure, however, increased output and the percentage of linoleic acid in the sebum. They discovered that sebum in cattle was not produced entirely by a holocrine mechanism, i.e. by the death of the secretory cells. Smith, with S. U. Ahmed, isolated sebaceous glands and showed (Smith & Ahmed, 1976) that freshly collected sebum on the surface was fairly representative of the composition in the gland. Poon Wai Yee and collaborators illustrated changes in the free fatty acid composition on the skin surface with time and P. L. Clarke, Jenkinson and Smith have determined the volatile fatty acid composition of the skin surface in preparation for a study of their role in insect attraction.

Epidermal structure

Lloyd and Jenkinson, in collaboration with visiting scientists, studied the stratum corneum of cattle and sheep and found it to be thicker than previously supposed. It varied in thickness at different parts of the hair pore and was, especially in sheep, permeated with lipid and aqueous secretions from the skin glands. The stratum corneum was almost completely removed by abrasion or by the application of lipid solvents.

Skin surface microflora

The normal microflora of the skin surface of cattle, which may itself contribute to body defence, was determined by Lloyd who also demonstrated the existence of mixed microcolonies of bacteria and fungi in the superficial layers of the stratum corneum. He is now testing the effects of the skin secretions on the growth of the normal inhabitants *in vitro* and the effects of environment and of infection by a pathogen on the stability of the normal skin microbial ecosystem.

The Department of Physiology of the Hannah Research Institute has over the past three decades not only contributed data of importance to cattle husbandry in developing countries but also provided comparative information relevant to mammary gland function and spotlighted some of the skin defence systems. The substantial amount of fundamental research on the structure and function of the skin that stemmed from the skin biology programme has provided the basis for applied studies at the Institute and elsewhere. It has provided a new method of treatment for demodectic mange; this method was developed by Jenkinson and McLean in conjunction with G. S. Walton of the University of Liverpool (Jenkinson *et al.*, 1974*b*).

In the future an even more comprehensive knowledge of skin biology is likely to be required as ectoparasites and bacteria become more resistant to pesticides and antibiotics and new methods of disease control are sought.

REFERENCES

Beakley, W. R. & Findlay, J. D. (1955) *J. Agric. Sci.* **45**, 353–364

Bianca, W. (1959) *J. Agric. Sci.* **52**, 380–383

Bianca, W. & Hales, J. R. S. (1970) *Br. Vet. J.* **126**, 45–53

Bligh, J. (1967) *Environ. Res.* **1**, 28–45

Burcev, A. V. (1937) *Usp. Zootekh. Nauk* **4**, 137–143

Day, N., Jenkinson, D. McE. & Walker-Love, J. (1971) *Anim. Prod.* **13**, 93–99

Ferguson, K. A. & Dowling, D. F. (1955) *Aust. J. Agric. Res.* **6**, 640–644

Findlay, J. D. (1950) *Hannah Dairy Res. Inst. Bull.* no. 9

Findlay, J. D. & Jenkinson, D. McE. (1964) *Res. Vet. Sci.* **5**, 109–115

Findlay, J. D. & Robertshaw, D. (1965) *J. Physiol (Lond.)* **179**, 285–297

Findlay, J. D. & Yang, S. H. (1948) *Nature (Lond.)* **161**, 1012–1013

Findlay, J. D. & Yang, S. H. (1950) *J. Agric. Sci.* **40**, 126–129

Findlay, J. D., Goodall, A. M. & Yang, S. H. (1950) *J. Dairy Res.* **17**, 22–36

Ingram, D. L., McLean, J. A. & Whittow, G. C. (1963) *J. Physiol. (Lond.)* **169**, 394–403

Jenkinson, D. McE. (1967) *Br. Vet. J.* **123**, 311–316

Jenkinson, D. McE. (1971) *Res. Vet. Sci.* **12**, 152–155

Jenkinson, D. McE. (1973) *Br. J. Dermatol.* **88**, 397–406

Jenkinson, D. McE. & Blackburn, P. S. (1967*a*) *J. Anat.* **101**, 333–341

Jenkinson, D. McE. & Blackburn, P. S. (1967*b*) *Res. Vet. Sci.* **8**, 306–312

Jenkinson, D. McE. & Blackburn, P. S. (1968) *Res. Vet. Sci.* **9**, 521–528

Jenkinson, D. McE. & Mabon, R. M. (1973) *Br. Vet. J.* **129**, 282–295

Jenkinson, D. McE. & Nay, T. (1968) *Aust. J. Biol. Sci.* **21**, 1001–1011

Jenkinson, D. McE. & Robertshaw, D. (1971) *J. Physiol. (Lond.)* **212**, 455–465

Jenkinson, D. McE., Sengupta, B. P. & Blackburn, P. S. (1966) *J. Anat.* **100**, 593–613

Jenkinson, D. McE., Blackburn, P. S. & Proudfoot, R. (1967) *Br. Vet. J.* **123**, 541–549

Jenkinson, D. McE., Thompson, G. E., Kenny, J. D. R. & Pearson, J. M. (1970) *Histochem. J.* **2**, 419–424

Jenkinson, D. McE., Mabon, R. M. & Manson, W. (1974*a*) *Res. Vet. Sci.* **17**, 75–80

Jenkinson, D. McE., McLean, J. A. & Walton, G. S. (1974*b*) *Vet. Rec.* **94**, 8–12

Lloyd, D. H., Mabon, R. M. & Jenkinson, D. McE. (1977) *J. Comp. Pathol.* **87**, 75–82

Lovatt Evans, C., Nisbet, A. M. & Ross, K. A. (1957) *J. Comp. Pathol. Ther.* **67**, 397–405

Mabon, R. M. & Jenkinson, D. McE. (1971) *Res. Vet. Sci.* **12**, 289–292

Mackie, A. M. & Nisbet, A. M. (1959) *J. Agric. Sci.* **52**, 376–379

McLean, J. A. (1963) *J. Agric. Sci.* **61**, 275–280

Nay, T. & Jenkinson, D. McE. (1964) *J. Dairy Res.* **31**, 53–58

Robertshaw, D. (1968) *J. Physiol. (Lond.)* **198**, 531–539

Robertshaw, D. & Taylor, C. R. (1969) *J. Physiol. (Lond.)* **205**, 79–89

Smith, M. E. & Ahmed, S. U. (1976) *Res. Vet. Sci.* **21**, 250–252

Smith, M. E. & Jenkinson, D. McE. (1975) *J. Agric. Sci.* **84**, 57–60

Thompson, G. E., Robertshaw, D. & Findlay, J. D. (1969) *Can. J. Physiol. Pharmacol.* **47**, 310–311

Whittow, G. C. (1962) *J. Agric. Sci.* **58**, 109–120

Yang, S. H. (1952) *J. Agric. Sci.* **42**, 465–467

INSTRUMENTATION

J. A. McLean

An important part of the programme of studies in environmental physiology at the Hannah Research Institute (see review by Bligh in this volume, p.73) has been the development of suitable instrumentation. Practical work on the subject depends on three types of instrumental requirement: the provision of climatic environments to which animals are exposed, the measurement of these environments and the measurement of the physiological reactions of animals to the environments. Instrumentation for these purposes must have special features in addition to accuracy and reliability. For example, if some sensing device is fixed on or near an animal it must cause minimal disturbance to the animal; or if it is enclosed with an animal inside a metabolic chamber it must either need no adjustment throughout an experiment or be capable of adjustment from outside. It must also be proof against all the wiles of an unco-operative experimental subject. The extraordinary ingenuity and persistence of cattle in finding ways of destroying experimental apparatus can only be appreciated after much bitter experience! Commercially available instrumentation seldom meets these requirements without modification. This chapter reviews the development of suitable instrumentation for environmental physiology at the Institute and discusses in greater detail some aspects of current methodology that may have wider application. Emphasis is deliberately placed on some of the mistakes that have been made in the hope that it will help to avoid repetition by others.

CLIMATIC CHAMBERS

The first psychrometric chamber at the Institute was constructed by modifying a loose box in old farm buildings. It was large enough to accommodate only one calf of about a year old, which had to leave and enter via a door opening directly to outside. Its economical and somewhat makeshift design can be judged from its description and photographs (Findlay, 1950). Nevertheless, much of the early work on environmental physiology was accomplished within it. The experience gained with this room was invaluable when it became necessary to draw up specifications for the main climatic laboratory, which was completed in 1959 with the aid of a grant from the Commonwealth Development and Welfare Fund. The climatic laboratory houses two psychrometric chambers, an operating theatre, engineering workshop and analytical laboratories. An extension to house a direct calorimeter was added in 1965. The psychrometric chambers, with ample room for one and four adult cattle respectively, have a temperature range from 0 to 60°C and a wide range of humidities. They have been fully described by Findlay *et al.* (1959). An important concept in the design of these rooms was thought to be that of flexibility. No firm structure was built into the rooms and a network of fixing points was included so that stalls of different designs could be erected and modified according to experimental needs. With the benefit of hind-sight it is clear that though this concept was excellent it was not carried nearly far enough. The rooms themselves should not have been built as firm structures. Heavy fixed structures are not required to restrain a large animal; it only needs to be contained on the platform upon which it is standing. The ideal would be to have plant of high capacity and ample storage for refrigerated fluid, steam, water and electricity, which could be piped or wired to all parts of a large working area. Individual specialized psychrometric chambers of lightweight construction could then be assembled and dismantled as required by the experimental programme. Climatic facilities intended for work on cattle should also be designed with greater emphasis on provision of cold rather than hot environments. Despite these criticisms the climatic rooms at the Institute provide an excellent facility which has been and still is used for a wide variety of studies in environmental physiology. As has been written in the chapter describing this work in detail (see review by Bligh in this volume, p. 73) it is right to emphasize that none of these facilities nor any of the work done in them would exist were it not for the imagination and drive of the late J. D. Findlay.

TEMPERATURE AND HUMIDITY

The temperature measuring systems in the climatic laboratory are based on copper/constantan thermocouples. This method was preferred for work with animals, where temperature sensors have to be disposable, on the grounds of cheapness and ease of replacement. Each psychrometric chamber and the direct calorimeter is equipped with a system consisting of a constant temperature reference bath, selector switches and a display or recording device. All thermocouples consist of three parts connected together; short lengths of 40 SWG copper and constantan wires twisted and joined together at one end to form the measuring junction; a longer length of twin 25 SWG copper/constantan cable joined at one end to the 40 SWG wires; and a length of 25 SWG twin copper wire joined at one end to the 25 SWG copper/constantan to form the reference junction and at the other end to a selector switch. The disposable part is the 40 SWG tip wire, which can be mounted in a variety of ways depending on its intended use. All joints between wires are made by first tightly twisting the cleaned wires together and then locking the joint with silver solder. The need to ensure complete uniformity of thermo-electric characteristics throughout the system and over many replacements of disposable elements was foreseen when the climatic laboratory was built; adequate stocks of matching 25 and 40 SWG wires were obtained at that time and are still in use. These precautions have proved adequate to eliminate stray thermoelectric effects and the resulting systems are accurate to $\pm 0{\cdot}1°C$.

Humidity measurements are made by means of 'wet- and dry bulb' thermometers or by Foxboro* dewcells, the basic measurement in each case being temperature. Despite the many principles that have been used in commercial instruments for measuring humidity, the 'wet bulb' is still preferred on grounds of accuracy and simplicity. The small size of a thermocouple 'wet bulb' often allows it to be used successfully without forced ventilation. 'Wet- and dry bulbs' have been applied to measure evaporation from the skin by means of sweat cups (McLean, 1963*a*). These are inverted ventilated capsules placed on the skin of animals, the humidity difference between air leaving and entering being measured by two opposing thermocouple 'wet bulbs'.

RESPIRATORY MEASUREMENT

Respiratory physiology provides an example of how work on instrumentation has been stimulated by the needs of the study and at the same time influenced the development of the experimental approach. In the late 1940s when environmental physiology at the Institute was in its infancy the equipment for respiratory studies consisted of a stethoscope and a stopwatch. One of the first instruments developed was the pneumotachometer (Nisbet, 1956). The principle, that of detecting minute pressure changes induced across a fine-mesh gauze in the air-flow path, had been used with human subjects, but the system had to be scaled up in size and speed of operation to make it suitable for cattle. Also, a suitable face-mask had to be developed before it could be used for a study of the respiratory behaviour of cattle in different environments (Findlay, 1957). The pneumotachometer could measure respiratory volume, frequency and ventilation rate. The next logical requirement was for the collection of respiratory gases for analysis. For this purpose W. Nisbet developed lightweight valves which he fitted to the face-mask to separate inspired from expired air, the pneumotachometer gauze being placed in the expiratory air-flow. With this mask, and associated gas analysis equipment, heat production and respiratory evaporative heat loss could be measured. In combination with

* A list of manufacturers' names and addresses is included as an Appendix to this chapter.

the sweat cups, the mask was used to assess the overall heat balance of calves and to make a rough partition of heat losses from the body (McLean, 1963*b*). A final development of the face-mask technique was the addition of a suction pump which drew ventilating air through the mask and past the animal's nose and mouth (Hales & Findlay, 1968*a*). This mask eliminated the resistance to respiration caused by valves, yet could be temporarily reconverted, by means of taps, to the original 'pneumotachometer' system for measurement of respiratory volume. Respiratory ventilation was partitioned into alveolar and dead space ventilation and the efficiency of panting was assessed (Hales & Findlay, 1968*b*). Face-masks were thus essential to the basic studies in respiratory physiology, but they undoubtedly imposed an undesirable restraint on the animal and their use was limited to periods of 2 to 3 hours.

For routine measurement of respiratory gas exchanges a simpler and less disturbing system was needed and this gave rise to the ventilated hood or head-cage which is in continual use today for work on cold exposure (see review by Thompson in this volume, p. 191). The hood (McLean & Calvert, 1972) consists of a polythene-covered framework placed over the animal's head. The animal has freedom of head-movement within the fixed cage. Air is pumped out of the cage continuously and passed to instruments for recording air-flow rate and gas concentrations. The neck-opening, which is gathered loosely by means of a draw-thread, also serves as air inlet. The main advantages of the head-cage lie in its simplicity in construction and operation. So long as the ventilation is adequate to ensure that the animal on expiration cannot cause gas to flow out of the hood then leaks are not important. Further improvements have since been made to the head-cage system but these are concerned mainly with methods of gas sampling and are discussed later. With the introduction of the gradient-layer (direct) calorimeter, also discussed later, the head-cage was effectively increased in size to contain the whole animal. Heat production can now be measured over prolonged periods during which the animal has freedom of movement and may be fed or watered when desired. Rapid response in the measurement of gas exchange is still possible, despite the slow wash-out time of the large enclosure, by making allowance for the rate of change of gas concentration in the system.

INDIRECT CALORIMETRY

Indirect calorimetry is the assessment of heat produced by an animal from quantitative estimations of the net chemical changes resulting from the processes of metabolism. In closed-circuit calorimetry air is circulated round an enclosed system containing the animal. The chemical changes are estimated from the weight or volume change of vessels that selectively supply or absorb gases consumed or produced (principally O_2 and CO_2). In open-circuit calorimetry the animal is confined within, or breathes into a ventilated space. Gas exchanges are estimated from the flow rate and changes of concentration in the ventilating air. The open-circuit system, which allows easier access to the experimental animal and can be made rapid in response, is the method used in the studies on environmental physiology.

It is desirable first to consider the calculation of heat production from the measurements made, as this has an important bearing on the choice of instrumentation. Following the suggestion of its sub-committee (Brouwer, 1965), the European Association for Animal Production recommended a formula for calculating heat production (M, kJ) from oxygen consumption (O_2, litres), carbon dioxide production (CO_2, litres), methane production (CH_4, litres) and urine nitrogen (N, g) as follows:

$$M = (16{\cdot}18 \times O_2) + (5{\cdot}02 \times CO_2) - (2{\cdot}17 \times CH_4) - (5{\cdot}99 \times N)$$

This formula is ideal for application to measurements by closed-circuit calorimetry.

In open-circuit calorimetry application of the Brouwer equation becomes tedious because the primary quantities measured are not gas consumption and production rates, but gas concentrations and ventilation rate. It has been shown that for open-circuit systems ventilated with fresh air the equation may be expressed in a different form (McLean 1972)

$$M = V_s(-20{\cdot}47X + 0{\cdot}73Y - 6{\cdot}64Z) - 5{\cdot}99\mathrm{N}$$

where V_s (litres) is the volume of air exhausted from the system and X, Y and Z are the differences in concentration (of O_2, CO_2 and CH_4) between the air leaving and entering the system (note that X is a negative quantity). This formula is ideal for application to measurements where the ventilation results in rapid wash-through of the system compared with the duration of the measurement period (i.e. face-masks and head-cages). In systems where the whole animal is confined in a chamber the last condition is not usually satisfied. There is a consequent slow build-up of gas concentrations towards a new equilibrium, following any change in metabolic activity. In such instances it is still possible to estimate metabolic rate with rapid response if account is taken of the rates of change of gas concentrations (McLean & Watts, 1976). The appropriate equation is

$$\dot{M} = \left(\dot{V}_s + V\frac{\mathrm{d}}{\mathrm{d}t}\right)(-20{\cdot}47X + 0{\cdot}73Y - 6{\cdot}64Z) - 5{\cdot}99\dot{N}$$

where V is the volume (litres) of the open-circuit system. $\dot{M}$ (watts), $\dot{V}_s$ (litres/sec) and $\dot{N}$ (g/sec) are rates of heat production, stale air ventilation and urinary nitrogen production. Similarly $\frac{\mathrm{d}X}{\mathrm{d}t}$, etc., are rates of change of differences in gas concentrations. In the latter two equations the terms in oxygen (X) are large compared with the other terms which tend to cancel each other out. A good approximation is possible by omitting all but the oxygen term.

The critical measurement for accuracy in open-circuit calorimetry is that of the product, ventilation rate × oxygen concentration difference ($\dot{V}_s X$); for fast response using animal chambers it is also desirable to measure the product, chamber volume × rate of change of oxygen concentration $\left(V\frac{\mathrm{d}X}{\mathrm{d}t}\right)$. The percentage accuracy with which a difference measurement can be made improves as the difference increases. It is therefore preferable to arrange the open-circuit system with a low ventilation rate and correspondingly high gas-concentration differences. However, if too great a differential is allowed between inlet and outlet gas concentrations, respiratory alkalosis or other adverse effects may be induced in the experimental subject. The limit is usually set at about 1% gas-concentration difference. For oxygen which has a fresh-air concentration of 20·95% this limits exhaust concentration to about 20%. If heat production is to be estimated with an accuracy of say ±1% then it follows that oxygen-concentration difference must be estimated to better than 1% or oxygen concentration itself to better than 1 part in 2000. Fortunately this is possible with modern paramagnetic analysers.

GAS ANALYSIS

Oxygen concentration is measured by means of a Servomex OA 184 paramagnetic analyser. The heart of this instrument is a pair of Munday cells both housed in the same temperature-controlled enclosure. They may be used either as two separate analysers or in differential mode, the claimed accuracy being 1 part in 2000. In our experience, three or four times better accuracy than this can be obtained if the gas flows are temporarily switched, once an hour, between the two cells to check on inter-cell drift. An accuracy of 1 part in 2000 is obtainable also with a single-cell analyser (Servomex OA 137) by means of gas switching. To achieve these levels of accuracy it is essential to ensure that the gas samples are completely dry before they enter the analysers, and to account for changes in barometric pressure. The outputs of the instruments, when calibrated in terms of oxygen concentration, are directly proportional to barometric pressure. Since oxygen concentration of fresh air is constant, hourly checking of the analysers against fresh air, effectively provides a very accurate indication of any change in barometric pressure.

Two further essential requirements to obtain maximum accuracy are a high-quality indicating instrument and a system of calibration or standardization. The panel-meter included in Servomex instruments provides only a rough indication, and a precision potentiometric recorder, or better still, a digital voltmeter must be used. For setting up the instruments gas mixtures of standard composition are needed, and there is no better standard available than dry fresh air. The actual value of the oxygen concentration of dry fresh air is debatable, different authorities giving values ranging from 20·93 to 20·96%, but the essential point is that it is invariant; we use the value 20·95%. The other standard is pure nitrogen. Each cell of the analyser is set so that it gives zero output for pure nitrogen and 20·95 mV for fresh air. On the assumption that the analyser output is linear over its whole range (0 to 21%), this gives a differential output of 1 mV for 1%. The assumption of linearity may be proved by means of a reduced-pressure calibration. A simple water manometer is connected to the outlet port of the analyser and a restriction applied to the inlet, so that when dry fresh air is drawn through, the resultant negative pressure in the cell is indicated by the manometer. If the analyser is linear the fall in output is proportional to the manometer reading. It must not, however, be expected that extrapolation of the resultant straight-line calibration will pass through the point corresponding to zero output at zero pressure. This is because the zero point of the analyser, as it is set up, corresponds to the cell being filled with nitrogen; whereas the zero point of the reduced pressure calibration corresponds to the cell being totally evacuated. Nitrogen is not inert magnetically; it is in fact slightly diamagnetic, as may be demonstrated by doing a pressure-calibration having the cell filled with nitrogen instead of fresh air. When this small effect is allowed for, we find that the outputs of our analysers have the expected sensitivity over the full range. A fuller explanation of most of the points discussed in this section has been published elsewhere (McLean & Watts, 1976).

For carbon dioxide and methane measurements the requirements for accuracy are less stringent. Several commercially available instruments are suitable. We use Grubb Parsons Infra Red Gas Analysers for both gases and standardize them against special gas mixtures obtained from BOC Ltd.

GAS SAMPLING

When a face-mask or head-cage is used the speed of response to gas exchanges by the animal can become an embarrassment. Fluctuations in gas concentration corresponding to minor positional or respiratory alterations by the animal can cause considerable fluctuation in the output record from the gas analysers. This can be prevented by passing the expiratory gas through a divided reservoir. The reservoir stores successive samples of the exhaust gas for a convenient period, before passing the mixed gas through the analysers whilst the next sample is being accumulated. Two such systems with differing degrees of versatility have been described (Kendall *et al.*, 1971; McLean, 1974).

For heat production measurements in the gradient-layer (direct) calorimeter, the animal chamber is effectively used as an open-circuit respiration chamber. Gas samples are drawn continuously from both fresh and exhaust air ducts and passed via a switching unit and $CaCl_2$ drying towers to the twin-channel oxygen analyser; the carbon dioxide and methane analysers normally monitor only exhaust air. For a short period once every hour the switching unit interchanges the two gas samples between analysers for the purpose of checking zero drifts.

AIR FLOW AND AIR MOVEMENT

Ventilation rate in open-circuit systems is measured by means of Rotameter flowmeters. Flow rate is indicated by the height of a float which is supported by the airstream passing up a tapered glass tube. To make the Rotameter compatible with automatic data logging a photoelectric detecting device has been fitted to it. This operates over a limited portion of the Rotameter scale giving an analogue voltage output. Details of this device are not yet published. It has been found that, following initial calibration, the Rotameter gives accurate and repeatable measurements of air flow rates. A disadvantage is that at any particular level of flow (volume per unit of time) its output reading varies in proportion to the square root of the density of the air passing through it. This means that the calibration is affected by temperature, humidity and barometric pressure, and calculation of corrected flow rates is complex. A great simplification would have been achieved if the flow measuring device had been one which records volumetric flow rate, such as a turbine flowmeter. It has been shown (McLean & Watts, 1976) that if a volumetric flow device is used for open-circuit calorimetry the effects of pressure and temperature on the flow device and gas analysers are compensatory and no corrections are needed. Unfortunately, the pressure drop required to operate a turbine flowmeter proved too great for this particular application. However, the tedium and potential source of errors involved in making corrections for pressure, temperature and humidity are virtually eliminated by the availability of a computer.

Localized levels of air movement near animals (i.e. draughtiness) are measured by means of Tinsley hot wire anemometers. The effectiveness of these and other instruments used for monitoring the animal environment is unfortunately limited by the necessity of enclosing the instruments in protective cages.

RECORDING SYSTEMS FOR INDIRECT CALORIMETRY

For the open-circuit head-cage system a potentiometric strip-chart instrument is used for recording. A separate measuring device, e.g. digital voltmeter, is needed for setting up the oxygen analyser. This is because the two setting points (0 and 20·95 %) are separated by more than the span of the recorder which covers a range of only 1 %. An alternative method of overcoming this problem is to prepare a simple battery-operated potential divider which provides voltages in a fixed ratio of 1 to 20. The recorder is first used to examine the smaller voltage which is adjusted to read 100 % of the recorder span. The larger voltage now corresponds to 20 times the recorder span. It may then be connected in series with the output of the oxygen analyser to back-off the equivalent of 20 % oxygen. The span of the analyser is now set to read 95 % of recorder span for fresh air which contains 20·95 % oxygen. The recorder is thus set up to cover the range 20 to 21 % oxygen.

For the gradient-layer calorimeter where many other parameters are measured in addition to those involved in indirect calorimetry a computer-controlled data logger is used for recording. This is described later.

DIRECT CALORIMETRY

Direct calorimetry is the measurement of the heat given off from an animal by the physical processes of conduction, convection, radiation and evaporation. The direct calorimeter at the Institute, designed for use with cattle, operates on the gradient-layer principle. The animal chamber is a fully water-jacketed aluminium enclosure, its inside surfaces being completely lined with a uniform layer of insulating material that incorporates a network of copper/constantan thermocouples. The thermocouples are positioned to record the average thermal gradient across the entire insulating layer. This is a measure of the integrated heat flow through the surfaces of the chamber, which in the main represents the sum of sensible, or non-evaporative, heat losses from the animal. The remaining, mainly evaporative, heat loss from the animal passes out of the chamber with the ventilating air. This heat loss is detected by cooled units (platemeters) which also operate on the gradient-layer principle.

The heat measured inside the platemeters is that of returning a sample of the exhaust air to its original (inlet) temperature and re-condensing the moisture out of it. The calorimeter, which can be operated over a variety of climatic conditions, has a high speed of response to changes in heat evolved within it and measures this heat with an accuracy of ±1 %. It has been fully described elsewhere (McLean, 1971). Since, as already mentioned, the direct calorimeter is also used as an open-circuit indirect calorimeter, it provides simultaneous and rapidly responding measurement of all heat produced and heat lost by an animal inside it. The calorimeter is thus used for studies on the heat balance of cattle with particular reference to short-term imbalances that result from feeding, activity and normal daily cycles.

There are two major causes of anomalous measurement that can occur in direct animal-calorimetry. Firstly, if water lies in the chamber it evaporates. The heat required to evaporate the water is given up by the contents of the chamber and is recorded as negative sensible heat. An equal and opposite positive amount of heat is recorded by the platemeters. There is thus no error in total heat loss but the partition between sensible and evaporative loss from the animal is affected. Urine and faeces must, therefore, be collected into containers and water must not be allowed to lie in drinking bowls. Secondly, after the animal has consumed food or water, the body gives up heat to the ingested materials and this heat loss by the animal is not recorded by the calorimeter. Similarly when the animal excretes urine, faeces or evaporated water, this material cools giving up heat to the chamber. This heat is recorded by the calorimeter. Over a long period the heats of warming ingesta and of cooling excreta cancel each other out but they do not occur at the same time and they represent false temporary imbalances between heat production and heat loss of the animal.

DATA RECORDING AND PROCESSING

All measurements associated with the gradient-layer calorimeter (including direct and indirect systems and temperatures) are monitored by a Solartron computer-controlled logging system. This consists of a digital voltmeter, scanner, central processing unit and various peripheral devices. Information can be fed to or from the computer via three alternative systems (a teletype, fast paper punch/reader and visual display unit). The computer controls and processes all measurements and gives output information in terms of calculated heat flows (watts), temperatures (°C). The computer also controls much of the operation of the calorimeter and experimental timetable (e.g. gas switching systems, availability of drinking water, illumination). It also checks results and gives warnings of any apparent malfunction (e.g. loss of temperature control, door improperly closed, unreasonably high or low results). At the present time it is being used to control and monitor heat balance experiments that normally last 10 to 12 days. During this time heat balance measurements are automatically recorded at 10-minute intervals. Operators have to enter the calorimeter briefly three times each day for renewal of faeces collectors and supply of food. The gas analysers are set up every 24 hours.

CALIBRATION AND PROVING

The accuracy of all measuring instruments depends upon their calibrations and the standards against which the calibrations are made. Accurate standards for basic physical measurements such as mass, temperature, voltage, pressure and time are obtained easily and calibration of many instruments depends in essence only on these. There remain, however, a few instruments for which accurate calibration standards are difficult to obtain. For calibrating gas analysers, cylinders containing accurately made up gas mixtures can be purchased or special mixing pumps may be used. But as stated earlier, for highly linear oxygen analysers, fresh air and nitrogen used as standards can give a more accurate calibration even though the range to be covered is small. The calibration of gas flowmeters presents a problem. Low levels of flow can be produced accurately by a displacement method such as weighing the amount of water displaced out of a rigid container by the calibration gas in a given time. For larger flows this is not practicable. Wet or dry gas meters are probably as accurate as any other commercially available instrument. We have calibrated our Rotameters against another, standardized Rotameter, at the National Engineering Laboratory. The combined system of flowmeters and gas analysers is tested by a recovery trial after injection of a known quantity of the gas to be measured. Nitrogen is used as the injection gas to simulate oxygen consumption. We find our nitrogen recovery trials are repeatable to $\pm 1\%$.

For calibration of the direct calorimeter an accurate source of 400 or 500 W of heat is necessary. This is inconveniently large to generate from direct current sources which would make its measurement simple. Our ultimate standard is a Sangamo Weston S67 wattmeter, but for routine checks we use an Ormandy & Stollery Digiwatt wattmeter which, having an analogue output, is compatible with the data logger. Calibration of all parts of the gradient-layer calorimeter is based on a dry heat load measured by the wattmeters. The platemeter sensitivity so determined requires a further correction. In normal usage, water evaporated in the chamber at temperature T_c is condensed in the platemeters at a lower temperature T_p. The heat recorded is thus the latent heat of vaporization λ plus the heat evolved on cooling the water from T_c to T_p. The output from the platemeters therefore requires reduction by a factor $[1-(T_c-T_p)\ c/\lambda]$ where c is the specific heat of water. An analogous correction will be necessary in most systems of direct calorimetry depending upon the conditions under which evaporative heat loss is measured.

It will be clear from the description given that instrumental development has played an essential part in the programme of work in environmental physiology at the Hannah Research Institute. More recently some of the same instrumental techniques have been employed in a joint project with the West of Scotland Agricultural College to monitor the environment of calf-rearing houses. A growing recognition of the importance of local environment to efficient livestock production has prompted this and other relevant studies elsewhere. Continued advances in instrumentation will be necessary to further both laboratory and field studies.

REFERENCES

Brouwer, E. (1965) in *Energy Metabolism* (Blaxter, K. L., ed.), pp. 441–443, Academic Press, London
Findlay, J. D. (1950) *Bull. Hannah Dairy Res. Inst.* no. 9, 163–168
Findlay, J. D. (1957) *J. Physiol. (Lond.)* **136**, 300–309
Findlay, J. D., McLean, J. A. & Bennet, R. D. (1959) *Heat. Vent. Eng.* **33**, 95–100
Hales, J. R. S. & Findlay, J. D. (1968*a*) *Respir. Physiol.* **4**, 333–352
Hales, J. R. S. & Findlay, J. D. (1968*b*) *Respir. Physiol.* **4**, 353–362
Kendall, P. P., McLean, J. A. & Smith, J. F. (1971) *J. Physiol. (Lond.)* **216**, 4P–5P
McLean, J. A. (1963*a*) *J. Physiol. (Lond.)* **167**, 417–426
McLean, J. A. (1963*b*) *J. Physiol. (Lond.)* **167**, 427–447
McLean, J. A. (1971) *J. Inst. Heat. Vent. Eng.* **39**, 1–8
McLean, J. A. (1972) *Br. J. Nutr.* **27**, 597–600
McLean, J. A. (1974) *Lab. Pract.* **23**, 14–15
McLean, J. A. & Calvert, D. T. (1972) *J. Agric. Sci.* **78**, 303–307
McLean, J. A. & Watts, P. R. (1976) *J. Appl. Physiol.* **40**, 827–831
Nisbet, W. (1956) *J. Sci. Instrum.* **33**, 154–158

APPENDIX – SUPPLIERS OF SOME OF THE INSTRUMENTS

BOC Special Gases Ltd, Deer Park Road, London SW19 3UF (Standard gas mixtures)
Fenlow Electronics Ltd, Whittlets Eyot, Jessamy Road, Weybridge, Surrey (Digital voltmeter)
Foxboro-Yoxall Ltd, Redhill, Surrey RH1 2HL (Dewcells)
GEC-Elliot Process Instruments Ltd, 330 Purley Way, Croydon CR9 4PG (Rotameter)
Sir Howard Grubb Parsons & Co. Ltd, Walkergate, Newcastle upon Tyne NE6 2YB (CO_2 and CH_4 analysers)
Ormandy & Stollery, Regal Works, Station Road, Brightlingsea, Essex (Digital wattmeter)
Sangamo Weston Ltd, Enfield, Middlesex (Wattmeter)
The Solartron Electronic Group Ltd, Farnborough, Hampshire (Computer-based data logger, digital voltmeters)
Taylor-Servomex Ltd, Crowborough, Sussex TN6 3DU (Oxygen analysers)
H. Tinsley & Co. Ltd, Werndee Hall, South Norwood, London SE25 (Anemometers)

TRIGLYCERIDE STRUCTURE AND BIOSYNTHESIS

W. W. CHRISTIE

VIRTUALLY all the commercially important fats of animal and vegetable origin consist largely of triglycerides. In animals, the main site of triglyceride deposition is in adipose tissue where they serve as an energy store. During lactation, milk fat, which consists mainly of triglycerides, is synthesized for the nutrition of the young calves. Fatty acids may be transported in plasma in the form of triglycerides to many tissues of the animal where they may be hydrolysed by the enzyme lipoprotein lipase and the constituent fatty acids utilized for the synthesis of other lipids including triglycerides. Although triglycerides may appear superficially to be rather simple molecules consisting of the trihydric alcohol glycerol, to which three molecules of fatty acids are esterified, the detailed structures can be complex. For example, in a natural triglyceride mixture that contains only five different fatty acid constituents, each triglyceride molecule can contain only three fatty acid molecules and these can be the same fatty acid or different fatty acids and they can be esterified in any order to the glycerol moiety. It can be calculated that in this example seventy-five different combinations of fatty acids or molecular species of triglycerides can exist (not including enantiomers). In bovine milk fat where 437 different constituent fatty acids have been characterized to date (Patton & Jensen, 1975), the number of possible combinations is astronomically high. When positions 1 and 3 of a triglyceride molecule are occupied by different fatty acids, a centre of asymmetry is created so that natural triglycerides may exist in enantiomeric forms so increasing the number of possible fatty acid combinations still further.

The physical properties such as melting point of triglycerides will depend largely on their fatty acid compositions but they will also be influenced by the manner in which the fatty acids are arranged within the molecules. This in turn will depend on the mechanism of triglyceride biosynthesis, which may vary somewhat from tissue to tissue and be related to the function of triglycerides in particular tissues. A knowledge of the detailed structures of triglycerides is therefore important from a practical standpoint in that the uses to which a fat can ultimately be put by the consumer will be to a large degree controlled by its physical properties. It is also important in that it may yield information about the method of biosynthesis. Such knowledge might eventually be used to control or modify the composition and properties of natural fats or indeed to control the rate of synthesis. It would be economically advantageous to be able to stimulate milk fat production or conversely to limit the rate of fat deposition in relation to protein synthesis in domestic animals. Information on the latter process would also be of value in controlling obesity in humans.

The profound influence that variations in the structures of triglycerides can have on their physical properties became apparent at the Hannah Research Institute during a study of the effects of dietary supplements of copper sulphate on the composition of the depot fats of pigs. Copper sulphate in the diet increased the growth rate of pigs and improved their feed conversion efficiency (Barber *et al.*, 1955). The triglycerides in the inner and outer back fats of pigs fed on a diet containing 250 p.p.m. copper sulphate had melting points 10 to 15°C lower than those of the corresponding control animals although the only significant difference in the fatty acid compositions of the fats was that the control animals contained 2% more stearic acid and correspondingly less oleic acid in their triglycerides than pigs fed on the copper-supplemented diets (Moore *et al.*, 1969). Apparently, differences in the structures of the triglycerides rather than the small differences in fatty acid compositions were responsible for the observed change in melting point, because when fats from pigs fed on the control or copper-supplemented diets were randomly interesterified, there was little difference between melting points. When the proportions of the principal molecular species of triglycerides in the backfat from pigs fed on the control and copper-supplemented diets were determined, it was demonstrated that only the compositions of the fatty acids in the primary positions of the triglycerides were altered and that a 2% change in the overall fatty acid composition of the triglycerides produced a 6% change in the amounts of the two most saturated molecular species of triglycerides, i.e. the tri-saturated and di-saturated—mono-unsaturated species which were believed to play a major role in governing the melting points of triglycerides (Christie & Moore, 1969*a*).

To study this and related problems further, it was necessary to develop a number of new analytical procedures. Foremost among these was a technique to differentiate between the two primary positions of the triglyceride molecule, i.e. for the stereospecific analysis of triglycerides. Brockerhoff (1965) had shown how this might be achieved but the procedure then described was time-consuming and required comparatively large amounts of material (approx. 1 g). It proved possible, however, to improve and adapt the procedure into a rapid and highly accurate method that required only 10 to 40 mg of triglycerides for a complete analysis (Christie & Moore, 1969*b*). A schematic representation of the procedure is shown in Fig. 1. α, β-Diglycerides (a mixture of *sn*-1,2- and *sn*-2,3-diglycerides) were prepared by partial hydrolysis of triglycerides with ethyl magnesium bromide and converted by chemical synthesis to phospholipids, which were in turn hydrolysed by the stereospecific phospholipase

A_2 (*EC* 3.1.1.4) of snake venom. The fatty acid composition of the resulting lysophosphatide accurately represented that of position *sn*-1 of the original triglyceride, that of position *sn*-2 was determined separately by means of pancreatic lipase hydrolysis and that of position *sn*-3 could then be calculated by difference. Alternative methods were available to confirm the compositions of positions *sn*-2 and *sn*-3 and various checks on the accuracy of the analysis were possible. A further advance in lipid methodology that facilitated a number of subsequent studies was the development of a new procedure for quantifying lipid classes. Lipids separated by thin layer chromatography were quantified by gas chromatography of their fatty acid constituents in the presence of an internal standard, i.e. a fatty acid that did not occur naturally in the sample analysed and usually an odd-chain component. The amount of lipid present was determined simply by relating the sum of the amounts of the various fatty acid constituents to the amount of the internal standard (Christie *et al.*, 1970; Christie, 1972). Related procedures that utilized internal standards proved useful in estimating molecular species of triglycerides separated according to degree of unsaturation by means of silver nitrate thin layer chromatography or according to molecular weight by means of high temperature gas chromatography (Christie, 1973).

The stereospecific analysis procedure was used to demonstrate that the structures of triglycerides from various sites in an animal differed markedly. For example in the pig (Christie & Moore, 1970*a*), triglycerides from adipose tissue, heart, kidney and adrenals were found to be similar in that

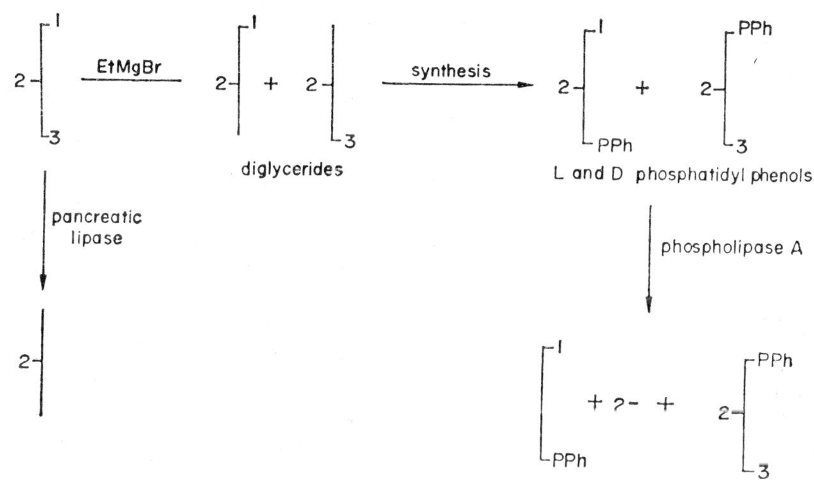

Fig. 1. *Stereospecific analysis of triglycerides.*

palmitic acid constituted a high proportion of the fatty acids in position *sn*-2, most of the remaining saturated fatty acids were in position *sn*-1, and position *sn*-3 contained mainly unsaturated fatty acids. Liver triglycerides, on the other hand, had very much less palmitic acid in position *sn*-2 as most of the saturated fatty acids were in position *sn*-1 whereas plasma triglycerides were intermediate in structure between those of liver and adipose tissue. Sow's milk triglycerides resembled those of adipose tissue except that there was much less difference between the compositions of positions *sn*-1 and *sn*-3 than was found in the adipose tissue. The physical properties of triglycerides from these various sites must also have differed markedly and would be related to their function in the animal, although more information is needed before these can be fully correlated. In order to study the arrangement of fatty acids in the triglycerides, simpler molecular species were obtained by fractionation according to degree of unsaturation and these were separately subjected to stereospecific analysis. The proportions of the various molecular species found were compared with those predicted from the positional distribution in the intact triglycerides on the basis of various distribution theories. Excellent agreement was obtained for pig adipose tissue and milk triglycerides if a 1-random, 2-random, 3-random arrangement was assumed, i.e. in biosynthetic terms if position *sn*-2 was acylated from a particular pool of fatty acids without regard to the composition of fatty acids already esterified to position *sn*-1 in the intermediate mono-acylglycerol derivative, and if position *sn*-3 was subsequently esterified without selectivity for particular diglyceride intermediates. In pig liver triglycerides, there appeared to be some specificity for the chain-lengths of fatty acids in the intermediates during acylation but not for degree of unsaturation (Christie & Moore, 1970*a*; Hunter *et al.*, 1973).

Adipose tissue triglycerides from the pig differed from those of most other mammals in the high content of palmitic acid in position *sn*-2. Those of the sheep (Christie & Moore, 1971) were more typical in that saturated fatty acids were concentrated in position *sn*-1, unsaturated in position *sn*-2 and C_{18} saturated and unsaturated in position *sn*-3. Sheep liver and plasma triglycerides differed from this in that they contained higher proportions of longer-chain polyunsaturated components which were concentrated in positions *sn*-2 and *sn*-3 as well as higher proportions of palmitic acid in position *sn*-2. In triglycerides from all tissues, the minor branched-chain fatty acid components were distributed among the three positions in much the same way as the corresponding straight-chain components but *trans*-18 : 1 was concentrated in positions *sn*-1 and *sn*-3 although *cis*-18 : 1 was found in greatest concentrations in position *sn*-2. Human adipose tissue and heart muscle triglycerides resembled those of the sheep in structure if not in composition, with some degree of asymmetry between positions *sn*-1 and *sn*-3, whereas those of human liver were highly asymmetric (Christie *et al.*, 1971). Triglycerides from aortic lesions, however, were rather different from those of the other tissues examined and in particular contained higher proportions of polyunsaturated fatty acids; these were present in greatest concentration in position *sn*-3.

Adipose tissue triglycerides from chickens (Christie & Moore, 1972*a*) were found to resemble those of mammals, in contradiction of an earlier report (Brockerhoff *et al.*, 1966). In this instance also, the proportions of the various molecular species found agreed closely with those predicted if a 1-random, 2-random, 3-random arrangement were assumed. In the triglycerides of the egg of the chicken, an exceptionally high degree of asymmetry was found as palmitic acid constituted over 70% of the fatty acids of position *sn*-1, but only 12% of position *sn*-3 and 4% of position *sn*-2; most of the linoleic acid was found in position *sn*-2 whereas position *sn*-3 contained nearly 70% oleic acid (Christie & Moore, 1970*b*). This unusual structure was later found to be characteristic of triglycerides from avian eggs in general as it was found in those of more than twenty quite disparate species of birds (Christie & Moore, 1972*c*). As triglycerides from the liver and plasma of laying hens were found to have virtually identical structures to those of the egg and indeed as the proportions and compositions of all the lipids in the plasma were identical to those in the egg, it was suggested that egg lipids were synthesized in the liver and transported in the plasma to the ovary where they were deposited intact in the egg follicles without hydrolysis and resynthesis at the avian equivalent of the placental barrier (Christie & Moore, 1972*b*). Direct support for this hypothesis was supplied by evidence that the activities of the various lipid hydrolytic and biosynthetic enzymes in the ovary were much too low to account for the amount of lipid deposited in the egg (Gornall *et al.*, 1972).

Prior to the development of stereospecific analysis procedures, pancreatic lipase hydrolysis was used widely to obtain the fatty acid composition of position *sn*-2 of natural triglycerides. As a consequence, there has been a tendency to assume that the composition of fatty acids esterified to the sole secondary hydroxyl group must have greater importance than those of the two primary positions. The composition of position *sn*-2 is of course important when triglycerides are digested by animals, since 2-monoglycerides may then be formed which can be absorbed by the intestine and utilized as such, and also when monoglycerides derived from triglycerides are re-utilized for triglyceride biosynthesis via the monoglyceride pathway elsewhere in the animal, but full stereospecific analyses from this and other laboratories have shown that the compositions of all three positions may be of interest. In pig adipose tissue triglycerides, for example, position *sn*-2 has been widely held to be unique as it contained 80% of the total palmitic acid but stereospecific analysis has shown that position *sn*-1 was also unusual in that it contained more than 70% of the stearic acid whereas position *sn*-3 contained two-thirds of the linoleic acid (Christie & Moore, 1970*a*). If any single position in natural triglycerides can be held to have any special significance, especially in terms of their biosynthesis, it could be argued that it is position *sn*-3. This is the last position to be esterified during triglyceride biosynthesis by the glycerol-*sn*-3-phosphate pathway (Weiss & Kennedy, 1956) and this step is potentially important in the cellular control mechanism for triglyceride biosynthesis.

Analyses of certain natural triglycerides have illustrated the importance of position *sn*-3 in some circumstances. The seed oil of *Sapium sebiferum*, for example, was found to contain two unusual fatty acids, 8-hydroxy-5, 6-octadienoic acid and *trans*-2, *cis*-4 decadienoic acid, which were joined by an estolide linkage (Sprecher *et al.*, 1965). By means of stereospecific analysis, it was shown that the estolide-linked fatty acids were located entirely on position *sn*-3 of the triglycerides (Christie, 1969). Other seed oils have been found that contained acetic acid exclusively esterified to position *sn*-3 of the triglycerides, e.g. that of *Impatiens edgeworthii* (Bagby & Smith, 1967). It has long been known that the triglycerides of milk fats from ruminant animals contained appreciable amounts of butyric acid and recently it was shown, by means of stereospecific analysis, that all of this was esterified exclusively to position *sn*-3 (Breckenridge & Kuksis, 1968). All the butyric acid was synthesized within the mammary gland and as position *sn*-3 was the last to be esterified, it might be suggested that butyric acid synthesis was involved intimately in controlling the rate of triglyceride biosynthesis in the mammary gland. Conversely, it could be argued that the enzymes of triglyceride biosynthesis exerted a key role in controlling butyric acid synthesis. In either circumstance, the two processes are inextricably linked. It has also been suggested, if not conclusively proved, that all of the oleic acid formed *de novo* by desaturation of stearic acid within the mammary gland of ruminant animals is incorporated into position *sn*-3 of triglycerides and that this process may have also a regulatory function in triglyceride biosynthesis (Kinsella, 1972). Related processes may operate in plant tissues as it was demonstrated that avocado mesocarp microsomes *in vitro* incorporated acyl-acyl carrier protein esters, that would normally be the product of the plant fatty acid synthetase, exclusively into position *sn*-3 of triglycerides (Shine *et al.*, 1976).

Evidence that fatty acids synthesized *de novo* within adipose tissue might be used for triglyceride biosynthesis in a different way from fatty acids originating in the diet or elsewhere in the animal and reaching the tissue via the plasma was obtained from a study of the variation of triglyceride structure with fatty acid composition in pig adipose tissue (Christie & Moore, 1970*c*). Pig adipose tissue triglycerides that varied widely in composition were subjected to stereospecific analysis. There was always a constant amount of palmitic acid and constant proportions of myristic, stearic, palmitoleic and oleic acids in position *sn*-1. Positions *sn*-2 and *sn*-3 contained constant proportions of myristic and palmitoleic acids but the amounts of palmitic, stearic and oleic acids varied in a linear manner with respect to each other. In none of the positions was there a constant proportion of linoleic acid. As the two fatty acids which must have been synthesized *de novo* largely within the tissues, i.e. myristic and palmitoleic acids, were found in constant proportions in all three positions, it was suggested that position *sn*-1 contained only fatty acids synthesized *de novo*, but that fatty acids derived from the plasma complicated the pattern in positions *sn*-2 and *sn*-3. The only fatty acid that was entirely of dietary origin, i.e. linoleic acid, exhibited a distribution pattern distinct from that of any of the other fatty acid components.

Analytical studies of the kind described above can provide valuable clues to the mechanism of triglyceride biosynthesis in tissues but they cannot prove unequivocally that a particular mechanism is operative. For the latter purpose, it is necessary to develop some system that uses part of the living tissue *in vitro* to study the incorporation of isotopically labelled fatty acids into triglycerides and any factors that might influence this process. Analytical studies often suggest potentially useful approaches to biosynthetic problems, however, and with modification, the analytical techniques developed for structural studies can also be used to advantage in biochemical experiments.

Initially, it was decided to study triglyceride biosynthesis in mammary tissue. As the use of biopsy samples of mammary tissue *in vitro* is beset by numerous practical difficulties, not the least of which is the removal of the large amount of fibrous connective tissue in the gland, the potential of freshly secreted milk as a source of mammary tissue enzymes was examined in the light of reports that it had the capacity to synthesize triglycerides and other lipids from added isotopically labelled fatty acids (McCarthy & Patton, 1964) and to desaturate stearic acid to oleic acid (McCarthy *et al.*, 1965). A preliminary study (Christie, 1974) confirmed that these reports were correct and that the reaction was enzymatic since boiling, brief ultrasonification or addition of detergents eliminated all activity. The pattern of the various products formed (i.e. triglycerides, diglycerides and phosphatidic acid, in addition to some phosphatidyl choline and phosphatidyl ethanolamine) was consistent with triglyceride biosynthesis by the glycerol-*sn*-3-phosphate pathway yet the reaction was not stimulated by the addition of any of the cofactors such as ATP, α-glycerophosphate or coenzyme A that are normally required for this mechanism to operate. Esterification was, however, stimulated by added glycerol. The enzymes of the monoglyceride pathway of triglyceride biosynthesis were also present since isotopically labelled 1-O- and 2-O-alkyl glycerols were incorporated into lipids by the milk. In addition to confirming that desaturation of stearic acid to oleic acid occurred, it was shown that all the newly formed oleic acid was esterified immediately, which suggested that the desaturase and the esterifying enzymes were in close physical proximity in milk.

By means of a simple precipitation and differential centrifugation procedure, it proved possible to obtain a preparation which contained all the enzyme activity but which was free of most of the other milk constituents (Christie & Wooding, 1975). Electron microscopy showed that it contained membrane-bounded pieces of cytoplasmic material, each with a lipid droplet but no nucleus. Rough endoplasmic reticulum in sheets, vesicles or swollen cisternae was usually present together with an occasional mitochondrion, but only rarely could pieces of the Golgi apparatus or casein-containing vesicles, that are typical of the lactating mammary epithelial cell, be identified. Autoradiography of sections of the pellet that had been incubated with a radioactive fatty acid prior to precipitation, showed significant incorporation into the lipid droplets contained within each cytoplasmic fragment. Used *in vitro* in short-term incubations, the particles were more akin to mammary epithelial cells in their properties than to subcellular preparations. The 'cytoplasmic droplets' obtained from milk were therefore analogous to some mammary tissue preparations obtained by surgical means yet they were obtained routinely from goats without harming or unduly exciting them in any way. Pooled samples have been taken from several goats, so minimizing biological variation, on almost a daily basis from the same animals for over a year with very little change in biosynthetic activity.

More recently (Christie *et al.*, 1976*b*), it was demonstrated that in addition to the metabolic activities already described, the cytoplasmic droplets would incorporate [1-^{14}C]acetate into the normal range of fatty acids known to be synthesized in goat mammary tissue and found in goat milk fat and also would oxidize [U-^{14}C]glucose to CO_2 and incorporate [^{3}H]leucine into protein.

The incorporation of labelled fatty acids into each of the three positions of the triglycerides was also studied (Christie, 1974), although it was necessary to use a modified stereospecific analysis procedure in this instance (Christie & Hunter, 1973; Christie, 1975). In a conventional stereospecific analysis, the relative proportions of several different fatty acids in each position are obtained; when triglycerides are formed biosynthetically and contain a single labelled fatty acid, it may be necessary to determine the absolute amount of that fatty acid incorporated into each position by correcting for all the losses that inevitably occur during a multi-step analysis. The problem was resolved by adding, prior to the analysis, an internal standard consisting of triglyceride with a single labelled odd-chain fatty acid in each position.

The same basic mechanisms for triglyceride biosynthesis that operate in mammary gland also function in adipose tissue although there are points of difference in detail. In preliminary studies with rat adipose tissue slices *in vitro* (Christie & Vernon, 1975), the manner in which [1-^{14}C]palmitic, stearic, oleic and linoleic acids were incorporated into triglycerides from an unesterified fatty acid pool of controlled composition was determined for comparison with the natural distributions of these components in rat adipose tissue triglycerides. Similar relative proportions of oleic and linoleic acids entered each of the three positions of the triglycerides as were found naturally, but less palmitic acid entered position *sn*-1 and much less stearic acid was found in position *sn*-2 than was anticipated from the natural distributions. In addition, considerably more oleic acid formed *de novo* from stearic acid in the tissue entered position *sn*-2 than was found in this position naturally or was found when [1-^{14}C]oleic acid was incubated as such.

The optimum cofactor requirements for triglyceride biosynthesis by the glycerol-*sn*-3-phosphate pathway in rat adipose tissue homogenates *in vitro* have been determined also (Christie *et al.*, 1976*a*). Some differences from the requirements for total fatty acid esterification were observed but in general the results provided further evidence for the key role of the enzyme, phosphatidate phosphohydrolase, in the regulation of triglyceride biosynthesis. A requirement for a critical concentration of albumin was also observed. In parallel experiments to those with adipose tissue slices, it was found that in this instance, the distributions of fatty acids incorporated into the three positions of the triglycerides were similar to those of the natural distributions. It appeared then that, in rat adipose tissue, the relevant acyl transferases have the major role in determining the positional distribution of fatty acids within triglycerides but that other factors may play a part. For example, transfer of newly synthesized fatty acids to triglycerides may be a factor as may compartmentation of the various enzymatic steps for triglyceride biosynthesis in the cell.

Triglycerides then are extremely important compounds commercially within the agricultural industry but also they are important to various metabolic processes in domestic animals and in humans. A knowledge of the structures of triglycerides and of the factors that control their synthesis may, therefore, have innumerable practical applications. Analytical techniques such as the stereospecific analysis procedure will continue to have an important part to play in studies of both facets of the problem.

REFERENCES

Bagby, M. O. & Smith, C. R. (1967) *Biochim. Biophys. Acta* **137**, 475–477

Barber, R. S., Braude, R., Mitchell, K. G. & Cassidy, J. (1955) *Chem. Ind.* (*Lond.*) **74**, 601

Breckenridge, W. C. & Kuksis, A. (1968) *J. Lipid Res.* **9**, 388–393
Brockerhoff, H. (1965) *J. Lipid Res.* **6**, 10–15
Brockerhoff, H., Hoyle, R. J. & Wolmark, N. (1966) *Biochim. Biophys. Acta* **116**, 67–72
Christie, W. W. (1969) *Biochim. Biophys. Acta* **187**, 1–5
Christie, W. W. (1972) *Analyst* **97**, 221–223
Christie, W. W. (1973) *Lipid Analysis*, Pergamon Press, Oxford
Christie, W. W. (1974) *Lipids* **9**, 876–882
Christie, W. W. (1975) *J. Chromatogr. Sci.* **13**, 411–415
Christie, W. W. & Hunter, M. L. (1973) *Biochim. Biophys. Acta* **316**, 282–287
Christie, W. W. & Moore, J. H. (1969*a*) *Lipids* **4**, 345–349
Christie, W. W. & Moore, J. H. (1969*b*) *Biochim. Biophys. Acta* **176**, 445–452
Christie, W. W. & Moore, J. H. (1970*a*) *Biochim. Biophys. Acta* **210**, 46–56
Christie, W. W. & Moore, J. H. (1970*b*) *Biochim. Biophys. Acta* **218**, 83–88
Christie, W. W. & Moore, J. H. (1970*c*) *Lipids* **5**, 921–928
Christie, W. W. & Moore, J. H. (1971) *J. Sci. Food Agric.* **22**, 120–124
Christie, W. W. & Moore, J. H. (1972*a*) *J. Sci. Food Agric.* **23**, 73–77
Christie, W. W. & Moore, J. H. (1972*b*) *Comp. Biochem. Physiol. B. Comp. Biochem.* **41**, 287–295
Christie, W. W. & Moore, J. H. (1972*c*) *Comp. Biochem. Physiol. B. Comp. Biochem.* **41**, 297–306
Christie, W. W. & Vernon, R. G. (1975) *Biochem. Biophys. Res. Commun.* **66**, 243–249
Christie, W. W. & Wooding, F. B. P. (1975) *Experientia* (*Basel*) **31**, 1445–1447
Christie, W. W., Noble, R. C. & Moore, J. H. (1970) *Analyst* **95**, 940–944
Christie, W. W., Moore, J. H., Lorimer, A. R. & Lawrie, T. D. V. (1971) *Lipids* **6**, 854–856
Christie, W. W., Hunter, M. L. & Vernon, R. G. (1976*a*) *Biochem. J.* **159**, 571
Christie, W. W., Vernon, R. G. & Wooding, F. B. P. (1976*b*) *Biochem. Soc. Trans.* **4**, 242–243
Gornall, D. A., Kuksis, A. & Morley, N. (1972) *Biochim. Biophys. Acta* **280**, 225–230
Hunter, M. L., Christie, W. W. & Moore, J. H. (1973) *Lipids* **8**, 65–70
Kinsella, J. E. (1972) *Lipids* **7**, 349–355
McCarthy, R. D., Ghiardi, F. L. A. & Patton, S. (1965) *Biochim. Biophys. Acta* **98**, 216–217
McCarthy, R. D. & Patton, S. (1964) *Nature* (*Lond.*) **202**, 347–349
Moore, J. H., Christie, W. W., Braude, R. & Mitchell, K. G. (1969) *Br. J. Nutr.* **23**, 281–287
Patton, S. & Jensen, R. G. (1975) *Prog. Chem. Fats Other Lipids* **14**, 163
Shine, W. E., Mancha, M. & Stumpf, P. K. (1976) *Arch. Biochem. Biophys.* **173**, 472–479
Sprecher, H. W., Maier, R., Barber, M. & Holman, R. T. (1965) *Biochemistry* **4**, 1856–1863
Weiss, S. B. & Kennedy, E. P. (1956) *J. Am. Chem. Soc.* **78**, 3550

MILK COMPOSITION

J. C. D. WHITE and D. T. DAVIES

INTRODUCTION

COWS' MILK, an extremely complex biological fluid, plays a very important role in human nutrition. Because of this complexity and importance, studies of milk composition have formed the basis of much research as can readily be seen from the extensive literature which now exists on this subject. Workers at the Hannah Research Institute have been closely engaged in this area of research, and in this review an outline is given of their contributions to present knowledge. The type of work involved in the study of milk composition can vary considerably and for convenience in this article the work done at the Institute is divided on the one hand into studies that were concerned with the development of methods for the analysis of milk, and on the other into investigations directly devoted to the composition of milk.

DEVELOPMENT OF ANALYTICAL METHODS

The examination of milk composition, for whatever purpose, clearly requires reliable and convenient analytical methods and consequently considerable effort has been and continues to be directed to the development of such methods at the Institute. One practical result of this effort, of direct use to the dairy industry, was the furtherance of milk recording and compositional quality payment schemes, which arose from the investigation of densitometric and gravimetric methods for the large-scale determination of milk solids (Waite & Abbot, 1956; Waite, 1959; Waite & McPhillips, 1965). The report by Waite & Paterson (1959) on the composition of bulk milk measured on milk-recorded farms in Scotland was one of the first to show the value of densitometric solids-not-fat (SNF) determinations for recording purposes. Also of value in developing milk quality schemes was the investigation by Waite & Taylor (1966) of the effectiveness of potassium dichromate in preserving milk for analysis, and the study by Waite & Smith (1972) of the influence of subclinical mastitis and stage of lactation on the accuracy of the rapid, amido-black dye-binding method for determining the protein content of milk. Although the above methods have now been partially superseded by fully automated, infra-red absorption techniques for the rapid determination of fat, protein and lactose in milk, their availability did encourage a reduced emphasis on fat content as the principal criterion of milk quality. This change of emphasis led to the adoption in the United Kingdom of compositional quality payment schemes based on the total solids content of milk, and the Scottish Milk Records Association (SMRA), on whose executive committee the Institute is represented, now includes the determination of the protein content of milk in addition to yield and fat content in its official recording scheme. The Institute has also for many years maintained close contact with the Scottish Milk Marketing Board as well as with the SMRA in helping to develop analytical procedures.

In addition to this work on rapid analytical methods for use principally in the dairy industry, the Institute has been involved in the development, modification and improvement of research methods for the detailed study of milk constituents, in particular the protein. The major pioneering work on the quantitative determination of the main components of milk protein was done in the 1930s by the late S. J. Rowland at the National Institute for Research in Dairying, Reading, where his nitrogen-partition method (Rowland, 1938) was improved subsequently by Aschaffenburg & Drewry (1959). A scaled-down variant of the latter method, devised at the Hannah, has proved extremely useful for many routine studies. However, increasing evidence of the complexity of the protein mixture in milk necessitated analytical methods of improved resolution and much work has been done at the Institute to meet this need. This work has shown that it is possible by ion-exchange chromatography to partition casein quantitatively into γ-casein, κ-casein, β-casein, major α_s-caseins and minor α_s-caseins (Rose *et al.*, 1969; Davies & Law, 1977) and by gel chromatography to partition milk serum protein (excluding proteose-peptone and γ-globulins) quantitatively into β-lactoglobulin, α-lactalbumin, serum albumin, and lactoferrin plus minor proteins (Davies, 1974). The detailed partition of milk proteins so achieved has already provided information on compositional variation that could have practical implications but the chromatographic methods are laborious and it is probable that their main use will be as standards for assessing the validity of results obtained by more rapid methods such as gel electrophoresis.

Because of their influence on the coagulation of milk protein by rennet and heat, the mineral constituents of milk, especially calcium and magnesium, together with phosphate and citrate, have also been of considerable interest to Institute workers, and for their studies on the heat stability of milk White & Davies (1958) first devised semi-micro, classical methods of high accuracy for the determination of calcium and magnesium, which were followed later by more rapid yet equally reliable methods which make use of ethylenediaminetetra-acetic acid (Davies & White, 1962). During this work it was also shown that a readily prepared TCA-filtrate was a satisfactory alternative to a solution of milk-ash for use in determining the total amounts of calcium and magnesium in milk. For the routine determination of these and other metals, however, atomic absorption spectrophotometry is now favoured and one use of this technique at the Institute was in determining the trace levels of copper in farm bulk milks (Hankinson, 1975).

The determination of citric acid in milk was a slow and laborious procedure based on the formation of pentabromoacetone until eventually more rapid methods utilizing the colour-forming reaction between pyridine, acetic anhydride and citric acid were developed. These methods, however, gave conflicting results and therefore White & Davies (1963) carried out a detailed study of the problem that resulted finally in a pyridine–acetic anhydride method which gave accurate and precise values for the citric acid content of milk and milk sera. This method is now used widely and it formed the basis of the reference method adopted by the International Dairy Federation (IDF), the International Standards Organization (ISO), the Association of Official Analytical Chemists (AOAC), the Food and Agriculture Organization and the World Health Organization, for determining the citric acid content of cheese and processed cheese products. Also adopted by the IDF was the semi-micro classical method referred to above for determining the calcium content of milk.

In addition to having satisfactory methods for determining the total amounts of the mineral elements in milk it is necessary when studying the physico-chemical properties of the system to know how these elements are distributed between the aqueous (soluble) and the disperse (colloidal) phases and in what form they exist in each. The essential first step in obtaining this information is to have a method of separating the two phases that causes the minimum disturbance to the existing equilibria so that a serum representative of the aqueous phase can be isolated. By determining the concentration of constituents in such a serum and knowing their total concentration in the milk, the colloidal fraction of each can then be obtained by difference. At the time when knowledge of the mineral partition in milk was required for heat stability studies in the Institute, there was a lack of information on how best to isolate a suitable milk serum. Accordingly, Davies & White (1960) made a detailed comparative study of milk sera obtained by ultrafiltration, dialysis, high-speed centrifuging and rennet addition. Analysis of these sera showed that high-speed centrifugate and rennet whey were not wholly satisfactory because both contained some residual colloidal caseinate. Ultrafiltrate and diffusate, on the other hand, could be prepared free of caseinate and were found to be similar in composition provided they were prepared at the same temperature. It was also demonstrated that much of the reported difference in composition between these two sera resulted from their customary preparation at different temperatures. Ultrafiltrate composition, however, was influenced to a certain extent by a membrane sieving phenomenon, which resulted in slightly low calcium, citrate and lactose values and it was concluded that diffusate was the most satisfactory serum. Results obtained for diffusate indicated that provided it was prepared at 20°C by dialysing 50 volumes of milk containing 0·25% of chloroform (as a preservative) against 1 volume of water for 48 hours in Visking tubing, then its composition after correction for the dilution and for bound water should approximate closely to that of the aqueous phase of milk at ambient temperature. The results reported in this investigation are still quoted widely and have been used to characterize the ionic environment of the caseinate micelles in milk (Waugh, 1971; Farrell & Thompson, 1974). Other analytical techniques required to establish a fairly complete picture of the forms in which calcium, magnesium, phosphorus and citric acid occur in milk, and which were developed by the modification of various reported methods, are summarized by White & Davies (1958).

As well as carrying out the above type of analytical investigations it has always been the policy of the Institute to co-operate with national and international organizations concerned with developing standard analytical methods for milk and milk products, not only for use by the dairy industry but also in research laboratories. This co-operation started with one or two members of staff representing the Institute on a few technical committees of the British Standards Institution (BSI) engaged in preparing British dairying standards. However, the Institute's participation increased gradually to encompass representation on working groups of ISO and of IDF and finally, with rationalization of international standardizing activities, on joint IDF/ISO/AOAC working groups. Further proliferation of the Institute's work in this field came with the advent of the European Economic Community (EEC) and its associated dairy trade organizations, e.g. ASSILEC (Association de l'industrie laitière de la CEE). Thus today the Institute participates actively in the work of most standardizing organizations in dairying with the principal objective being the preparation of IDF/ISO/AOAC analytical standards, which can be endorsed by all other standardizing bodies for universal use. This work involves not only assistance in drafting standards but also taking part in collaborative investigations to develop methods and to assess their precision.

The individual standardization projects which the Institute has aided are too numerous for all to be mentioned here but assistance has been given in the preparation of many British Standards for the sampling and analysis of milk and milk products. More recently, the Institute has been involved with BSI in preparing standard methods for determining the solubility of dried milk, the protein content of cheese, the composition of caseins and caseinates, and extraneous water in milk. In the international field, much work has been done in preparing methods for determining the fat content of milk and milk products, the dispersibility, wettability and solubility of instant dried milk, the heat-treatment classification of dried milk and the repeatability and reproducibility of analytical methods. This selection of projects illustrates the wide range of practical standardization work in which the Institute has taken part.

STUDIES ON MILK COMPOSITION

Because of its location most of the studies on milk composition which have been carried out at the Institute have naturally involved milk from the Ayrshire breed of cattle but this position will no doubt change in the future with the increasing proportion of British Friesian and British Friesian × Ayrshire cows in the Scottish national herd. In this article, however, all the data refer to milk from Ayrshire cows, and of the factors that influence milk composition, only heredity, stage of lactation, season of year and subclinical mastitis are discussed here. Other factors such as feeding and processing treatments, e.g. heating, are dealt with in other reviews in this publication.

One of the most extensive investigations on the chemical compositior of milk carried out at the Institute was that described by Waite *et al.* (1956) and Robertson *et al.* (1956). Although this investigation, undertaken jointly with the SMRA and the Institute of Animal Genetics, Edinburgh, was designed specifically to obtain information on the inheritance of the levels of the major constituents of the SNF fraction of milk from the Ayrshire breed, and on the genetic interrelationships of these constituents, it also provided valuable information on the influence on milk composition of various non-genetic factors including stage of lactation, season of year and age of the cow. The experimental work involved the determination of total solids, fat, total nitrogen, non-casein nitrogen, lactose and 'ash' (by difference) in over 6000 milk samples from about 800 cows comprising 500 dam–daughter pairs on thirty-three farms in south-west Scotland. From the dam–daughter correlations within herds, heritabilities of about 0·50 were obtained for SNF, crude protein, casein and 'ash', about 0·35 for fat and lactose, 0·25 for milk yield and 0·13 for non-casein nitrogen. From these it was concluded that a breeding policy based entirely on SNF content should lead to an increase of about 0·02% SNF per year and that selection for yield alone would be unlikely to cause any decline in SNF content provided some attention was paid to fat content. Of the phenotypic (overall) within-herd correlations among individual constituents, the most noteworthy were positive correlations between yield and lactose, fat and casein, and casein and lactose. Statistically significant, purely genetic correlations were found between fat and SNF, fat and casein, fat and lactose, and casein and lactose.

As to the influence of non-genetic factors on milk composition, it was found that stage of lactation (effect of season eliminated) had a marked effect on the level of the major constituents in addition to the well known effect on milk yield. With advancing lactation, the initially high levels of total solids, SNF, fat, crude protein and casein all decreased rapidly during the first 45 days, with the decrease in total solids and fat continuing for the next 30 days, whereas thereafter, levels gradually increased for the remainder of lactation with the increase being more rapid after about 200 days. The lactational changes in lactose content were the opposite of those just described and closely resembled those for yield. Seasonal effects (lactational effect eliminated) on milk composition were less marked but more irregular and appeared to be associated mainly with the dietary changes in going from byre feeding to pasture and *vice versa*. The development of maximum yield in May–June following the start of grazing coincided with a depressed fat content, and an increased SNF content due to an increase in crude protein, whereas the decline in yield to near-minimum values in October was accompanied by high levels of fat and crude protein, and low levels of lactose. These purely seasonal effects, which are reflected in the changes in the composition of herd-bulk milks, are not usually so unequivocally demonstrated as in this study and they still remain to be explained fully. It was also found, generally speaking, that increasing age of the cow resulted in increasing yields of poorer quality milk, with the average difference in composition between milk from the first lactation and that from the ninth and higher lactations, being for fat 0·19%, SNF 0·34%, crude protein 0·08%, casein 0·21%, and lactose 0·25%. These results emphasize the need to avoid having herds containing many old cows, particularly when the herds are small. In a subsequent investigation of the composition of the milk of young and old cows, Waite (1968) concluded that the effect of age of the cow on milk composition was due mainly in fact to a higher incidence of subclinical mastitis in older cows but he also found some evidence for a genuine age-effect which led to a decrease in the lactose content of the milk of old cows after about 80 to 90 days in lactation. Other results obtained in this investigation for quarter-milk samples, taken during almost 300 days of the first lactation of four cows that were free of subclinical mastitis, showed that the milk varied little in composition except for fat content. Whether this absence of the accepted changes in milk composition with advancing lactation means that with healthy, first-lactation cows they do not in fact occur, or only take place in a very extended lactation, is uncertain and requires further study.

Subclinical mastitis has been the subject of much work at the Institute. Many of the studies (e.g. Blackburn, 1956*a*, *b*, 1966, 1968; Sweetsur & Phillips, 1976) have not been related directly to milk composition and will not be discussed further here, but others have involved an examination of the effect of the disease on composition. One of the earliest of the latter was by Chanda (1953) who studied the effect on the carotenoids, vitamin A and phosphorus compounds in quarter-milk samples from Ayrshire cows, with the presence of *Streptococcus agalactiae* together with high chloride levels in the milk being taken to indicate subclinical mastitis. Among the effects associated with the disease were decreased levels in the milk of total phosphorus, ester phosphorus, thiamin and vitamin A, increased levels of phosphatase and β-carotene, and an increased nitrogen : phosphorus ratio in both the milk and the casein. These effects were interpreted as indicating decreased metabolic activity of the mammary gland, changes in the proportions of the casein components and leakage of blood serum into the milk through damaged epithelium, conclusions that still hold today. Subsequently, using total and differential cell counts in milk as criteria of the severity of subclinical mastitis, Waite & Blackburn (1957) correlated cell count and milk composition using over 2000 milk samples from individual cows and over 800 samples of farm bulk milk. Total cell counts varied from less than 10 000/ml to 5 000 000/ml, and as counts increased from low levels to about 500 000/ml, SNF content decreased by an average of 0·25 percentage unit solely as the result of a decrease in lactose content. Casein content was not reduced until total cell counts approached 1 000 000/ml and fat content appeared to be unrelated to cell count; in a later study however, Waite *et al.* (1963) found that fat content could be lowered considerably in milks containing large numbers of cells. These results on the influence of udder disease on milk composition are of considerable practical importance, and Waite (1961) emphasized the need for regular examination of milk from each quarter of the udder individually, when assessing the effects of feeding treatments on milk composition (Waite *et al.*, 1963) and for special studies on individual cows with subclinical mastitis (Waite & Blackburn, 1963; Blackburn & Castle, 1968).

As already indicated, in addition to studies on the gross composition of milk much work has been done at the Institute on the forms in which various constituents occur in milk and the extent of variation in the amounts of the various forms. An example of this type of work is the investigation of White & Davies (1958) which involved the comprehensive analysis of 132 samples of milk comprising herd-bulk samples and samples from individual healthy cows in early-, mid- and late lactation, as well as some from cows with subclinical mastitis. The analysis partitioned various milk constituents into soluble and colloidal forms and the results showed that whereas nearly all the sodium, potassium, citrate and chloride were present in solution, an appreciable amount of the magnesium and a large amount of the calcium and phosphate were present in colloidal form. The more detailed partition for calcium showed, for example, that in the herd-bulk milks, on average, 42% was colloidal inorganic, 27% was bound to casein, 21% was soluble un-ionized and 10% was as free ions. The results showed also that the level of some of these fractions was affected by stage of lactation, e.g. soluble un-ionized and ionized calcium were high in early lactation whereas sodium and chloride were high in late lactation. In addition, it was found that as the pH of milk decreased the ratio of soluble to colloidal forms of calcium and phosphorus increased and *vice versa*. In this study, the nitrogen was also partitioned into various fractions and these results showed that lactational changes in casein, heat-denaturable serum protein and non-protein nitrogen were the same as for total nitrogen, i.e. very high values at the beginning of lactation rapidly falling to below average values and then a gradual increase until near the end of lactation when very high values similar to those for colostrum were attained; the lactational changes in proteose-peptone differed from those for the other nitrogen fractions only in that the initial levels were below rather than above average. The information obtained from this detailed study proved useful in contemporary attempts at the Institute and elsewhere to identify factors that control the stability of milk protein to heating and to the addition of rennet or ethanol.

The method of partitioning nitrogen used in the above investigation was rather limited, in that the casein and the heat-denaturable serum protein were treated as single entities, and it does not supply the detailed information required for modern protein studies. Some of this information however, has now been obtained using the newer methods of fractionation developed at the Institute (Davies, 1974; Davies & Law, 1977). These have shown for example that in the milk of the Institute herd the casein has a relatively constant composition and consists of about 38% major α_s-caseins, 36% β-casein, 13% κ-caseins, 11% minor α_s-caseins and 2% γ-caseins, whereas the mixture of serum proteins now collectively termed 'albumin', although somewhat more variable in composition, consists on average of 67% β-lactoglobulin, 22% α-lactalbumin, 5% serum albumin and 6% lactoferrin plus other minor proteins. Application of the new casein-fractionation method to samples from healthy individual cows on the other hand has indicated a more variable composition, with the casein in early and late lactation being poorer in β-casein and richer in κ- and γ-casein than in mid-lactation. While it is too early to judge the implications of these findings, they may provide a partial explanation of the variable behaviour of the caseinate complex during the manufacture of milk products at certain times of the year and they may also be useful in future attempts to improve milk quality through breeding.

In this review article an attempt has been made to give a résumé of some of the work on milk composition at the Hannah Research Institute. There is no doubt this topic will remain of perennial interest to research workers and members of the dairy industry alike since new views on the feeding of cows, the utilization of milk and milk products and even on the suitability of milk as a staple part of the human diet will continue to be propounded and tested, as in the past.

REFERENCES

Aschaffenburg, R. & Drewry, J. (1959) *Proc. 15th Int. Dairy Congr.* **3**, 1631–1637
Blackburn, P. S. (1956*a*) *Br. Vet. J.* **112**, 366–369
Blackburn, P. S. (1956*b*) *J. Dairy Res.* **23**, 225–228
Blackburn, P. S. (1966) *J. Dairy Res.* **33**, 193–198
Blackburn, P. S. (1968) *J. Dairy Res.* **35**, 59–65
Blackburn, P. S. & Castle, M. E. (1968) *Br. Vet. J.* **124**, 27–35
Chanda, R. (1953) *Biochem. J.* **54**, 68–77
Davies, D. T. (1974) *J. Dairy Res.* **41**, 217–228
Davies, D. T. & Law, A. J. R. (1977) *J. Dairy Res.* **44**, 213–221
Davies, D. T. & White, J. C. D. (1960) *J. Dairy Res.* **27**, 171–190
Davies, D. T. & White, J. C. D. (1962) *J. Dairy Res.* **29**, 285–296
Farrell, H. M. & Thompson, M. P. (1974) in *Fundamentals of Dairy Chemistry* (Webb, B. H., Johnson, A. H. & Alford, J. A., eds), 2nd ed., chap. 9, Avi, Westport, Connecticut
Hankinson, D. J. (1975) *J. Dairy Sci.* **58**, 326–336
Robertson, A., Waite, R. & White, J. C. D. (1956) *J. Dairy Res.* **23**, 82–91
Rose, D., Davies, D. T. & Yaguchi, M. (1969) *J. Dairy Sci.* **52**, 8–11
Rowland, S. J. (1938) *J. Dairy Res.* **9**, 42–46
Sweetsur, A. W. M. & Phillips, J. D. (1976) *J. Dairy Res.* **43**, 53–62
Waite, R. (1959) *Proc. 15th Int. Dairy Congr.* **3**, 1559–1565
Waite, R. (1961) *J. Dairy Res.* **28**, 75–79
Waite, R. (1968) *J. Dairy Res.* **35**, 407–422
Waite, R. & Abbot, J. (1956) *Proc. 14th Int. Dairy Congr.* **3**(2), 699–710
Waite, R. & Blackburn, P. S. (1957) *J. Dairy Res.* **24**, 328–339
Waite, R. & Blackburn, P. S. (1963) *J. Dairy Res.* **30**, 23–33
Waite, R. & McPhillips, J. (1965) *J. Soc. Dairy Technol.* **18**, 27–31
Waite, R. & Paterson, J. A. (1959) *J. Soc. Dairy Technol.* **12**, 117–122
Waite, R. & Smith, G. M. (1972) *J. Dairy Res.* **39**, 195–201
Waite, R. & Taylor, M. M. (1966) *Proc. 17th Int. Dairy Congr.* **B**, 87–93
Waite, R., White, J. C. D. & Robertson, A. (1956) *J. Dairy Res.* **23**, 65–81
Waite, R., Abbot, J. & Blackburn, P. S. (1963) *J. Dairy Res.* **30**, 209–215
Waugh, D. F. (1971) in *Milk Proteins: Chemistry and Molecular Biology* (McKenzie, H. A., ed.), vol. 2, chap. 9, Academic Press, New York
White, J. C. D. & Davies, D. T. (1958) *J. Dairy Res.* **25**, 236–255
White, J. C. D. & Davies, D. T. (1963) *J. Dairy Res.* **30**, 171–189

MILK PROTEINS

W. MANSON

THE protein constituents of bovine milk amount in all to about 3·25% of the total weight of milk itself and the annual production of milk protein in the United Kingdom's milk supply now amounts to 425 000 tonnes. This is sufficient to supply everyone in the country with 25% of his total requirement for protein and with about 65% of his animal protein intake. Since these last two figures do not differ greatly from those of other Western European countries or of the United States of America it is apparent that in the affluent areas of the world milk protein in one form or another contributes significantly to the human diet. For this reason and in terms of output, milk protein may be regarded as an agricultural product of considerable value and importance.

As may be seen from Table 1 bovine milk protein is composed of several different groups of proteins, the most plentiful of which is the mixture of phosphoproteins which constitutes the casein complex and which amounts to about 80% of the total milk protein or to about 2·6% by weight of milk itself. It is with the investigation of the chemistry and the biochemistry of this, the major component of bovine milk protein, that work performed at the Hannah Research Institute has been concerned.

TABLE 1

The main constituents of milk protein expressed as a percentage of the total protein

Protein	Percentage of total milk protein
Caseins	78
β-Lactoglobulin	14
α-Lactalbumin	3
Immunoglobulins	2
Blood albumin	3

In its native state in bovine milk, casein occurs in the form of complex colloidal particles which contain, in addition to individual phosphoproteins, the positive ions of calcium and magnesium together with the negatively charged phosphate and citrate ions. It is fundamental to the understanding of the chemistry of these micellar particles to appreciate that they exist in equilibrium with the milk serum in which they are dispersed. They will react to changes produced in the concentrations of ionic constituents of milk serum by making constitutional changes within their own structures. Many of the problems encountered in dairy technology are connected with the induction in casein micelles of changes of this type as a result of variations produced in their immediate environment during manufacturing processes. For this reason the study of the physical chemistry of the casein micelles of bovine milk is now being actively pursued in this Institute. Micellar structure is also, and to a considerable extent, determined by the primary structures of the individual constituent phosphoproteins. The elucidation of the chemical characteristics and primary structures of these proteins was consequently recognized early as being a first step towards the attainment of two important objectives, namely (1) the understanding of the processes involved in the biosynthesis of the casein micelle, and (2) the utilization of dietary phosphoprotein. The investigation of milk proteins at this Institute has contributed significantly to the advances made in both fields.

Study of milk protein in the Institute began early in its existence and has continued steadily since. At that time, little was known of the chemistry of milk proteins, and casein was generally considered to be a single denatured protein of uncertain chemical composition whose main value was associated with its use in the manufacture of plastics or as a constituent of adhesives, a role first suggested by the ancient Egyptians. Recognition of its great potential value as a nutrient or as a food additive led to efforts in the Institute to encourage its use in non-traditional ways. Thus, the role of milk constituents in baking and breadmaking was an early interest of the Institute (Allen & Bell, 1931). At the same time, it was realized that for such purposes the preservation of milk constituents in the form of an edible powder was desirable. Research was therefore undertaken on technical problems associated with the preparation of dried milk powders (Scott, 1931, 1932; Wright, 1932), a subject which became a major interest of the Institute and which is reviewed elsewhere in this volume (p. 20).

If these investigations can be considered as falling within the field described by the second of the objectives mentioned above, further studies were undertaken which may be considered to lie within the scope of the first of these. They were concerned with the understanding of the mechanisms involved in the production of milk protein by the dairy cow or to be more precise, the defining of factors influencing the synthesis and yield of milk protein. Attempts were made to determine the uptake of amino acids and phosphate from blood during its passage through the lactating mammary gland of the cow and to relate this to milk and protein yield (Blackwood, 1932; Blackwood & Stirling, 1932). These experiments are particularly noteworthy since they were serious scientific attempts at understanding protein production in the dairy cow at a time when little effort was being expended elsewhere on this subject. They were of course severely circumscribed by the knowledge of protein chemistry and biochemistry then available and also by a lack of methods suitable for the accurate analysis of amino acids and peptides. These difficulties rendered virtually impossible any detailed characterization of individual milk proteins although some of the reactions of casein such as its susceptibility to attack by the digestive enzymes trypsin and pepsin (Stirling & Wishart, 1932) and the release of phosphate from it during heat coagulation (Howat & Wright, 1934, 1936) were investigated. These studies underlined the complex nature of the milk proteins, which was further confirmed when physico-chemical experiments in Sweden (Pedersen, 1936; Mellander, 1939) finally proved that casein consisted not of a single protein but of a family of phosphoproteins.

Further advances in this field could not be made until techniques employed in protein chemistry had themselves progressed sufficiently for the separation of these casein components to be achieved on a preparative scale. This required some time and was accomplished initially by Warner (1944) in the United States of America. Two of the three fractions prepared using this procedure and designated α- and β-casein were later studied in this Institute by T. Hofman. In experiments using a phosphoprotein phosphatase he concluded that only one type of phosphate linkage existed in both proteins (Hofman, 1958*a*) a conclusion now known to be correct since all the phosphorus of the caseins is present in the form of monoesters of orthophosphoric acid and the hydroxyl group of hydroxyamino acid residues. When made, however, this claim was at variance with the accepted thinking, which was based on the findings of Perlmann (1954*a*, *b*) that pyrophosphate, phosphoamide and phosphomonoester linkages were all present in these proteins.

The rapid increase in the number of techniques available for the separation of proteins taking place at this time made possible a further fractionation of R. C. Warner's bovine α-casein into α_s- and κ-casein (von Hippel & Waugh, 1955; Waugh & von Hippel, 1956) a finding which was also made for goat casein almost simultaneously by Hofman although not reported until later (Hofman, 1958*b*). The number of components recognized in the casein complex was therefore increased to four, α_s-, β-, κ- and γ-caseins. While the identifiable components were increasing in number attempts were also being made to characterize them chemically. Much of this effort was directed towards the most abundant fraction, α-casein, which had been reported to consist of a variable number of polypeptide chains that terminated at their amino ends in lysyl or arginyl residues (Mellon *et al.*, 1953; Wissman & Nitschmann, 1957). Estimates of the molecular weight varied between 31 000 and 250 000. In a series of papers, Manson (1959, 1960, 1961) showed that α-casein contained only a single polypeptide chain terminating at the amino end in arginine. From the amounts of arginine present the molecular weight was estimated at about 30 000, a value subsequently shown to be somewhat high.

Despite the apparent purity of these four casein fractions doubts existed as to their homogeneity and efforts were continued to evolve more effective fractionation procedures. Most of these relied on the use of strong solutions of urea to suppress the aggregation of the individual casein. Unfortunately these conditions were shown to cause chemical modification of the proteins (Manson, 1962*a*) and attention here was turned towards fractionation procedures that would yield small amounts of pure caseins without the involvement of urea. This was achieved by electrophoresis of whole casein in solution in a column stabilized by a density gradient (Manson, 1962*b*, 1965) which yielded all four casein components again apparently in a high degree of purity. However, when these preparations were examined using modifications of the recently introduced technique of starch electrophoresis it was found that whereas the β-casein behaved as a single homogeneous protein the others exhibited signs of heterogeneity. The α_s-casein fraction was further investigated and six proteins were finally separated from it and partially characterized by Annan & Manson (1969). These were designated, α_{s0} to α_{s5}-casein in decreasing order of electrophoretic mobility at pH 9·2, the main component being designated α_{s_1}-casein.

The number of known and partially characterized members of the casein complex had therefore risen considerably with improvements in analytical and preparative techniques and the way was now open for the total characterization of these proteins. At the Institute this centred on gaining an understanding of the reactivity of the phosphate groups associated with the two most plentiful members of the casein complex, the α_{s_1}- and β-caseins. Both reacted with the basic dyestuffs Safranine and Acridine Orange, through intervention of their phosphate groups to form insoluble dye-protein complexes (Manson, 1967*a*). When the reaction was analysed, however, it was found that the dye bound for a fixed weight of phosphorus differed between that located in α_{s_1}-casein and that in β-casein. Similarly, from the study of a reaction between Fe^{2+} and α_{s_1}- and β-casein it was found that the reactivity of equal amounts of phosphate in the two proteins with a fixed weight of Fe^{2+} was not the same and thus not all of the phosphate groups were equally reactive (Manson, 1967*b*). Since they all shared a common chemical structure it was deduced that their reactivities differed because of differences in grouping of these phosphoseryl residues between the α_{s_1}- and β-casein structures. That such differences in arrangement do exist in fact became apparent when the chemical structure of the region of β-casein containing its complement of phosphate was elucidated by Manson & Annan (1971) and confirmed by Ribadeau Dumas *et al.* (1972). From this it could be seen that all five phosphorus atoms of β-casein were accommodated in a short region of the protein's polypeptide chain. A similar disposition was shown to exist in α_{s_1}-casein (Mercier *et al.*, 1971) but in addition a further three phosphorus atoms were accommodated outside this grouping. Thus although all the phosphorus is present in the caseins as phosphomonoesters of hydroxyamino acid residues as deduced by Hofman (1958*a*) the grouping of these phosphoseryl residues in clusters confers on them properties not possessed when they occur singly. The possibility that those groupings fit α_{s_1}- and β-casein for specialized biochemical roles is presently under consideration and further studies have been made of the stability of those phosphate residues towards both chemical (Manson & Carolan, 1972) and enzymic attack (West & Towers, 1976). In addition the possible role of these residues in micelle formation and ion-binding is being pursued currently.

For most of its existence this Institute has concerned itself with a study of the chemistry of milk proteins and particularly of the caseins. It has contributed consistently to the accumulating knowledge of their chemistry and of techniques for their study, which has culminated in the recent complete elucidation of their primary structures. The determination of the sequence of amino acid residues present in α_{s_1}-casein was completed in 1971 (Mercier *et al.*, 1971) that of β-casein in 1972 (Ribadeau Dumas *et al.*, 1972) and that of κ-casein in 1973 (Mercier *et al.*, 1973). The γ-casein fraction has now been recognized as being composed of fragments of β-casein formed by the action of proteolytic enzymes in milk. The structure of minor components of the casein complex has been investigated also and one of them, α_{s_0}-casein, has been found to possess a sequence of amino acid residues identical to that of α_{s_1}-casein but to contain one additional phosphate residue (Manson *et al.*, 1977).

The mechanisms involved in the biosynthesis of the caseins, however, have not been disregarded. It has been shown recently at this Institute (Mackinlay *et al.*, 1977) that the phosphate attached to casein molecules is incorporated into preformed polypeptide synthesized in the lactating mammary gland by a phosphorylating system present in the same tissue. The enzyme system has been isolated and partially purified. Thus one of the goals of the work initiated with the foundation of this Institute, namely the understanding of the processes involved in the biosynthesis of milk phosphoprotein and of the nature of these phosphoproteins themselves has been attained. Similar objectives have been achieved for the main non-phosphorylated milk proteins with the elucidation of the primary structures of β-lactoglobulin (Frank & Braunitzer, 1967, 1968) and of α-lactalbumin (Brew *et al.*, 1970). While much work remains to be done particularly with respect to elucidating the structure of casein micelles it cannot be emphasized too strongly that the completeness of the knowledge of the chemistry of milk protein is unequalled by that of any other protein foodstuff and provides a unique opportunity for its exploitation in non-traditional forms in new food and pharmaceutical products.

REFERENCES

Allen, L. A. & Bell, J. (1931) *J. R. Tech. Coll. (Glasgow)* **2**, 550–564

Annan, W. D. & Manson, W. (1969) *J. Dairy Res.* **36**, 259–268

Blackwood, J. H. (1932) *Biochem. J.* **26**, 772–777

Blackwood, J. H. & Stirling, J. D. (1932) *Biochem. J.* **26**, 778–784

Brew, K., Castellino, F. J., Vanaman, T. C. & Hill, R. L. (1970) *J. Biol. Chem.* **245**, 4570–4582

Frank, G. & Braunitzer, G. (1967) *Hoppe-Seyler's Z. Physiol. Chem.* **348**, 1691–1692

Frank, G. & Braunitzer, G. (1968) *Hoppe-Seyler's Z. Physiol. Chem.* **349**, 1456–1462

Hofman, T. (1958*a*) *Biochem. J.* **69**, 139–145

Hofman, T. (1958*b*) *Nature (Lond.)* **181**, 633–634

Howat, G. R. & Wright, N. C. (1934) *Biochem. J.* **28**, 1336–1345

Howat, G. R. & Wright, N. C. (1936) *Biochem. J.* **30**, 1413–1415

Mackinlay, A. G., West, D. W. & Manson, W. (1977) *Eur. J. Biochem.* **76**, 233–243

Manson, W. (1959) *Nature (Lond.)* **184**, 1393–1394

Manson, W. (1960) *Biochim. Biophys. Acta* **44**, 182–184

Manson, W. (1961) *Arch. Biochem. Biophys.* **95**, 336–342

Manson, W. (1962*a*) *Biochim. Biophys. Acta* **63**, 515–517

Manson, W. (1962*b*) *J. Electroanal. Chem.* **3**, 203–208

Manson, W. (1965) *Biochem. J.* **94**, 452–457

Manson, W. (1967*a*) *Arch. Biochem. Biophys.* **122**, 370–377

Manson, W. (1967*b*) *Proc. Nutr. Soc.* **26**, xvii–xviii

Manson, W. & Annan, W. D. (1971) *Arch. Biochem. Biophys.* **145**, 16–26

Manson, W. & Carolan, T. (1972) *J. Dairy Res.* **39**, 189–194

Manson, W., Carolan, T. & Annan, W. D. (1977) *Eur. J. Biochem.* **78**, 414–417

Mellander, O. (1939) *Biochem. Z.* **300**, 240–245

Mellon, E. F., Korn, A. H. & Hoover, S. R. (1953) *J. Am. Chem. Soc.* **75**, 1675–1678

Mercier, J.-C., Grosclaude, F. & Ribadeau Dumas, B. (1971) *Eur. J. Biochem.* **23**, 41–51

Mercier, J.-C., Brignon, G. & Ribadeau Dumas, B. (1973) *Eur. J. Biochem.* **35**, 222–235

Pedersen, K. O. (1936) *Biochem. J.* **30**, 948–960

Perlmann, G. E. (1954*a*) *Biochim. Biophys. Acta* **13**, 452–453

Perlmann, G. E. (1954*b*) *Nature (Lond.)* **174**, 273–274

Ribadeau Dumas, B., Brignon, G., Grosclaude, F. & Mercier, J.-C. (1972) *Eur. J. Biochem.* **25**, 505–514

Scott, A. W. (1931) *J. R. Tech. Coll. (Glasgow)* **2**, 456–461

Scott, A. W. (1932) *J. R. Tech. Coll. (Glasgow)* **2**, 620–630

Stirling, J. D. & Wishart, G. M. (1932) *Biochem. J.* **26**, 1989–1999

von Hippel, P. H. & Waugh, D. F. (1955) *J. Am. Chem. Soc.* **77**, 4311–4319

Warner, R. C. (1944) *J. Am. Chem. Soc.* **66**, 1725–1731

Waugh, D. F. & von Hippel, P. H. (1956) *J. Am. Chem. Soc.* **78**, 4576–4582

West, D. W. & Towers, G. E. (1976) *Biochim. Biophys. Acta* **453**, 383–390

Wissmann, H. & Nitschmann, H. (1957) *Helv. Chim. Acta* **40**, 356–363

Wright, N. C. (1932) *J. Dairy Res.* **4**, 122–141

THE HEAT STABILITY OF THE CASEIN COMPLEX OF MILK

D. D. MUIR

INTRODUCTION

THE processing of milk has been the subject of continuing study throughout the history of the Hannah Research Institute and particular interest has been focused on the production of dried milk. Over the past 50 years marked improvements have been achieved in the quality of commercial dried milk, mainly through a reduction in the severity of the heat treatment applied to the milk during processing. Although milk protein is extremely stable at temperatures up to 100°C at the concentrations found in raw milk, at higher temperatures and concentrations the protein becomes progressively more susceptible to denaturation. Heat treatment during processing can cause milk protein to become insoluble and can lead to defects such as excessive thickening of evaporated milk and poor reconstitution of milk powder. To further improve the quality of fully and partially dehydrated milk products, a sound understanding is required of factors governing the heat stability of milk protein.

The central role which protein stability plays in determining product quality has been recognized since the inception of work on processing at the Institute. In response to changes in the extent of contemporary knowledge and in manufacturing practice, however, several differences in emphasis have occurred in research over the last 50 years. Historically, three separate groups of experiments can be identified and these will be described in the context of the practical problems that were outstanding at the time the experiments were undertaken.

EARLY EXPERIMENTS CONCERNING THE MECHANISM OF HEAT DENATURATION

The period following the First World War saw a general expansion in the production of sweetened-condensed and evaporated milks. As these products increased in commercial importance, it was realized that considerable variation existed in the stability of milk protein since serious production problems arose at certain seasons and with milk from certain herds. Studies were therfore initiated to relate the composition of milk to the heat stability of the protein during manufacture (see review by Hunziker, 1935).

Three principal factors were identified in this early work for their effect on the heat stability of milk. First, it was recognized that the 'acidity' of milk had an important bearing on protein stability when heated, but it was suggested that only acidity resulting from bacterial growth was important and that the natural acidity of milk had little effect, e.g. Sommer & Hart (1919), Rogers *et al.* (1921) and Rice & Markley (1924).

Secondly, the proportion of 'albumin' – that is, whey protein – in the milk was shown to influence heat stability and it was found that denaturation of the albumin by heat treatment prior to concentration had a pronounced effect on the heat stability of the concentrate (Rogers *et al.*, 1920, 1921). In addition, Hunziker (1935) showed that there were substantial differences in milk albumin levels between cows and that these differences were reflected in the heat stability of the milk.

Finally, the mineral composition of milk was thought to be a determinant of heat stability. Sommer & Hart (1919) proposed that, for maximum heat stability in raw and concentrated milk, a balance must exist between the levels of calcium + magnesium and phosphate + citrate. The destabilizing effect of acidity associated with bacterial growth was then attributed to the effect of acid on the salt balance. Benton & Albery (1926) supported the salt balance theory, but work by Rogers *et al.* (1921) and Holm *et al.* (1932) cast serious doubts upon this mechanism.

Despite the strong body of evidence to the contrary, the salt balance theory was attractive to many workers in the milk industry since it rationalized the still current practice of adding phosphate or citrate to evaporated milk to improve the heat stability. This practice, however, is carried out on an *ad hoc* basis, and, whilst often effective, does not justify the salt balance theory as proposed by Sommer & Hart (1919).

Against this background, work at the Institute suggested an alternative mechanism for the destabilization by heat of milk protein. Howat & Wright (1934, 1936) showed that, on heating, milk protein was dephosphorylated and that the temperature coefficient of the dephosphorylation reaction was similar to that of the coagulation process. A mechanism was then proposed whereby the coagulation of milk protein on heating was a result of the loss of colloidal stability caused by the cleavage of bound phosphate groups. This mechanism found general acceptance as an alternative to the salt balance theory, but it has only recently been tested. Belec & Jenness (1962*a*, *b*) considered the dephosphorylation reaction in some detail and, although they confirmed the results of Howat & Wright (1934, 1936) with respect to the temperature dependence of the dephosphorylation reaction, they also showed that over the pH range 6 to 7 the rate of dephosphorylation was largely unchanged. Contrary to the suggestions of early workers and as will be detailed later, Rose (1961*a*, *b*) showed clearly that within the pH range 6 to 7 large differences in heat stability occurred in milk unaffected by bacterial degradation. Therefore, the mechanism proposed by Howat & Wright (1934, 1936) can no longer be regarded as the rate limiting step in the destabilization of milk protein by heat.

Following this work in the 1930s, a hiatus occurred in work on heat stability and at the Institute the emphasis of research on milk processing changed to a more applied nature in support of the war effort. After 1945, however, a new generation of research workers, recognizing the deficiencies of the various proposed mechanisms for the control of protein stability during heating set about exploring alternative mechanisms.

POST-WAR RESEARCH – THE DEVELOPMENT OF NEW METHODOLOGY

As work on heat stability regained impetus, substantial advances were made in the development of new methodology (see p. 229). Two developments in analytical methods were of particular relevance to studies on heat stability. First, at the National Institute for Research in Dairying, Reading, Aschaffenburg & Drewry (1959) developed a reliable and comprehensive method for the separation and analysis of milk protein. Secondly, at the Hannah Research Institute the work of White & Davies (1958*a*) produced reliable methods for the detailed analysis of the mineral composition of milk.

Using these new techniques, White & Davies (1958*b*), undertook a detailed study of milk from individual animals at various stages of lactation and of herd bulk milk over a period of a year and, in a separate series of experiments, Davies & White (1959) also studied the heat-induced changes in protein stability and chemical composition of milk over the temperature range 50 to 120°C. From these studies (White & Davies, 1958*b*; Davies & White, 1959) no general relation was found between milk composition and heat stability. In certain circumstances, the mineral composition of the milk appeared to influence heat stability but the results did not support the salt balance theory of Sommer & Hart (1919), nor the simplification of this theory by Seekles & Smeets (1947), which suggested that calcium ion concentration was the controlling factor in protein coagulation.

The next major advance occurred when Rose (1961*a*, *b*) demonstrated that the time taken for milk protein to coagulate (CT) depended strongly on pH in the range 6·4 to 7·4 and that two distinct types of CT-pH profile were obtained. These types, named A and B were shown to be related to the β-lactoglobulin content of the milk (Rose, 1962) and are typified by the CT-pH profiles shown in Fig. 1. Type A milks are characterized by a CT maximum which may occur between pH 6·6 and 6·9 and a corresponding CT minimum which occurs between pH 6·7 and 7·1, whereas type B milks show an approximately linear increase across this pH range.

The differences in heat stability between cows giving milks with different albumin levels reported by Hunziker (1935) could now be rationalized, for Aschaffenburg & Drewry (1957) had identified genetic variants of β-lactoglobulin and had shown that the ratio of β-lactoglobulin to casein was related to genotype. Therefore, the differences in albumin level observed by Hunziker (1935) were reflexions of genotype and would, on the basis of the work of Rose (1962) lead to differences in heat stability. Tessier & Rose (1964) went on to show that not only was the minimum in the type A CT-pH profile induced by the addition of β-lactoglobulin but also that κ-casein had the reverse effect. Hence, by alteration of the relative concentrations of these components the CT-pH profile of milk could be interconverted between type A and type B.

Whilst acknowledging the importance of β-lactoglobulin, Morrissey (1969) suggested that the main cause of type A behaviour was the depo-

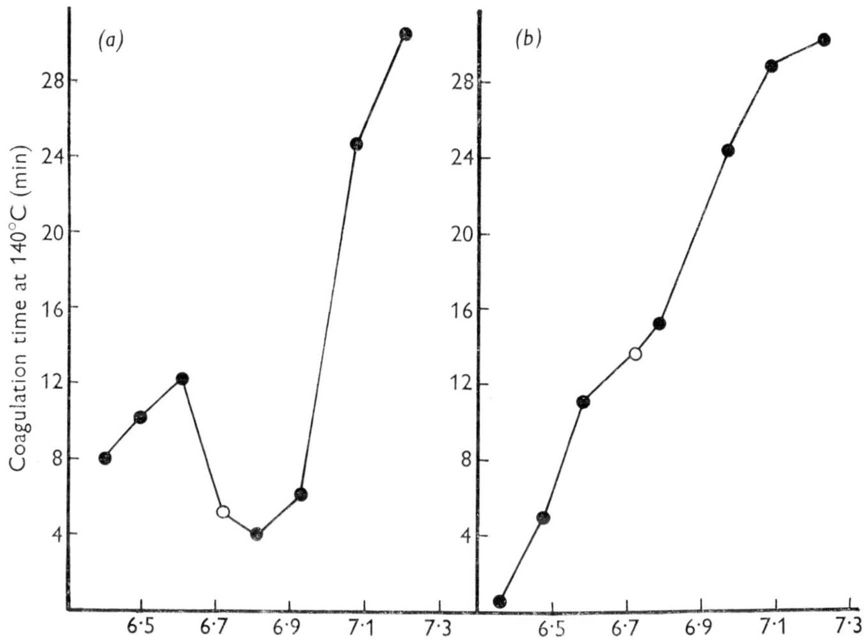

Fig. 1. *Type A and type B coagulation time-pH profiles.* (a) Curve for milk from an Ayrshire cow homozygous for the β-lactoglobulin A allele. The concentration of β-lactoglobulin was 300 mg/100 ml milk. (b) Curve for milk from an Ayrshire cow homozygous for the β-lactoglobulin B allele. The concentration of β lactoglobulin was 165 mg/100 ml milk. ○ – pH of milk unadjusted; ● – pH of milk adjusted with either HCl or NaOH.

sition of calcium phosphate on the caseinate-β-lactoglobin complex which, in the presence of calcium ions, was now less stable to heat. However, despite this work – and the earlier studies by Pyne & McHenry (1955) and Pyne (1958, 1959) concerning the relation between soluble salt concentration and heat stability – the nature of the type A CT-pH profile remained unresolved.

During the 1960s the main effort at the Institute was in the development of tests for the estimation of heat stability. This resulted in the publication of subjective and objective heat stability tests by Davies & White (1966) and White & Davies (1966). These tests showed that the observation of visible coagulation in milk at a single pH value was an ambiguous measure of the heat stability. By measurement of the rate of protein precipitation, White & Davies (1966) identified two distinct types of heat coagulation for, in one type, milk protein coagulates as a single, in the other as a two-stage process (Fig. 2).

These newly developed techniques were used (Sweetsur & White, 1970,

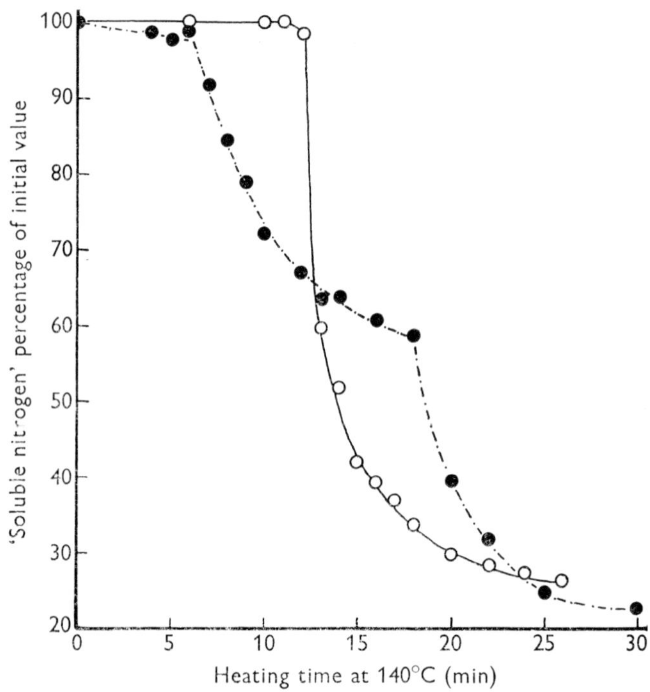

FIG. 2. *Curves typical of single-step* (○) *and two-step* (●) *coagulation of milk protein.* Single-step behaviour occurs outside the minimum in type A coagulation time-pH profiles and at all pH values for type B coagulation time-pH profiles.

1974, 1975*a*, *b*) to show that, within the minimum of the CT-pH profile, coagulation invariably took place in a two-stage process and outside the minimum a single-stage mechanism appeared to operate. Furthermore, Sweetsur & White (1974) demonstrated that the interconversion of type B with type A behaviour could be achieved by increasing the temperature of heating.

In other work, Fox & Hoynes (1975) re-assessed the influence of colloidal calcium phosphate and β-lactoglobulin on the shape of CT-pH profiles and concluded that the effect of β-lactoglobulin in milk is to shift the entire CT-pH profile to lower pH values and to sensitize the caseinate complex to heat-induced calcium phosphate precipitation at pH values greater than about 7·0 (cf. Morrissey, 1969).

It is apparent that the mechanism of heat coagulation within the CT-pH minimum is unresolved and that further work will be required before the appearance and extent of a minimum in a CT-pH profile can be predicted from an analysis of milk composition.

CURRENT RESEARCH ON HEAT STABILITY

Recent work on heat stability at the Institute has been concerned with the difficulties that occur in the commercial production of heat-stable skim-milk powder. This area of research was suggested by senior representatives of the United Kingdom dairy industry on the Institute Consultative Panel, who wished to find new outlets for the surplus of dried skim-milk which occurred on the entry of the United Kingdom into the European Economic Community.

In a preliminary study Sweetsur (1976) examined the stability of 'instant' dried skim-milk when added to hot coffee and showed that the formation of insoluble material was related directly to the casein number of the milk protein (i.e. the percentage of the milk protein precipitated at pH 4·6) and related inversely to the maximum CT measured at 140°C. As the history of the milk from which the powders were made was unknown, it was impossible to differentiate between natural variations in heat stability of the raw milk supply and subsequent changes brought about by denaturation of the whey protein (leading to an apparent increase in casein number). Therefore, studies were initiated to examine seasonal changes in the commercial supply of bulk milk and to examine the effect of processing variables on the heat stability of milk.

Previous studies on the seasonal variations in heat stability (e.g. White & Davies, 1958*b*; De Koning *et al.*, 1974) had failed to identify any close association between composition and heat stability, but neither investigation had encompassed two factors which have recently been shown to influence heat stability – that is, the concentration of non-protein nitrogen

(NPN) in milk (Robertson & Dixon, 1969) and the degree of hydration of the casein micelles (Thompson *et al.*, 1969).

Robertson & Dixon (1969) found a weak but statistically significant correlation between the NPN level of milk and CT, and in addition demonstrated that adding small quantities of urea – the major component of milk NPN – increased CT of milk at 120°C. Work at the Institute subsequently confirmed these observations and demonstrated that the destruction of urea by the enzyme urease substantially reduced CT (Muir & Sweetsur, 1976). Subsequently, Muir & Sweetsur (1977) found that additions of urea progressively increased the CT of milk outwith the CT-pH minimum, but that within the minimum larger amounts of urea were required before CT increased. It was shown also that urea affects CT over the temperature range from 120 to 150°C and in the presence of sulphydryl-blocking agents.

Following the report of Thompson *et al.* (1969), which showed a strong correlation between the solvation of casein micelles in milk and CT, C. Holt (unpublished work) observed substantial seasonal variation in the casein micelle size and apparent specific volume in the milk of individual animals of the Institute herd.

Therefore, in a 15-month study of the heat stability of milk from creamery silos in south-west Scotland, C. Holt, D. D. Muir & A. W. M. Sweetsur (unpublished work) considered the effect on heat stability of changes in urea content and micelle size in addition to other compositional factors. This survey has shown that, with the exception of a short period in the early summer, over 80% of the seasonal variation in heat stability of the commercial milk supply could be explained by changes in urea level (Fig. 3). It was noted, however, that a form of instability, not related to urea content, occurred in early summer when the natural pH of the milk fell within the CT-pH minimum. No rationalization of this effect has yet been possible in terms of milk composition.

A statistically significant inverse relation was also found between micelle size and CT over the period of the creamery survey, but as the addition of urea to milk caused large changes in CT but not in micelle size, the apparent association between micelle size and CT does not appear to be causal.

As a result of fundamental studies, Parker & Dalgleish (1977*a*, *b*) proposed a new model for the heat coagulation process. They suggested that the reaction that results in coagulation is a polymerization of the individual units of the casein complex which must contain two or more reactive sites on their surfaces. By applying this theoretical model to milk with type B CT-pH profiles, Parker & Dalgleish (1977*b*) have shown that the observed rate of nitrogen depletion (White & Davies, 1966; White & Sweetsur, 1977) may be predicted by a mathematical relation derived from the theory of polyfunctional condensation. Current studies by T. G.

Parker and D. G. Dalgleish have developed the application of their treatment to include type A CT-pH profiles and the model is now being tested.

The commercial implications of the recent work on heat stability at the Institute are now being evaluated by an examination of the role that urea plays in determining heat stability in concentrated milk and milk powder. Preliminary experiments by D. D. Muir and A. W. M. Sweetsur have shown that, during the concentration of skim-milk, the influence of urea diminishes until at a total solids level of 25% no effect on CT is observed. Thus urea appears to have no potential use in the stabilization of evaporated milk. These studies, however, reveal that the heat stability of concentrated milk bears no apparent relation to that of skim-milk from which it was made (cf. Rogers *et al.*, 1921). Current evidence suggests that, in concentrated milk, coagulation occurs by a mechanism similar to that

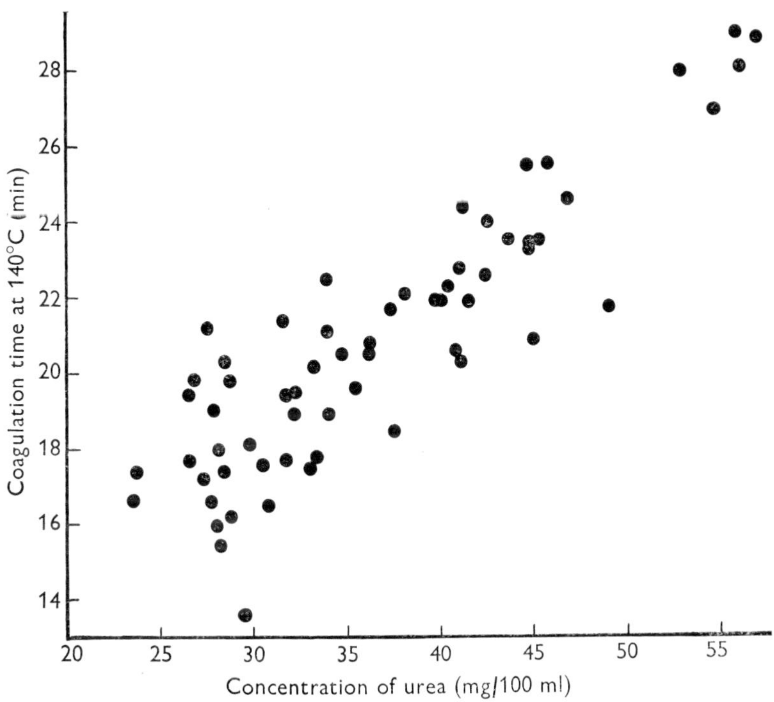

Fig. 3. *The relation between coagulation time at the natural pH of milk and urea level for samples of milk collected over a 15-month period from creamery silos in south-west Scotland.* Five points in which the natural pH fell within the CT-pH minimum have been excluded.

within the CT-pH minimum of type A milks, urea is ineffective, and the coagulation occurs in a two-step sequence (cf. Sweetsur & White, 1975*a*, *b*).

Urea has a similar effect on the heat stability of reconstituted dried skim-milks as on that of the original milk but its effect on the stability of powder when added to hot beverages has yet to be studied.

CONCLUSIONS

This paper has demonstrated that workers at the Institute have made significant and useful contributions towards an understanding of the mechanisms controlling the heat stability of milk. Notwithstanding this effort, many problems remain. The most pressing need is for an understanding of the chemistry of the coagulation reaction, since the ideas proposed by the earliest workers in the field, based on the loss of colloidal stability of the micelles during heating, no longer appear to be adequate. The alternative hypothesis of T. G. Parker and D. G. Dalgleish is attractive but the chemical nature of the reactive groups of the casein molecules, which result in polymerization, is unknown. Similarly, though it is now clear that urea plays a major role in determining the heat stability of the commercial milk supply, the mechanism by which it acts is obscure.

REFERENCES

Aschaffenburg, R. & Drewry, J. (1957) *Nature (Lond.)* **180**, 376–378
Aschaffenburg, R. & Drewry, J. (1959) *Proc. 15th Int. Dairy Congr.* **3**, 1631–1637
Belec, J. & Jenness, R. (1962*a*) *J. Dairy Sci.* **45**, 12–19
Belec, J. & Jenness, R. (1962*b*) *Biochim. Biophys. Acta* **63**, 512–514
Benton, A. G. & Albery, H. G. (1926) *J. Biol. Chem.* **68**, 251–263
Davies, D. T. & White, J. C. D. (1959) *Proc. 15th Int. Dairy Congr.* **3**, 1677–1685
Davies, D. T. & White, J. C. D. (1966) *J. Dairy Res.* **33**, 67–81
De Koning, P. J., Koops, J. & van Rooijen, P. J. (1974) *Neth. Milk Dairy J.* **28**, 186–202
Fox, P. F. & Hoynes, M. C. T. (1975) *J. Dairy Res.* **42**, 427–435
Holm, G. E., Webb, B. H. & Deysher, E. F. (1932) *J. Dairy Sci.* **15**, 331–343
Howat, G. R. & Wright, N. C. (1934) *Biochem. J.* **28**, 1336–1345
Howat, G. R. & Wright, N. C. (1936) *Biochem. J.* **30**, 1413–1415
Hunziker, O. F. (1935) in *Condensed Milk and Milk Powder*, 5th ed., Published by the Author, La Grange, Illinois
Morrissey, P. A. (1969) *J. Dairy Res.* **36**, 343–351
Muir, D. D. & Sweetsur, A. W. M. (1976) *J. Dairy Res.* **43**, 495–499
Muir, D. D. & Sweetsur, A. W. M. (1977) *J. Dairy Res.* **44**, 249–257
Parker, T. G. & Dalgleish, D. G. (1977*a*) *J. Dairy Res.* **44**, 79–84
Parker, T. G. & Dalgleish, D. G. (1977*b*) *J. Dairy Res.* **44**, 85–92
Pyne, G. T. (1958) *J. Dairy Res.* **25**, 467–474
Pyne, G. T. (1959) *Proc. 15th Int. Dairy Congr.* **3**, 1673–1676
Pyne, G. T. & McHenry, K. A. (1955) *J. Dairy Res.* **22**, 60–68
Rice, F. E. & Markley, A. L. (1924) *J. Dairy Sci.* **7**, 468–483
Robertson, N. H. & Dixon, A. (1969) *Agroanimalia* **1**, 141–144

Rogers, L. A., Deysher, E. F. & Evans, F. R. (1920) *J. Dairy Sci.* **3**, 468–485
Rogers, L. A., Deysher, E. F. & Evans, F. R. (1921) *J. Dairy Sci.* **4**, 294–309
Rose, D. (1961*a*) *J. Dairy Sci.* **44**, 430–441
Rose, D. (1961*b*) *J. Dairy Sci.* **44**, 1405–1413
Rose, D. (1962) *J. Dairy Sci.* **45**, 1305–1311
Seekles, L. & Smeets, W. T. G. M. (1947) *Neth. Milk Dairy J.* **1**, 7–18
Sommer, H. H. & Hart, E. B. (1919) *J. Biol. Chem.* **40**, 137–170
Sweetsur, A. W. M. (1976) *J. Soc. Dairy Technol.* **29**, 157–160
Sweetsur, A. W. M. & White, J. C. D. (1970) *Proc. 18th Int. Dairy Congr.* **1E**, 53
Sweetsur, A. W. M. & White, J. C. D. (1974) *J. Dairy Res.* **41**, 349–358
Sweetsur, A. W. M. & White, J. C. D. (1975*a*) *J. Dairy Res.* **42**, 57–71
Sweetsur, A. W. M. & White, J. C. D. (1975*b*) *J. Dairy Res.* **42**, 73–88
Tessier, H. & Rose, D. (1964) *J. Dairy Sci.* **47**, 1047–1051
Thompson, M. P., Boswell, R. T., Martin, V., Jenness, R. & Kiddy, C. A. (1969) *J. Dairy Sci.* **52**, 796–798
White, J. C. D. & Davies, D. T. (1958*a*) *J. Dairy Res.* **25**, 236–255
White, J. C. D. & Davies, D. T. (1958*b*) *J. Dairy Res.* **25**, 281–296
White, J. C. D. & Davies, D. T. (1966) *J. Dairy Res.* **33**, 93–102
White, J. C. D. & Sweetsur, A. W. M. (1977) *J. Dairy Res.* **44**, 237–247

THE PRESENT ORGANIZATION AND RESEARCH INTERESTS OF THE INSTITUTE

J. A. F. Rook

In 1971 the scientific staff of the Institute was reorganized into four departments, three specialist departments, Biochemistry, Chemistry and Physiology, and a department of Applied Studies. The following is a brief description of the structure and research interests of the four departments.

THE BIOCHEMISTRY DEPARTMENT

The present Biochemistry Department may be said to have originated with the appointment of Dr J. H. Moore as head of the Nutrition Department in 1966. Dr Moore had worked previously with Professor S. K. Kon at the National Institute for Research in Dairying, Reading. Under Dr K. L. Blaxter, the research work of the Nutrition Department had been concerned mainly with the energy metabolism of ruminant animals, but when Dr Moore succeeded Dr Blaxter, it was agreed that the work of the department should be directed further towards the biochemical aspects of nutrition and this change was recognized in the re-naming of the department in 1971.

In 1966, the graduate staff of the department consisted of Dr J. W. Czerkawski, Dr J. L. Clapperton and Dr A. K. Martin. Together with Dr Clapperton, Dr Czerkawski continued an investigation into the influence of dietary supplements of different fatty acids, including those with branched carbon chains, on the production of methane by sheep. In addition, Dr Czerkawski began a longer-term detailed research project on the mechanism of and factors affecting methane synthesis and hydrogen transfer by rumen bacteria; this project involved the development of new techniques for the investigation of the metabolism of rumen microorganisms *in vitro*.

From earlier nutritional experiments with cattle and sheep, Dr Martin had shown that large quantities of aromatic acids were excreted in the urine of ruminant animals, and that these quantities varied considerably with the type of diet given to the animals. Consequently, Dr Martin began a systematic search for dietary precursors of these urinary aromatic acids, a project that was subsequently augmented by an investigation into the metabolic transformations involved in the conversion of such precursors into the aromatic compounds that are ultimately excreted in the urine.

Since Dr Moore's recent interest and experience had been in the field of lipid biochemistry and nutrition, it was decided that this line of investigation should form a major part of the research work of the department. This necessitated the recruitment of additional staff with experience in lipid research, and a lipid group was established with the appointment in 1966 of Dr R. C. Noble who had worked in Dr Moore's laboratory at the National Institute for Research in Dairying, and in 1967 of Dr W. W. Christie, a graduate of the University of St Andrews, who had worked with Professor R. T. Holman at the Hormel Institute, in the University of Minnesota, USA.

Between 1966 and 1969 the research programme included several lipid topics that had originated in Dr Moore's laboratory at Reading; these included studies on the interrelationships between diet, lipid metabolism and cardiovascular degeneration in the rabbit, the transport and metabolism of lipids by the developing chick embryo and the effects of dietary fat on blood lipids and milk fat secretion in the dairy cow. An additional research project initiated by Dr Noble and Dr Moore was a study of certain aspects of the absorption and transport of lipids in ruminant animals; the results of this investigation raised the question of how the ruminant animal is able to maintain its essential fatty acid status if hydrogenation of dietary unsaturated fatty acids in the rumen is as effective as had previously been supposed. It was clear that this question could be answered only by a detailed examination of the metabolism of polyunsaturated fatty acids by rumen micro-organisms. Before such a research project could be undertaken it was evident that the department required some strengthening in the field of microbiological chemistry, and in 1967, Dr C. G. Harfoot from the Department of Microbiology at the University of Reading was appointed to work with Dr Noble and Dr Moore on factors affecting the biohydrogenation of linoleic acid by rumen micro-organisms. In collaboration with Dr Czerkawski, Dr Harfoot also began work on the isolation of methanogenic bacteria from the rumen.

Later expansion of the department's researches into the metabolism of polyunsaturated fatty acids by the ruminant animal and by rumen microorganisms resulted in the appointment to the staff in 1973 of Dr M. L. Crouchman, a graduate of Loughborough University of Technology. Dr Crouchman worked with Drs Noble, Harfoot and Moore on the mechanism of the microbial hydrogenation of linoleic acid and also on the lecithin-cholesterol-acyl transferase system in ruminant plasma. After 2 years, Dr Crouchman resigned from his position at the Hannah Research Institute to take up a post with the Unilever Research Laboratory; he was succeeded by Dr J. H. Shand from the Department of Biochemistry in the University of Aberdeen. Since joining the lipid group, Dr Shand has worked with Dr Noble and Dr Moore on the metabolic interconversion of polyunsaturated fatty acids in the tissues of the neonatal lamb.

Dr Christie commenced working with Dr Moore on the relationships between triglyceride structure and the physical properties of animal fats, principally those present in pig adipose tissue; this was a continuation and extension of a project already begun by Dr Moore before leaving the National Institute for Research in Dairying. Rapid progress was made in this field of investigation and in 1968 Dr Christie was joined by a senior research assistant, Miss M. L. Hunter, a graduate of the University of Strathclyde. In 1972, the lipid group was enlarged by the appointment of Dr R. G. Vernon, a graduate of the University of Birmingham, who had been working at the Banting and Best Department of Medical Research in the University of Toronto. By this time Dr Christie had extended his work on triglyceride structure to include a detailed study of the mechanism of triglyceride synthesis in mammary and adipose tissues, and Dr Vernon was appointed so that more emphasis could be placed on the metabolic aspects of this research topic. A considerable contribution to the group's work on triglyceride biosynthesis in adipose tissues was made also by Dr R. J. Henderson, a biochemistry graduate from the University of Strathclyde, who worked in Dr Christie's laboratory from 1974 to 1977 whilst in receipt of a Postgraduate Scholarship from the Agricultural Research Council. Early in 1977, Dr Vernon left the Biochemistry Department for a more senior post in the Physiology Department.

Dr Harfoot resigned from the staff of the Institute in 1975 to take the post of Senior Lecturer in Microbiology in the University of Waikato, New Zealand, and was succeeded by Dr R. A. Clegg, a biochemistry graduate from the University of Cambridge, who had worked with Professor Garland in the University of Bristol and with Professor Borst in the University of Amsterdam. Dr Clegg joined the lipid group of the department in 1976 and is investigating the factors that determine the activity of lipoprotein lipase in milk and in mammary tissues.

In 1971, two changes to the staff of the department were made as part of the Institute reorganization. Dr Clapperton, whose work had been mainly in the field of applied nutrition, was transferred to the new Applied Studies Department and Dr I. M. Morrison, who had been a member of the old Technical Chemistry Department since 1967, was transferred to the new Biochemistry Department. In the Technical Chemistry Department, Dr Morrison, who had worked originally in Sir Edmund Hirst's group of carbohydrate chemists in the University of Edinburgh, was investigating the nature of the chemical linkages that are involved in the formation of complexes between the lignin polymer and the structural carbohydrates in grasses. On joining the Biochemistry Department, Dr Morrison was encouraged to continue this research project together with investigations into the detailed structures of the hemicellulosic polysaccharides and studies of the changes that occur in these structures as the maturity of the herbage increases.

The excellent progress made by Dr Morrison in the elucidation of the structures of lignin-carbohydrate complexes in herbage encouraged him to begin an investigation into the mechanisms involved in the microbial degradation of these and other constituents of the plant cell wall. To participate in this research project, Dr A. G. Williams, a microbiologist from University College, Cardiff, was appointed to the staff of the department in 1973; in addition Dr Williams began a study of the metabolism of carbohydrates by rumen protozoa. The following year, Dr Williams was joined by a senior research assistant Mrs S. E. Withers, a graduate in microbiology from the University of Edinburgh, and in 1975 Mr R. E. Brice, a graduate of the University of East Anglia, began working as a senior assistant in Dr Morrison's laboratory; Mr Brice had worked previously on carbohydrate metabolism in plant tissues at the University of St Andrews. Recent evidence suggested that phenolic cinnamic acids may be involved in the cross-linkages of the hemicelluloses in plant cell walls, and that the disruption of these cross-linkages may be a rate-limiting step in the degradation of hemicelluloses by rumen micro-organisms. Since it is probable that the liberated phenolic acids are important precursors of the aromatic acids that are excreted in the urine of ruminant animals, there has been some convergence of the research interests of Dr Martin and Dr Morrison. This facet of the microbial degradation of plant cell wall material is being studied by Mr G. R. Tanner, a biochemistry graduate from the University of Liverpool, who began working with Dr Martin and Dr Morrison in 1976; Mr Tanner is in receipt of a Post-graduate Scholarship from the Agricultural Research Council. Also in 1976, Dr Martin was joined by a senior research assistant, Mrs P. J. Poole who is a graduate of the University of Nottingham.

Until recent years, there had been few facilities for the use of radioactive isotopes at the Institute, but in 1974 an isotope section was formed in the Biochemistry Department. This section, which is under the control of Dr Noble, provides a service for all the staff of the Institute. To assist Dr Noble in the provision of this service, Mr J. T. Drummond, a graduate of the University of Strathclyde, was appointed to the Biochemistry Department in 1975.

THE CHEMISTRY DEPARTMENT

The Chemistry Department was created in 1971 by the transfer of Dr W. Manson, Dr W. D. Annan, Dr D. W. West and Mr T. Carolan from the former Biochemistry Department and Dr D. T. Davies from the former Technical Chemistry Department. In 1972 Dr D. G. Dalgleish, after a period as Research Officer (Medical Research Council) in the Nuffield Department of Clinical Biochemistry, University of Oxford, took up the appointment of head of the department and he was joined in 1972 by Dr C. Holt from the Department of Biophysics, University of Leeds, in 1973 by Dr T. G. Parker from the Department of Pure and Applied Chemistry, University of Strathclyde and in 1975 by Dr D. S. Horne from the Research and Development Department of Laporte Industries Ltd.

The interests of Drs Manson and Annan were in the chemistry and biochemistry of the caseins. They had recently completed the characterization of the phosphopeptide of β-casein which contains four of the five phosphoseryl residues of the original protein and appears to be resistant to attack by proteolytic enzymes of the gastro-intestinal tract. Work on the biochemistry of the phosphate centre was begun by Dr Manson in collaboration with Dr West, Miss G. E. Towers, who was in receipt of a Postgraduate Studentship of the Agricultural Research Council, and Dr A. G. Mackinlay, a visiting worker from the University of New South Wales. Studies were made of the differences in sensitivity to enzyme attack of the various phosphoseryl residues and of the possibility of phosphate transfer between the phosphopeptide and adenosine diphosphate. In collaboration with Dr Clegg of the Department of Food Science at the University of Strathclyde, Dr Manson also examined the use of casein digests in the nutrition of patients with cystic fibrosis. At that time details of the complete amino acid sequences of the major casein fractions, α_{s_1}-casein, β-casein and κ-casein, were available but information was lacking for the minor caseins. Dr Manson, in collaboration with Dr Annan and Mr Carolan, devised a technique for the isolation of α_{s_0}-casein and has since determined its amino acid sequence. Following the publication of the recommendations of the Joint Consultative Organisation, the decision was taken that work on the phosphopeptide should be discontinued and that Dr Manson and his colleagues should investigate the chemical effects of processing on the protein of milk. Heating has profound effects on the behaviour of milk, particularly the protein fraction, and work has now started on the chemical changes induced in the proteins during heating, and their heat-induced interaction with other milk constituents.

Dr Davies had, for some time, been concerned with the development of accurate methods for the analysis of the milk serum protein and casein fractions. A suitable method for the quantitative analysis of serum proteins was completed in 1974 and in collaboration with Mr A. J. R. Law, a graduate in chemistry of the University of Strathclyde, a column-chromatographic method for analysis of whole casein in 1977. These techniques are being used to study variations in the protein composition of milk as affected by season, stage of lactation and breed of the animal and in the composition of creamery bulk milks. The object of these studies is to determine the influence of casein composition on the processing characteristics of milk. As the column-chromatographic method for casein analysis is relatively slow and is suitable for the thorough analysis of only a few samples, a search is being made for a reliable but more rapid method of casein analysis. The chosen technique is to use electrophoresis on polyacrylamide gels or cellulose acetate strips followed by a densitometric scanning of the electrophoresis patterns.

The aim of the group of workers appointed with Dr Dalgleish was to study the physical properties of the caseins. They have approached the problem in two ways. They have considered the properties of the individual caseins which aggregate together to form a large proportion of the framework of micelles and of the micelles themselves, and from studies on whole milk and separated milk have attempted to explain the formation and behaviour of these structures. The spectrometry, fluorimetry and thermochemistry of α_s-casein were investigated, as were the light scattering properties of micellar suspensions of casein. This latter work was later extended to allow simple turbidimetric analyses of average micellar radii and molecular weights, a technique which was subsequently adopted as a routine analytical method in survey work on creamery bulk milks.

Quantitative information was also acquired on the interaction between calcium ions and α_s-casein and studies of the kinetics of the aggregation of α_s-casein caused by Ca^{2+} demonstrated that α_s-casein did not show the linear flocculation kinetics which were to be expected from current models of the precipitation process. Immediately after the mixing of solutions of casein and calcium ions, molecular weight increased only slowly, and this stage was succeeded by a very much faster, linear growth of molecular weight. These and certain other aspects of casein behaviour have been explained by Dr Parker and Dr Dalgleish according to theories based on polymerization by a mechanism embodying polyfunctional condensation, on the assumption that α_s- and β-caseins aggregate through Ca^{2+}-induced reactive areas on their surfaces. The wider applications of this theory are now being investigated, with special emphasis on the explanation of the kinetics of aggregation processes involving both individual caseins, and micelles which have been treated with rennin or subjected to high temperatures. In addition, the effect of chemically modifying the casein is under investigation by Dr Horne, to allow definition of the particular sites of the caseins by which aggregation occurs.

The studies on naturally occurring casein micelles have been extended by Dr Holt to include the comparison of micellar size distribution as determined by light-scattering and electron microscopy; it was found that adjustment of electron microscopic results was necessary to take account of artefacts produced in the measurement. The main emphasis over the last 2 years, however, has been on the relationship between naturally occurring differences in micelle size and the processing characteristics of milk, and on the effect of concentrations of ionic and non-ionic constituents of milk on micelle size. In a survey of creamery bulk milks conducted by Dr Muir of the Applied Studies Department, a correlation between micelle radius, coagulation time and the concentration of urea in milk was established but micellar size and stability may also be affected by the concentration in milk of free Ca^{2+}, Na^+, K^+, HPO_4^{2-}, citrate, and complexed calcium, magnesium phosphate and citrate, and these relationships are now being studied.

Of particular interest is the relation between free Ca^{2+}, precipitated calcium phosphate, and the calcium bound to the caseins in micelles. It appears that the micelles depend for their integrity upon the presence of colloidal calcium phosphate, and its removal causes disruption of the micellar structure. There is a need to determine how micelles are synthesized so as to contain calcium phosphate in addition to calcium caseinate and to characterize the ionic balances which exist in milk and allow the micellar complex to remain stable under a surprisingly wide range of conditions.

THE PHYSIOLOGY DEPARTMENT

A department of Physiology was established in 1945 under Dr J. D. Findlay who, following a reorganization of the Institute, had been appointed Research Assistant in Physiology. His initial brief was to undertake research into the housing of cattle and ventilation of byres but soon it became apparent that there would be little headway in the absence of detailed study of animal acclimatization. This was a subject with wide ramifications since it was of application not only to local agriculture but to the whole question of the suitability of livestock for export to overseas territories where the climatic conditions differ markedly from those in the United Kingdom.

From its inception, therefore, the department was concerned with the physiology of thermoregulation in farm animals, with particular reference to the effect of high temperatures and humidities on the ability of animals to withstand adverse climatic stress. In the course of the work a great deal of the basic physiology of farm animals, for example the effect of heat stress on the cardiorespiratory and cardiovascular systems, the morphology of sweat glands and the nervous control of sweating, had to be studied. The work rapidly moved from an *ad hoc* approach to a fundamental one and later concentrated on the central nervous control of body temperature and the quantitative measurement of heat balance in cattle under various climatic conditions. When, however, the British colonies became independent some of the impetus for the work diminished, though not its relevance for developing countries, and by 1971 the balance of effort had been altered substantially to include investigations of the effect of cold, more relevant to United Kingdom conditions. The work of Dr Findlay and his colleagues received international recognition and many facets of it have been reviewed in earlier chapters.

With the reorganization of the Institute in 1971, the organizational structure of the department was unaltered but its work, whilst remaining in environmental physiology, was reorientated to encompass the effects of environment on metabolism. The work of Dr Findlay and of Dr D. T. Calvert, a graduate in physiology of the University of St Andrews who had joined the staff in 1967, on the role of the brain in thermoregulation continued but the emphasis of the work of other groups changed. Dr G. E. Thompson, when he joined the Institute in 1966 from the Department of Physiology of the Royal Free Hospital School of Medicine, London, had begun an investigation of the physiological regulation of fat mobilization in response to exposure to acute cold. For a while he continued work on the role of catecholamines with Dr D. P. Clough, on the role of neurotransmitters with Mrs K. F. Darling and on the pharmacology of fat mobilization with Dr A. R. Marshall. Progressively, however, greater emphasis was given to the metabolic consequences of cold and work is proceeding currently on the changes that take place in the liver, in muscle and in the mammary gland. He was joined in this work by Dr A. W. Bell, a graduate in rural science of the University of New England, Australia, who, before he was appointed to the staff in 1972, had been working with Dr G. Alexander at the Ian Clunies Ross Animal Research Laboratory, Prospect, New South Wales, on the effects of environment on metabolism in the lamb. Dr Thompson was also joined by Mr J. W. Gardner from the University of Manchester, by Dr P. L. Clarke from the University of Aberdeen and more recently by Mr E. M. Thomson and Mrs S. M. Butler, who are in receipt of Postgraduate Studentships of the Agricultural Research Council and the Medical Research Council respectively.

Dr J. A. McLean, a physicist who had made a major contribution to the earlier studies on heat loss and thermoregulation, had recently designed and constructed a gradient-layer calorimeter and had used it initially in support of the investigations in the department on the role of the brain and spinal cord in thermoregulation. With the help of Mr P. R. Watts, a graduate in physics of the University of Wales, who joined the staff in 1972, he began to use the instrument to study heat balance and heat storage in cattle. It was calculated that variation in the heat content of the body of cattle between night and day could be equivalent to 2 hours of heat production by the animal. Changes in heat storage are also brought about by the animal drinking water and eating food. These changes in heat storage are large enough to affect productive efficiency but at present there is no simple method for measuring changes in body heat content. The objects of the work on heat balance are to analyse diurnal changes in heat storage by simultaneous direct and indirect calorimetry, to use calorimetric methods as a basis for the development of indirect thermometric methods of measuring heat storage and to assess the energetic significance of heat storage in relation to daily environmental fluctuations. The results of these studies will be applied to the evaluation and design of livestock housing.

From 1958, work in the department on histology was the responsibility of Dr D. McE. Jenkinson. In the late 1960s he had begun, in collaboration with Dr Miranda E. Smith and Mr R. M. Mabon, a study of the chemical composition and physiological significance of the secretions of the sweat and sebaceous glands. This work was broadened to encompass the effect of skin secretions of cattle and sheep on the viability of benign and pathogenic organisms, work greatly aided by the appointment in 1973 of Mr D. H. Lloyd, a veterinary surgeon, as an Animal Health Trust Wooldridge Farm Livestock Research Fellow. The work made good progress but unfortunately, in the research priorities established through the Joint Consultative Organisation, insufficient emphasis was given to work on skin ecology to justify the additional resources needed for the proper development of the programme. A narrower scheme, concerned with specific aspects of skin biochemistry and physiology has been introduced.

Following the untimely death of Dr Findlay in 1974, there was a further review of the work of the department. It was decided that studies of central temperature regulation should cease, and that additional effort should be given to the investigation of metabolic changes in lactating animals in response to environmental and dietary factors. Responsibility for this new effort was given partly to Dr Thompson and his colleagues, who were joined by Dr Calvert when his work on central temperature regulation was concluded, but to give direct endocrinological support a new group was established under Dr R. G. Vernon, formerly of the Biochemistry Department. He was joined by Dr Anne Faulkner, from the Nuffield Institute for Medical Research where she had studied the biochemical development of foetal liver, by Mr D. J. Flint on the completion of a period of postgraduate study in the Physiology Department of the University of Liverpool as a Science Research Council scholar and by Miss Shona Park, a graduate of the University of Glasgow. Research topics already under investigation are the hormonal regulation of adipose tissue metabolism in lactating animals, the role of the liver in the supply of substrates to the mammary gland and the role of the autonomic nervous system in the regulation of substrate supply.

The new work will have as its objective a better understanding of the way in which the endocrine changes which take place throughout the lactation cycle or in response to environmental stimuli, regulate or modify the use of food for milk production. It is designed to support the work of the Applied Studies Department on improving the efficiency of utilization of food for milk production and to complement that of the Biochemistry Department on the biochemistry of the mammary gland and other tissues.

THE APPLIED STUDIES DEPARTMENT

When the Institute was reorganized in 1971, staff whose research interests were in applied science were grouped together in a single department, which was given responsibility for applied work in both milk production and milk utilization. Dr R. Waite, formerly head of the Technical Chemistry Department, became head of the new department and he was joined by nine other senior members of staff. Dr M. E. Castle, Dr D. Reid and Mr J. N. Watson transferred from the former Grass and Dairy Husbandry Department, Dr J. L. Clapperton from the former Nutrition Department, Dr Margaret M. Taylor and Mr J. D. Phillips from the former Bacteriological Laboratory and Dr J. C. D. White, Dr A. W. M. Sweetsur and Mr J. Abbot from the former Technical Chemistry Department. Three additional senior posts were created and these were filled by the appointment of Dr P. C. Thomas and Mr D. G. Chamberlain in 1971 and Miss Morag E. Ferrie (Mrs Kelly) in 1972. Dr Thomas was a graduate of the University of Wales and had worked with me in the Department of Agricultural Sciences at the University of Leeds, Miss Ferrie was a graduate of the University of Edinburgh who had obtained experience in applied research during a period with Beechams Products Ltd, and Mr Chamberlain was a recent graduate from the Department of Biochemistry and Physiology at the University of Reading.

The staff was organized into four working groups, dairy husbandry, nutrition and metabolism, milk utilization and bacteriology, and a central analytical section which serviced the needs of the research sections of the department and of other departments when required. By the time that Dr Waite retired early in 1973 a varied and balanced research programme had been established.

Dr Waite was succeeded by Dr W. Banks who had worked previously with Professor Greenwood at the Flour Milling and Baking Research Association on the properties of starch and starch-degrading enzymes. Shortly after his appointment the administrative structure of the department was formalized and Dr Castle was given overall responsiblity for work in Dairy Husbandry, Dr Thomas for work in Nutrition and Metabolism, and Dr D. D. Muir, who was a new appointee, for work in Milk Utilization. Dr Muir, a graduate of the University of Edinburgh, had obtained considerable experience of applied research whilst working on the role of starch in baked goods for the Flour Milling and Baking Research Association. Dr White, who up to that time had been responsible for much of the work on milk utilization and for the analytical section, had for many years been heavily involved in specialist analytical work sponsored by the International Dairy Federation, the British Standards Institution and the International Standards Organization. In view of the increasing importance and scale of this work, it was decided that he should work on it exclusively.

For many years the Institute had undertaken research into the efficient production, conservation and utilization of herbage for milk production and this general theme was continued. Dr Castle and Mr Watson completed a systematic comparison of simplified and complex grazing systems and expanded their work on the production and use of silage to consider the problems and advantages of ensiling material of high digestibility. They were joined in this work in 1975 by Mr W. C. Retter, a graduate of the University of Leeds, in receipt of a postgraduate award from Imperial Chemical Industries Ltd.

Dr Reid's special interest was in the response of grass to the application of nitrogenous fertilizer; following the large increase in the price of such fertilizer in 1973/74, he began an investigation into the use of the new, upright varieties of white clover capable of withstanding competition from grasses. He is placing particular emphasis on defining the optimum conditions for high yields of nutrients from mixed grass/clover swards when these are used for silage production.

The principal research interests of Dr Thomas, when at the University of Leeds, were in the effects of fermentation of food within the rumen on energy and nitrogen metabolism in the ruminant animal. He continued to work on this subject, in collaboration with Mr J. C. Hodgson and Mr M. R. Stokes, postgraduate students who had transferred with him from Leeds, and later with Mr Chamberlain. Initially, the diets that were studied were based on hay and were rich in concentrated foods as diets of this type are associated with great variability in rumen fermentation. The work of Dr Castle, however, highlighted a number of problems in the use of high-digestibility silage for milk production. The more important of these were a loss of digestibility of herbage constituents during ensilage, a comparatively low voluntary intake of the silage which was depressed further by a supplement of cereals, and a depression of the protein content of milk when the silage was included in the diet of lactating cows. The emphasis of the work of the Nutrition and Metabolism section was therefore changed and is now concerned with diets which contain silage as the sole forage. Work on the evaluation of the energy content of silage diets is the specific responsibility of Mr N. C. Kelly, who initially was in receipt of a Meat and Livestock Commision Postgraduate Studentship but has now transferred to the staff; work on protein utilization is the responsibility of Mr Chamberlain and that on the dietary factors affecting the protein content of milk is the responsibility of Mr J. S. Chalmers, in receipt of a Department of Agriculture and Fisheries for Scotland Postgraduate Studentship. New work is planned by Dr Thomas on the use of body reserves to support milk secretion in the early part of lactation, by Dr Clapperton on methods of improving the nutritive value of ensiled materials and by Dr Steele on suitable supplementary foods for diets of high-digestibility silage. Dr Steele who spent 1970 to 1973 on sabbatical leave at the Hormel Institute in the University of Minnesota, USA, joined the Applied Studies Department in 1975 after a short period in the Biochemistry Department.

In addition to the above lines of work, the section has been involved in evaluating the application of research ideas that have arisen from investigations by the specialist departments. Dr Clapperton has had prime responsibility for this work but others have collaborated, including Miss Ferrie of the Milk Utilization section. The first topic considered was the suppression of methanogenesis and in 1975, Mr S. Basmaeil, a graduate of the Universities of Riyadh and Aberdeen, joined Dr Clapperton to continue this work as a postgraduate study. A second topic arose from the demonstration by the Biochemistry Department that the physical properties of milk fat could be controlled by dietary fat composition. It has proved technically feasible to produce by dietary manipulation a butter having good spreadability at low temperature and adequate stand-up properties at high temperature. A third recent example is an assessment by Dr Steele of the practical benefits to be obtained from increasing the essential fatty acid status of the neonatal lamb, a project developed from the basic studies of Dr Moore and Dr Noble.

The Institute has a long-standing interest in heat treatment of milk and in 1971 Drs White and Sweetsur were undertaking an investigation of the stability to heat of the proteins of liquid milk. This work is being continued by Dr Muir and Dr Sweetsur in conjunction with members of the Chemistry Department, and present effort is being concentrated on the role of urea. After discussion with the Institute's Consultative Panel it was decided to extend the work on heat stability to include work on milk powder, and with the assistance of Unigate Ltd, a new Tall-Form spray drier was designed and built, and installed in the Dairy School of the West of Scotland Agricultural College. Technological facilities are now shared by the two institutions and much of the dairy technology research of the Institute is done in collaboration with the College. The installation and commissioning of the plant and of an instantizer and fluidized-bed drier were the responsiblity of Mr Abbot and Dr Muir.

The Consultative Panel also advised on the need for work on the chemical and bacteriological quality of milk going to manufacture in south-west Scotland and this project became the joint responsibility of Dr Muir, Dr Taylor, Miss Ferrie and Mr Phillips. Miss M. Cristina Vazquez, a graduate of the University of Madrid, joined Dr Taylor as a Postgraduate Scholar to work on one aspect of this problem, factors affecting the growth of psychrotrophs in milk. Unhappily, ill health forced Dr Taylor to retire in 1976 and Mr Phillips has assumed responsibility for the day-to-day management of the bacteriology laboratory.

A further project, introduced recently because of the high priority attached to it by the Milk and Milk Products Committee of the Joint Consultative Organisation, is the isolation of a heat-stable β-galactosidase for commercial use in the hydrolysis of the lactose of whey. This is the responsibility of Dr M. W. Griffiths who, prior to his appointment in 1974, had studied with Professor Kornberg at the University of Leicester. Initially, he was appointed to carry out with Dr Banks, and in collaboration with Dr R. L. Spooner of the Animal Breeding Research Organisation, an investigation of the amylase systems of bovine serum as an index of milk yield potential but the results were not promising and less emphasis is now being given to this work.

A feature of the recent work of the department has been the extent to which the work of the different sections has become interrelated. It is now common practice to examine an experimental treatment in terms both of efficiency of milk production and the consequences for milk utilization. As one example, in the study of dietary factors affecting milk protein content the importance of the observed changes in protein and non-protein nitrogen content to the manufacturing quality of the milk is also being assessed. Links with the specialist departments have also grown, as have those with the West of Scotland Agricultural College. The first topic selected for development at the Crichton Royal Development Farm is a comparison of systems based on medium- and high-quality silage, as an extension of the work being done within the Institute. The department now has the ability to carry out co-ordinated work on problems of practical importance, it has access to the advice of the specialist departments, the necessary working contacts with the West of Scotland Agricultural College for the development and extension of promising ideas and, through the Institute's Consultative Panel, good links with industry itself.

For the sake of brevity, in this account of the present work of the Institute, mention has been made only of those senior scientific staff with academic qualifications. The scientific work of the Institute could not proceed, however, without the active support of the many other members of the scientific and farm staffs, of the administrative staff under the control of the Secretary of the Institute, a position held by Miss Margaret M. Hughes, MBE, until 1975 and since then by Mr N. A. D. McCance, or of the Engineering and Maintenance Services section under the control of Mr J. Ross. Four members of the present senior scientific support staff have been associated with the Institute for the greater part of their working lives and deserve especial mention. Miss Agnes G. Wilson joined the staff in

1938 and for many years worked as a personal assistant to Dr Waite and now has responsibility for the analytical section of the Applied studies Department. Mr D. R. Paterson joined the staff in 1948 as an assistant to Dr Blaxter with responsibility for animal experimental work and has continued with this role in the Biochemistry Department under Dr Moore. Miss Grace H. McG. Breckenridge and Mr W. Nisbet both joined the staff in 1949. Miss Breckenridge was in charge of analytical work in the Nutrition Department when Dr Blaxter was the head and since Dr Moore's appointment has worked as a personal assistant to Dr Czerkawski. Mr Nisbet, a specialist in electronics, made a unique contribution to the development of the environmental facilities within the Physiology Department and now offers a specialist service to all departments.

MEMBERS OF THE JOINT COMMITTEE OF MANAGEMENT AND OF THE COUNCIL OF THE INSTITUTE, 1927 to 1978

Sir Donald MacAlister of Tarbert, Bt, KCB, MA, MD, DCL, LLD, DSc, PhD, Chairman	1927 to 1934
Professor Sir Robert Muir, MA, MD, ScD, LLD, DCL, FRS	1927 to 1936
Professor D. Noel Paton, FRS	1927 to 1928
Thomas C. Lindsay Esq.	1927 to 1931
W. G. R. Paterson, BSc, NDA	1927 to 1945
John Speir Esq.	1927 to 1933
The Right Hon. Lord Weir of Eastwood, LLD	1927 to 1930
Mrs Houison-Craufurd	1927 to 1930
Robert Brough Esq.	1927 to 1930
Professor E. P. Cathcart, CBE, MD, DSc, LLD, FRS, Vice-Chairman	1929 to 1954
John M. Hannah Esq.	1930 to 1938
Thomas Henderson Esq., BSc	1930 to 1935
J. Harling Turner Esq., CBE	1932 to 1939
T. W. Gibson Esq., BCom, Secretary	1930 to 1935
Sir Robert S. Rait, CBE, MA, LLD, Chairman	1934 to 1936
Sir George Wilson, KBE, MA, LLD, Chairman Farm Committee	1934 to 1972
F. A. Bell Esq.	1935 to 1940
J. McNair Esq., BCom, CA, Secretary	1935 to 1942
Sir Hector J. W. Hetherington, GBE, MA, LLD, Chairman	1936 to 1965
Professor Carl H. Browning, MD, LLD, DPH, FRS	1936 to 1939
The Right Hon. Lord Rowallan, MC	1938 to 1940
Professor G. M. Wishart, BSc, MD	1939 to 1948
James Dunlop Esq.	1939 to 1946
A. W. Montgomerie Esq.	1940 to 1949
J. S. Stevenson Esq.	1940 to 1958
Miss Margaret M. Hughes, MBE, Secretary	1942 to 1975
J. J. M. Hannah Esq., CBE, NDA, CDA, Vice-Chairman and Chairman House & Estate Committee	1945 to present
Sir James Denby Roberts, Bt, MA	1946 to 1952
Professor R. C. Garry, OBE, MB, DSc, FRSE, Vice-Chairman	1948 to 1972
J. R. Barron Esq.	1949 to 1952
J. Stevenson Esq., OBE, FRAgS	1952 to 1975
Professor J. W. (later Sir James) Howie, MD	1953 to 1964
Professor T. Ferguson, CBE, MD, DSc, DPH	1955 to 1964
W. G. M. Lohoar Esq.	1958 to 1962
W. W. Watson Peat Esq., CBE	1963 to present
Professor R. G. White, MA, DM	1964 to present
Professor R. B. Fisher, CBE, MA, PhD	1964 to 1974
Sir Charles H. Wilson, MA, LLD, Chairman	1965 to 1976
R. H. U. Stevenson Esq.	1971 to 1972
Professor J. S. Gillespie, MB, ChB, PhD	1972 to present
J. Caldwell Esq.	1972 to present
N. A. D. McCance Esq., AMBIM, Secretary	1975 to present
Professor W. W. Fletcher, BSc, PhD, FLS, FIBiol, FRSE	1976 to 1977
Professor F. D. Gunstone, BSc, PhD, DSc, FRIC, FRSE	1976 to present
Alwyn Williams, PhD, DSc, FGS, FRS, Chairman	1976 to present
G. L. Wilson Esq.	1977 to present

SENIOR STAFF OF THE INSTITUTE, 1928 to 1978

Abbot, James	*Dairy Technologist*	1955 to present
Alexander, Hazel	*Bacteriologist*	1950 to 1951
Allen, L. A.	*Bacteriologist*	1929 to 1930
Anderson, Agnes M.	*Chemist*	1958 to 1959
Annan, W. Douglas	*Chemist*	1964 to present
Armstrong, David G.	*Animal Nutritionist*	1954 to 1963
Banks, William	*Chemist*	1973 to present
Beakley, W. Robert	*Physicist*	1948 to 1954
Bell, Alan W.	*Physiologist*	1972 to 1977
Bianca, Wolfgang	*Physiologist*	1955 to 1967
Blackburn, Philip S.	*Veterinary Pathologist*	1951 to 1968
Blackwood, Janet H. (Mrs S. J. Edwards)	*Biochemist*	1929 to 1935
Blaxter, Kenneth L.	*Animal Nutritionist*	1948 to 1965
Bligh, John	*Physiologist*	1952 to 1956
Boyd, John	*Chemist*	1950 to 1952
Breckenridge, Grace McG.	*Biochemist*	1949 to present
Brown, Fred	*Biochemist*	1950 to 1952
Calvert, David T.	*Physiologist*	1967 to present
Campbell, Jean (Mrs W. Holmes)	*Chemist*	1947 to 1949
Carolan, Thomas	*Chemist*	1967 to present
Castle, Malcolm E.	*Agriculturalist*	1955 to present
Chamberlain, David G.	*Animal Nutritionist*	1971 to present
Christie, William W.	*Biochemist*	1967 to present
Clapperton, John L.	*Animal Nutritionist*	1957 to present
Clarke, Peter L.	*Chemist*	1974 to present
Clegg, Roger A.	*Biochemist*	1976 to present
Clough, David P.	*Physiologist*	1968 to 1972
Crawford, Jean	*Director's Secretary and Librarian*	1967 to 1974
Crouchman, Malcolm L.	*Biochemist*	1973 to 1974
Czerkawski, Julian W.	*Biochemist*	1963 to present
Dalgleish, Douglas G.	*Chemist*	1972 to present
Darling, Kathryn F.	*Physiologist*	1971 to 1974
Davidson, Elizabeth T.	*Director's Secretary and Librarian*	1953 to 1966
Davies, D. Thomas	*Chemist*	1952 to present
Dempster, Sheila A. R.	*Biochemist*	1969 to 1971
Dittmar, C. Julie	*Bacteriologist*	1939 to 1940
Drummond, James T.	*Chemist*	1975 to present
Drysdale, A. Douglas	*Agriculturalist*	1958 to 1966
Edwards, D. Caird	*Biochemist*	1952 to 1955
Edwards, S. John	*Veterinary Pathologist*	1934 to 1938
Faulkner, Anne	*Physiologist*	1976 to present
Ferrie, Morag E. (Mrs N. Kelly)	*Chemist*	1972 to 1977
Fergusson, Donald L.	*Agriculturalist*	1948 to 1951
Findlay, James D.	*Physiologist*	1942 to 1974

Flint, David J.	*Endocrinologist*	1977 to present
Fowler, Alexander B.	*Agriculturalist*	1929 to 1947
Gibson, Thomas W.	*Secretary*	1930 to 1935
Goodall, A. Myfanwy (Mrs W. Nisbet)	*Histologist*	1950 to 1957
Gorrod, Alan R. N.	*Chemist*	1953 to 1957
Graham, Norman McC.	*Animal Nutritionist*	1953 to 1958
Griffiths, Mansel W.	*Microbiologist*	1974 to present
Grindlay, Maurice	*Animal Nutritionist*	1948 to 1950
Harfoot, Christopher G.	*Microbiologist*	1967 to 1975
Hart, Leigh I.	*Biochemist*	1960 to 1964
Hay, William	*Agriculturalist*	1939 to 1940
Higginbottom, Constance	*Bacteriologist*	1939 to 1962
Hofman, Theodore	*Biochemist*	1952 to 1956
Holmes, William	*Agriculturalist*	1944 to 1955
Holt, Carl	*Chemist*	1972 to present
Horne, David S.	*Chemist*	1975 to present
Howat, George R.	*Chemist*	1932 to 1938
Howie, Janet	*Administrative Assistant*	1955 to present
Hughes, Margaret M.	*Administrative Assistant*	1933 to 1942
	Secretary of Institute	1942 to 1975
Hunter, Margaret L.	*Chemist*	1968 to present
Ingram, Douglas L.	*Physiologist*	1958 to 1961
Jackson, Christopher J.	*Chemist*	1935 to 1938
Jenkinson, David McE.	*Physiologist*	1958 to present
Johnston, Margaret J.	*Chemist*	1960 to 1963
Jordan, Leslie	*Veterinary Pathologist*	1929 to 1934
Kelly, Norman C.	*Animal Nutritionist*	1976 to present
Law, Andrew J. R.	*Chemist*	1971 to present
Mabon, Richard M.	*Chemist*	1957 to present
McCallum, E. Sheila	*Biochemist*	1957 to 1961
McCance, Neil A. D.	*Secretary of Institute*	1975 to present
Mackie, Alex	*Veterinary Physiologist*	1950 to 1952
McLean, John A.	*Physicist*	1952 to present
McLeod, Douglas H.	*Bacteriologist*	1943 to 1947
MacLusky, Donald S.	*Agriculturalist*	1950 to 1957
McNair, James	*Secretary of Institute*	1935 to 1941
McNaught, Mary L. (Mrs J. H. Moore)	*Biochemist*	1944 to 1952
Manson, William	*Chemist*	1952 to present
Marshall, Alan R.	*Physiologist*	1971 to 1974
Martin, A. Keith	*Animal Nutritionist*	1958 to present
Mitchell, I. Winifred	*Bacteriologist*	1934 to 1935
Montgomery, James P.	*Chemist*	1958 to 1959
Moore, John H.	*Biochemist*	1966 to present
	Deputy Director	1971 to present
Morris, Samuel	*Animal Nutritionist*	1930 to 1939
Morrison, Ian M.	*Biochemist*	1967 to present
Muir, D. Donald	*Chemist*	1974 to present
Murphy, Lilian	*Veterinary Pathologist*	1948 to 1949
Myatt, John E.	*Agriculturalist*	1948
Nettlefold, Brian A.	*Librarian*	1975 to present

Nichols, Agnes A.	*Bacteriologist*	1935 to 1938
		1940 to 1942
Nisbet, William	*Electronics Engineer*	1949 to present
Noble, Raymond C.	*Biochemist*	1966 to present
Owen, Edwin C.	*Biochemist*	1939 to 1970
Parker, Thomas G.	*Chemist*	1973 to present
Paterson, Donald R.	*Senior Animal Technician*	1948 to present
Philips, John D.	*Bacteriologist*	1967 to present
Reid, David	*Agronomist*	1953 to present
Roberts, P. Christopher V.	*Senior Animal Technician*	1973 to 1974
Robertshaw, David	*Physiologist*	1963 to 1969
Rook, John A. F.	*Animal Nutritionist*	1950 to 1954
	Director	1971 to present
Ross, John	*Services Engineer*	1972 to present
Shand, John H.	*Biochemist*	1974 to present
Smith, James A. B.	*Biochemist*	1936 to 1946
	Deputy Director	1948 to 1950
	Director	1950 to 1970
Smith, Miranda E.	*Physiologist*	1969 to 1975
Snow, Douglas	*Botanist*	1940 to 1944
Steele, William	*Animal Nutritionist*	1967 to present
Stirling, Anna C.	*Bacteriologist*	1942 to 1943
Stirling, John D.	*Biochemist*	1931 to 1933
Storie, George A. T.	*Climatic Engineer*	1957 to 1971
Strachan, Norman H.	*Chemist*	1955 to present
Sweetsur, A. W. Maurice	*Chemist*	1967 to present
Taylor, Margaret M.	*Bacteriologist*	1952 to 1976
Thom, Charles L.	*Agriculturalist*	1940 to 1944
Thomas, Philip C.	*Animal Nutritionist*	1971 to present
Thompson, Gordon E.	*Physiologist*	1966 to present
Vernon, Richard G.	*Biochemist*	1972 to present
Wainman, F. William	*Animal Nutritionist*	1954 to 1965
Wait, Martin K.	*Senior Animal Technician*	1975 to present
Waite, Robert	*Chemist*	1939 to 1973
Watson, John N.	*Farm Manager*	1952 to present
Watts, Paul R.	*Physicist*	1972 to 1977
Watts, Percival S.	*Veterinary Pathologist*	1938 to 1950
Webster, A. John F.	*Veterinary Physiologist*	1963 to 1966
West, David W.	*Biochemist*	1964 to present
Wheater, David W. F.	*Bacteriologist*	1946 to 1952
White, James C. D.	*Chemist*	1944 to present
Whittow, G. Causey	*Physiologist*	1959 to 1965
Wilkie, James	*Agriculturalist*	1935 to 1937
Williams, Alan G.	*Microbiologist*	1973 to present
Wilson, Agnes G.	*Chemist*	1938 to present
Wood, William A.	*Animal Nutritionist*	1950 to 1952
Wright, Norman C.	*Physiologist*	1928 to 1929
	Assistant Director	1929 to 1930
	Director	1931 to 1950
Yang, Shin-Hsing	*Histologist*	1947 to 1951

PUBLICATIONS,
1928 TO 1978

BULLETINS

No.

1. MacNeilage, A., Jr (1929) *Surplus milk and milk residues*
2. Fowler A. B. & Wright, N. C. (1931) *Reactors in tuberculin-tested (licensed) herds*
3. Allen, L. A. (1932) *The condensing and drying of whole milk, separated milk and whey. A review of existing knowledge*
4. Scott, A. W. (1932) *The engineering aspects of the condensing and drying of milk*
5. Stirling, J. D. & Blackwood, J. H. (1933) *The nutritive properties of milk in relation to pasteurisation*
6. Scott, A. W. & Wright, N. C. (1935) *An inquiry into the design, operation and efficiency of pasteurising plants*
7. Wright, N. C. (1936) *An inquiry into the drinking habits of children of school age, with special reference to milk drinking*
8. McNair, J. & Fowler, A. B. (1942) *The production of artificially dried grass*
9. Findlay, J. D. (1950) *The effects of temperature, humidity, air movement and solar radiation on the behaviour and physiology of cattle and other farm animals*

SPECIAL REPORTS

No.

1. Department of Health for Scotland (1933) *Tuberculous infection in milk. Med. Res. Counc. (G.B.) spec. Rep. Ser.* no. 189
2. Jordan, L. (1933) *The eradication of bovine tuberculosis. Med. Res. Counc. (G.B.) Spec. Rep. Ser.* no. 184
3. Wright, N. C. (1937) *Report on the development of the cattle and dairy industries in India.* Government of India Press, Simla
4. Agricultural Research Council (1944) *Modes of spread of* Streptococcus agalactiae *infection in dairy herds. Imp. Bur. Anim. Health Rev. Ser.* no. 2
5. Taylor, F. H., Coop, I. E. & Smith, J. A. B. (1945) *Some developments in dairying in Germany. A report of the British Intelligence Objectives Sub-committee.* His Majesty's Stationery Office, London
6. Department of Health for Scotland (1946) *The freezing point (Hortvet) test of milk.* His Majesty's Stationery Office, London
7. Wright, N. C. (1946) *Report on the development of cattle breeding and milk production in Ceylon. Sessional Paper XX.* Ceylon Government Press, Colombo
8. Food and Agriculture Organization (1946) *Report of the FAO Mission for Greece.* Food and Agriculture Organization, Washington, DC

NUMBERED SCIENTIFIC PUBLICATIONS

Institute reprint no.

1. Wright, N. C. (1929) The inorganic constituents of milk. *Science (Wash. DC)* **69**, 78
2. Wright, N. C. (1929) Recent advances in dairy science. *Agric. Prog.* **6**, 7–22
3. Wright, N. C. (1929) Membrane equilibria and selective absorption. *Biochem. J.* **23**, 352–357
4. Wright, N. C. (1929) The incidence of tuberculous infection in the milk supplies of Scottish cities. *Brit. Med. J.* **2**, 452–454

Institute reprint no.

5. Hannah Dairy Research Institute (1931) An inquiry into the length of life of dairy cattle. *Rep. Scott. Milk Rec. Ass.*
6. Jordan, L. (1930) Can tuberculosis be eliminated? *Farm. Stock Breed.* **44**, 768
8. Hannah Dairy Research Institute (1931) *The Hannah Dairy Research Institute. Its Origin and Work*
9. Wright, N. C. (1931) The Hannah Dairy Research Institute, Ayr. *Ayrsh. Cattle Soc. J.* **3**, 5–9
10. Jordan, L. (1930) Eradication of bovine tuberculosis. *Rep. Cent. Counc. Milk Rec. Soc.*
12. Scott, A. W. (1931) Small scale research on spray drying of milk. *J. R. Tech. Coll. Glas.* **2**, 456–461
13. Allen, L. A. & Bell, J. (1931) The role of milk constituents in breadmaking. *J. R. Tech. Coll. Glas.* **2**, 550–564
14. Fowler, A. B. (1931) Renewal of old pasture. *Scott. J. Agric.* **14**, 439–442.
15. Morris, S., Watson, A. M. & Morris, N. (1931) Metabolism studies in tetany. *Biochem. J.* **25**, 786–798
16. Allen, L. A. (1931) The mineral constituents and citric acid content of milk. *J. Dairy Res.* **3**, 1–51
17. Scottish National Milk and Health Association (1931) *Milk: Striking Facts Regarding Nature's Perfect Food.* Scottish National Milk and Health Association, Glasgow
18. Scott, A. W. (1932) The cooling effect of a hot air stream on a water surface. *J. R. Tech. Coll. Glas.* **2**, 620–630
19. Blackwood, J. H. & Stirling, J. D. (1932) An iodimetric method for the determination of iron in blood. *Biochem. J.* **26**, 353–356
20. Morris, N., Rennie, J. B. & Morris, S. (1932) The effect of intravenous injection of glycine on the serum calcium. *Br. J. Exp. Path.* **13**, 132–134
21. Blackwood, J. H. & Stirling, J. D. (1932) The absorption of milk precursors by the mammary gland. I. The Kaufmann-Mange technique. *Biochem. J.* **26**, 357–361
22. Blackwood, J. H. & Stirling, J. D. (1932) The absorption of milk precursors by the mammary gland. II. The relation of blood-sugar absorption to lactose secretion. *Biochem. J.* **26**, 362–368
23. Blackwood, J. H. (1932) The absorption of milk precursors by the mammary gland. III. The relation of amino-acid absorption to protein synthesis. *Biochem. J.* **26**, 772–777
24. Blackwood, J. H. & Stirling, J. D. (1932) The absorption of milk precursors by the mammary gland. IV. Aspects of the phosphorus metabolism of the mammary gland. *Biochem. J.* **26**, 778–784
25. Blackwood, J. H. & Stirling, J. D. (1932) The absorption of milk precursors by the mammary gland. V. Physico-chemical aspects of milk secretion. *Biochem. J.* **26**, 1127–1137
26. Williams, R. S. & Wright, N. C. (1932) Dairying in Finland. *J. Br. Dairy Farm. Assoc.* **44**, 1–8
27. Wright, N. C. (1932) Factors affecting the solubility of milk powders. I. The effect of heat on the solubility of milk proteins. *J. Dairy Res.* **4**, 122–141
28. Fowler, A. B. (1932) The Ayrshire breed of cattle. A genetic study. *J. Dairy Res.* **4**, 11–27
29. Wright, N. C. (1933) Wastage in dairy cows. *Scott. J. Agric.* **16**, 31–40
30. Morris, S. & Morris, N. (1932) Further observations on the volatile chlorine of blood. *Biochem. J.* **26**, 2015–2020

Institute reprint no.

31. Stirling, J. D. & Wishart, G. M. (1932) The hydrolysis of caseinogen by pepsin and by trypsin-kinase. *Biochem. J.* **26**, 1989–1999
32. Scott, A. W. (1933) The boiling point elevation of concentrated milk solutions. *J. R. Tech. Coll. Glas.* **3**, 116–120
33. Fowler, A. B. (1933) The construction and equipment of cattle byres. I. Byre flooring. *Scott. J. Agric.* **16**, 191–196
34. Morris, S. & Wright, N. C. (1933) The nutritive value of proteins for milk production. I. A comparison of the proteins of beans, linseed and meat meal. *J. Dairy Res.* **4**, 177–196
35. Howat, G. R. & Wright, N. C. (1933) Factors affecting the solubility of milk powders. II. The influence of temperature of reconstitution on protein solubility. *J. Dairy Res.* **4**, 265–272
36. Jordan, L. (1933) Diseases of young calves. *Vet. J.* **89**, 202–215
37. Morris S. (1933) The influence of the pituitary gland on parturition. I. Metabolism immediately prior to and immediately following parturition. *J. Obstet. Gynaecol. Br. Emp.* **40**, 230–239
38. Morris, S. (1933) The influence of the pituitary gland on parturition. II. Metabolism studies during injections of extracts of the posterior lobe of the hypophysis. *J. Obstet. Gynaecol. Br. Emp.* **40**, 580–585
39. Morris, S. (1933) The influence of the pituitary gland on parturition. III. The posterior lobe of the hypophysis and the water balance. *J. Obstet. Gynaecol. Br. Emp.* **40**, 598–605
40. Fowler, A. B. (1933) The construction and equipment of cattle byres. II. Automatic drinking bowls. *Scott. J. Agric.* **16**, 314–320
41. Wright, N. C. (1933) Some implications of compulsory pasteurisation. *Proc. Br. Assoc. Adv. Sci., Leicester*
42. Morris, S. & Wright, N. C. (1933) Nutritive value of proteins for milk production. II. A comparison of the proteins of blood meal, pea meal, decorticated earth-nut cake and a mixture of decorticated earth-nut cake and flaked maize. *J. Dairy Res.* **5**, 1–14
43. Bell, G. H. & Morris, S. (1934) The oxytocic property of the blood of the cow. *J. Physiol.* (*Lond.*) **81**, 63–69
44. Blackwood, J. H. & Wishart, G. M. (1934) The hydrolysis of lecitho-vitellin by pepsin and by trypsin-kinase. *Biochem. J.* **28**, 550–558
45. Morris, S. (1934) The nutritive value of proteins for milk production. III. The determination of the amino-acids of various feeds. *J. Dairy Res.* **5**, 108–112
46. Bell, G. H. & Morris, S. (1934) The oxytocic property of the blood and its relation to milk fever in cows. *Br. J. Exp. Pathol.* **15**, 143–147
47. Howat, G. R. & Wright, N. C. (1934) Factors affecting the solubility of milk powders. III. Some physico-chemical properties of concentrated solutions of milk solids. *J. Dairy Res.* **5**, 236–244
48. Fowler, A. B., Morris, S. & Wright, N. C. (1934) The quality of protein in relation to milk production. *Scott. J. Agric.* **17**, 261–269
49. Blackwood, J. H. (1934) The absorption of milk precursors by the mammary gland. VI. The relation of phosphorus to the fat metabolism of lactation. *Biochem. J.* **28**, 1346–1354
50. Blackwood, J. H. (1934) A technique for the determination of lactose in milk by the Hagedorn-Jensen method. *J. Dairy Res.* **5**, 245–249
51. Howat, G. R. & Wright, N. C. (1934) The heat-coagulation of caseinogen. I. The rôle of phosphorus cleavage. *Biochem. J.* **28**, 1336–1345

Institute reprint no.

52. Blackwood, J. H. (1935) The chemical constitution of sex hormones. *Glas. Med. J.* **124**, 113–125
53. Blackwood, J. H. & Wishart, G. M. (1936) The effect of salivary activity on the composition of bovine blood. *J. Physiol.* (*Lond.*) **86**, 37–45
54. Morris, S. & Wright, N. C. (1935) The nutritive value of proteins for maintenance. *J. Dairy Res.* **6**, 289–302
55. Edwards, S. J. (1936) Mastitis in dairy cows. *Scott. J. Agric.* **19**, 28–35
56. Nichols, A. A. (1936) The bacteriology of canned milk products. *J. Soc. Chem. Ind.* (*Lond.*) **55**, 78T–80T
57. Edwards, S. J. (1936) Mastitis in dairy cattle. *Vet. Rec.* **16**, 275–281
58. Edwards, S. J. (1936) The eradication of bovine tuberculosis. Further results. *Vet. Rec.* **16**, 561–566
59. Morris, S., Wright, N. C. & Fowler, A. B. (1936) The nutritive value of proteins for milk production. IV. A comparison of the proteins of (a) spring and autumn grass, (b) grass conserved as silage (A.I.V. acid treated, molasses treated and ordinary untreated) and (c) grass conserved by drying, with notes on (i) the effect of heat treatment on the nutritive value, and (ii) the supplementary relations of food proteins. *J. Dairy Res.* **7**, 98–121
60. Howat, G. R. & Wright, N. C. (1963) The heat-coagulation of caseinogen. II. The rate of phosphorus cleavage. *Biochem. J.* **30**, 1413–1415
61. Wright, N. C. (1936) The action of hypochlorites on amino-acids and proteins. The effect of acidity and alkalinity. *Biochem. J.* **30**, 1661–1667
62. Hutchinson, J. C. D. & Morris, S. (1936) The digestibility of dietary protein in the ruminant. I. Endogenous nitrogen excretion on a low nitrogen diet and in starvation. *Biochem. J.* **30**, 1682–1694
63. Hutchinson, J. C. D. & Morris, S. (1936) The digestibility of dietary protein in the ruminant. II. The digestibility of protein following a prolonged fast, with a detailed study of the nitrogen metabolism. *Biochem. J.* **30**, 1695–1704
64. Blackwood, J. H., Morris, S. & Wright, N. C. (1936) The nutritive value of raw and pasteurized milk for calves. The assimilation and retention of nitrogen, phosphorus and calcium. *J. Dairy Res.* **7**, 228–237
65. Aylward, F. X. & Blackwood, J. H. (1936) Fasting and realimentation in the ruminant. I. The effect of food and fasting on certain blood constituents. II. Calcium and phosphorus metabolism during fasting and during realimentation followed by fasting. *Biochem. J.* **30**, 1819–1832
66. Jackson, C. J., Howat, G. R. & Hoar, T. P. (1936) Discoloration and corrosion in canned cream. *J. Dairy. Res.* **7**, 284–290
67. Nichols, A. A. & Edwards, S. J. (1936) The comparative values of the plate count and the modified methylene-blue reduction test as routine methods for grading milk. *J. Dairy Res.* **7**, 258–270
68. Aylward, F. X., Blackwood, J. H. & Smith, J. A. B. (1937) Lipaemia and milk fat secretion in the ruminant. *Biochem. J.* **31**, 130–137
69. Wright, N. C. (1935) Methods of reducing infection in the milk supply. *Trans. R. Med. Chir. Soc. Glas.* **29**, 30–41
70. Wright, N. C. (1936) A century's progress in animal nutrition, in *Ayrshire Agricultural Assoc. Centenary Handbook*, pp. 73–78, Ayrshire Agricultural Association, Ayr
71. Wilkie, J., Edwards, S. J., Fowler, A. B. & Wright, N. C. (1937) The relative values of raw and pasteurized milk in the feeding of calves. *J. Dairy Res.* **8**, 311–323

Institute reprint no.

72. Nichols, A. A., Howat, G. R. & Jackson, C. J. (1937) Bitterness and thinning in canned cream. *J. Dairy Res.* **8**, 331–346
73. Jackson, C. J., Howat, G. R. & Hoar, T. P. (1937) Discoloration and corrosion in canned cream. Part II. *J. Dairy Res.* **8**, 324–330
74. Fowler, A. B. (1937) Grass drying. *Scott. Farm.* **45**, 1566
75. Nichols, A. A. (1938) Faults in canned milk products. Bitterness and thinning in canned cream. *Milk Trade Gaz.* **10**(195), 4
76. Edwards, S. J. (1937) The pathogenesis of tuberculosis in calves. *J. Comp. Pathol. Ther.* **50**, 377–382
77. Jackson, C. J. (1938) Faults in canned milk products. Discoloration and corrosion in canned cream. *Milk Trade Gaz.* **10**(201), 6
78. Wright, N. C. (1938) The importance of home-produced feeding stuffs. *Trans. High. Agric. Soc. Scotl.* **50**, 1–37
79. Howat, G. R. (1938) Faults in canned products. Some common defects in canned cream. *Milk Trade Gaz.* **10**(203), 6
80. Fowler, A. B. (1938) The construction and equipment of cattle byres. III. The use of rubber and rubber plastics for byre flooring. *Scott. J. Agric.* **21**, 153–157
81. Smith J. A. B., Howat, G. R. & Ray, S. C. (1938) The composition of the blood and milk of lactating cows during inanition with a note on an unidentified constituent present in certain samples of abnormal milk. *J. Dairy Res.* **9**, 310–322
82. Smith, J. A. B. (1938) Studies in the secretion of milk fat. I. The effect of inanition on the blood lipoids of the lactating cow. *Biochem. J.* **32**, 1856–1867
83. Smith, J. A. B. & Dastur, N. N. (1938) Studies in the secretion of milk fat. II. The effect of inanition on the yield and composition of milk fat. *Biochem. J.* **32**, 1868–1876
84. Edwards, S. J. (1938) The diagnosis of streptococcus mastitis by cultural methods. *J. Comp. Pathol. Ther.* **51**, 250–263
85. Smith, J. A. B. (1939) Fat metabolism in the animal body. *Chem. Ind.* (*Lond.*) **58**, 213–220
86. Wright, N. C. & Morris, S. (1939) Feeding standards for farm animals. I. The dry matter requirements. *J. Min. Agric. Fish.* **46**, 21–27
87. Wright, N. C. & Morris, S. (1939) Feeding standards for farm anaimals. II. The energy requirements. *J. Min. Agric. Fish.* **46**, 155–162
88. Strain, W. B. (1939) The agricultural value of screened dust. *Proc. Ann. Conf. Inst. Public Cleansing, Scarborough*
89. Wright, N. C. (1939) Feeding standards for farm animals. III. The protein requirements. *J. Min. Agric. Fish.* **46**, 251–257
90. Morris, S. & Ray, S. C. (1939) The nutritive value of proteins for milk production. V. The effect of high temperature and of season on the nutritive value of grass proteins, the supplementary effect of the maintenance ration on the production ration, and the effect of feeding a high-protein ration. *J. Dairy Res.* **10**, 165–185
91. Nicols, A. A. (1939) Bacteriological studies of canned milk products. *J. Dairy Res.* **10**, 231–249
92. Nichols, A. A. (1939) Bacteriological studies of spray-dried milk powder. *J. Dairy Res.* **10**, 202–230
93. Smith, J. A. B. (1939) The rapid determination of peroxide values for the fat in milk powders. *J. Dairy Res.* **10**, 294–299
94. McNair, J. & Watts, P. S. (1939) A pictorial record of orchitis due to *Br. abortus* in a bull. *Vet. Rec.* **51**, 758–760

Institute reprint no.

95. Wright, N. C. (1939) Feeding standards for farm animals. IV. The quality of protein. *J. Min. Agric. Fish.* **46**, 330–336
96. Wright, N. C. (1939) Feeding standards for farm animals. V. The mineral requirements. *J. Min. Agric. Fish.* **46**, 473–479
97. Morris, S. & Ray, S. C. (1939) The effect of a phosphorus deficiency on the protein and mineral metabolism of sheep. *Biochem. J.* **33**, 1209–1216
98. Morris, S. & Ray, S. C. (1939) The fasting metabolism of ruminants. *Biochem. J.* **33**, 1217–1230
99. Nichols, A. A. & Jackson, C. J. (1938) A comparison between the methylene blue reduction test and the plate count with changing conditions of milk production and with changing conditions of storage of the milk samples before testing. *Abstr. Proc. Soc. Agric. Bacteriol.* 22–25
100. Wright, N. C. & Smith J. A. B. (1939) Feeding standards for farm animals. VI. The vitamin requirements. *J. Min. Agric. Fish.* **46**, 559–565
101. Howat, G. R., Smith, J. A. B., Waite, R. & Wright, N. C. (1939) Factors affecting the solubility of milk powers. IV. The influence of speed and duration of stirring on solubility with a description of a rapid method for solubility determinations. *J. Dairy Res.* **10**, 498–514
102. Ray, S. C. (1939) Studies of the physico-chemical state of calcium in the serum of the ruminant. *Biochem. J.* **33**, 1599–1605
103. Smith, J. A. B. (1940) Chemistry of acyclic constituents of natural fats and oils. *Annu. Rev. Biochem.* **9**, 93–114
104. Wright, N. C. (1940) Britain's supplies of feeding-stuffs. *Emp. J. Exper. Agric.* **8**, 231
105. Smith, J. A. B. & Dastur, N. N. (1940) Studies on the secretion of milk fat. III. The effect of thyroxine administration on the blood lipoids and on the nature of the milk fat. *Biochem. J.* **34**, 1093–1107
106. Wright, N. C. (1940) The supply of animal feeding-stuffs. *Nature* (*Lond.*) **146**, 251–253
107. Wright, N. C. (1940) Britain's feeding-stuffs gap: can it be closed? *Agric. Prog.* **17**, 62–74
108. Nichols, A. A. (1940) The effect of variations in the fat percentage and in the reaction (pH) of milk media on the heat resistance of certain milk bacteria. *J. Dairy Res.* **11**, 274–291
109. Waite, R. (1940) A laboratory spray-drying machine. *Chem. Ind.* (*Lond.*) **59**, 659–662
110. Wright, N. C. (1941) The storage of artificially dried grass. *J. Agric. Sci.* **31**, 194–211
111. Watts, P. S. (1941) Contagious streptococcal mastitis of cattle. *Vet. Rec.* **53**, 61–66
112. Wright, N. C. (1941) The milk supply in time of war. *J. R. Agric. Soc. Engl.* **101**, 44–56
113. Waite, R. (1941) The drying of cheese whey and of acid casein whey by the roller process. *J. Dairy Res.* **12**, 71–77
114. Adair, W. (1941) Farm self-sufficiency via grass silage. *Farm. News* **93**, 10–11
115. Hannah Dairy Research Institute (1941) *How to Make Good Silage*, Boots, Nottingham
116. Watts, P. S. (1941) Contagious abortion provides its own control. *Farm. Week.* **14**, 29
117. Wright, N. C. (1941) Scottish dairying in war-time. *Scott. J. Agric.* **23**, 231–246
118. Wright, N. C. (1941) Wheatmeal bread or milk? *Chem. Ind.* (*Lond.*) **60**, 623–627

Institute reprint no.

119. Waite, R. (1941) The keeping quality of milk powders. Part I. Addition of antioxidants. *J. Dairy Res.* **12**, 178–183
120. Higginbottom, C. (1941) The examination of raw milk by the methylene-blue and resazurin tests and by the plate count. *Abstr. Proc. Soc. Agric. Bacteriol.*, 32
121. Watts, P. S. (1942) The milking machine as a factor in the incidence of mastitis as revealed by the plating method. *Vet. Rec.* **54**, 181–184
122. Higginbottom, C. (1944) Bacteriological studies of roller-dried milk powders, roller-dried buttermilk and of roller- and spray-dried whey. *J. Dairy Res.* **13**, 308–323
123. Wright, N. C. & Owen, E. C. (1941) Foods. *Rep. Prog. Appl. Chem.* **26**, 415–432
124. Wright, N. C. (1942) What the loss of milling offal means. *Farm. Stock Breed.* **56**, 547
125. Wright, N. C. (1942) Digestibility of national wheatmeal. *Lancet* **2**, 165–166
126. Watts, P. S. (1942) Field observations on the control of contagious mastitis. *Abstr. Proc. Soc. Agric. Bacteriol.*, 35
127. Watts, P. S. (1943) The practitioner and mastitis control under the national scheme. *Vet. Rec.* **55**, 175–179
128. Owen, E. C., Smith, J. A. B. & Wright, N. C. (1943) Urea as a partial protein substitute in the feeding of dairy cattle. *Biochem. J.* **37**, 45–53
129. Stableforth, A. W., Hignett, S. L., Watts, P. S., Paterson, J. D., Cunningham, A., Edwards, S. J., Malcolm, J. & Roach, R. W. (1942) Sulphanilamide in the treatment of bovine mastitis due to *Str. agalactiae*, with coincident observations on the value of irrigation with euflavine in lactose solution. *Vet. Rec.* **54**, 539–542
130. Wright, N. C., Fowler, A. B. & Thom, C. L. (1943) The relation of farm self-sufficiency to the attainment of milk targets. *Trans. Highl. Agric. Soc. Scotl.* **55**, 54–69
131. Lea, C. H., Moran, T. & Smith, J. A. B. (1943) The gas-packing and storage of milk powder. Parts I-V. *J. Dairy Res.* **13**, 162–215
132. Wright, N. C. (1944) The rival claims of animals and man for food. *Proc. Nutr. Soc.* **1**, 23–31
133. Wright, N. C. (1944) The potential contribution of milk, meat and eggs to the British diet. *Proc. Nutr. Soc.* **1**, 67–75
134. Pearson, R. M. & Smith, J. A. B. (1943) The utilisation of urea in the bovine rumen. I. Methods of analysis and preliminary experiments *in vivo*. *Biochem. J.* **37**, 142–148
135. Pearson, R. M. & Smith, J. A. B. (1943) The utilisation of urea in the bovine rumen. 2. The conversion of urea to ammonia. *Biochem. J.* **37**, 148–153
136. Pearson, R. M. & Smith, J. A. B. (1943) The utilisation of urea in the bovine rumen. 3. The synthesis and breakdown of protein in rumen ingesta. *Biochem. J.* **37**, 153–164
137. Baker, F. (1943) Direct microscopical observations upon the rumen population of the ox. Part I. Qualitative characteristics of the rumen population. *Ann. App. Biol.* **30**, 230–239
138. Stirling, A. C. & Watts, P. S. (1943) Notes on the bacteriological diagnosis of contagious mastitis. *Abstr. Proc. Soc. Agric. Bacteriol.*, 67–68
139. Wright, N. C. (1943) Trends and problems in the dairy industry–Research. *Dairy Ind.* **8**, 346–350
140. Smith, J. A. B. (1944) The effect of diet on lactation in animals. *Proc. Nutr. Soc.* **2**, 46–50
141. Owen, E. C. (1944) The excretion of borate by the dairy cow. *J. Dairy Res.* **13**, 243–248

Institute reprint no.

142. Higginbottom, C. (1944) The effect of storage on the plate counts of milk and whey powders. *J. Dairy Res.* **13**, 324–328
143. Watts, P. S. & Stirling, A. C. (1944) Mastitis and the resazurin test. *Vet. Rec.* **56**, 83–85
144. Watts, P. S. (1944) Mastitis and machines. *Dairyman* **1**, 13–15
145. Snow, D., Crichton, M. H. G. & Wright, N. C. (1944) Mould deterioration of feeding-stuffs in relation to humidity of storage. Part I. The growth of moulds at low humidities. *Ann. Appl. Biol.* **31**, 102–110.
146. Snow, D., Crichton, M. H. G. & Wright, N. C. (1944) Mould deterioration of feeding-stuffs in relation to humidity of storage. Part II. The water uptake of feeding-stuffs at different humidities. *Ann. Appl. Biol.* **31**, 111–116
147. Snow, D. (1944) Mould deterioration of stored feeding stuffs. *Fert. Feed. Stuffs* **30**, 407–413
148. McFarlane, D., Garside, J. S., Watts, P. S. & Stamp, J. T. (1944) An outbreak of bovine tuberculosis due to udder irrigation. *Vet. Rec.* **56**, 369–371
149. Mattick, A. T. R., Hiscox, E. R., Crossley, E. L., Thompson, S. Y., Kon, S. K., Findlay, J. D., Lea, C. H. & Smith, J. A. B. (1945) The effect of temperature of pre-heating, of clarification and of bacteriological quality of the raw milk on the keeping properties and on the residual bacterial flora of whole milk powder dried by the Kestner spray process. *J. Dairy Res.* **14**, 116–159
150. Findlay, J. D., Smith, J. A. B. & Lea, C. H. (1945) Experiments on the use of antioxidants in spray-dried whole milk powder. *J. Dairy Res.* **14**, 165–175
151. Smith, J. A. B. & Baker, F. (1945) The utilisation of urea in the bovine rumen. 4. The isolation of the synthesised material and correlation between protein synthesis and microbial activities. *Biochem. J.* **38**, 496–506
152. Snow, D., Smith, J. A. B. & Wright, N. C. (1945) The manufacture and storage of cereal feeding-stuffs with incorporated non-protein nitrogen compounds. *J. Agric. Sci.* **35**, 65–71
153. Owen, E. C., Snow, D. & Thom, C. L. (1945) The effect of borax dressing on the growth and yield of field beans (*Vicia faba*). *J. Agric. Sci.* **35**, 119–122
154. Snow, D. (1945) Mould deterioration of feeding-stuffs in relation to humidity of storage. Part III. The isolation of mould species from feeding-stuffs stored at different humidities. *Ann. Appl. Biol.* **32**, 40–44
155. Snow, D. & Watts, P. S. (1945) The effect of sulphanilamide and other bacteriostatic drugs on the growth of moulds. *Ann. Appl. Biol.* **32**, 102–112
156. Higginbottom, C. (1945) The technique of the bateriological examination of dried milks. *J. Dairy Res.* **14**, 184–194
157. Snow, D. & Wright, N. C. (1945) The loss of dry matter from stored bran. *J. Agric. Sci.* **35**, 126–132
158. Watts, P. S. (1945) The effect of the relative humidity on the survival of dried cultures of *Streptococcus Agalactiae*. *J. Pathol. Bacteriol.* **57**, 191–197
159. Owen, E. C. (1946) The determination of boron. *Analyst* (*Lond.*) **71**, 210–217
160. Watts, P. S. & Paterson, J. D. (1944) An attempt to control 'White Scour' with sulphaguanidine. *Vet. Rec.* **56**, 498
161. Snow, D. & Watts, P. S. (1945) Storage of oilcakes. Experiments with sulphonamide E.O.S. and propamidine as preservatives. *Food* (*Lond.*) **14**, 16–18
162. Smith, J. A. B. (1945) The formation of protein. *Proc. Nutr. Soc.* **3**, 203–216
163. McLeod, D. H. & Watts, P. S. (1945) The modified Hotis test for diagnosis of contagious bovine mastitis. *Abstr. Proc. Soc. Appl. Bacteriol.*, 37–40
164. Holmes, W. (1946) Prickly comfrey. *Agriculture* (*Lond.*) **52**, 515–516
165. Findlay, J. D., Higginbottom, C., Smith, J. A. B. & Lea, C. H. (1946) The effect

Institute reprint no.

of the pre-heating temperature on the bacterial count and storage life of whole milk powder spray-dried by the Krause process. *J. Dairy Res.* **14**, 378–399

166. Watts, P. S. & McLeod, D. H. (1946) The estimation of penicillin in blood serum and milk of bovines after intramuscular injection. *J. Comp. Pathol. Ther.* **56**, 170–179
167. Holmes, W. (1946) Self-sufficiency on the dairy farm. *Farming (Norwich)* **1**, 107–113
168. Henry, K. M., Kon, S. K., Lea, C. H., Smith, J. A. B. & White, J. C. D. (1946) Deterioration on storage of dried skim milk. *Nature (Lond.)* **158**, 348
169. Fowler, A. B. (1946) Cropping for milk. *Agriculture (Lond.)* **53**, 287–295
170. Owen, E. C. (1946) Digestion in the ruminant. *Proc. R. Soc. Med.* **39**, 804–805
171. Wright, N. C. (1947) Milk in the tropics and sub-tropics. *Dairyman* **64**, 101–107
172. Wright, N. C. (1946) Relation of animal husbandry to human nutritional needs. *Proc. Nutr. Soc*, **5**, 24–28
173. Watts, P. S. (1947) The importance of cowshed hygiene in the transmission of milk-borne disease. *J. R. Sanit. Inst.* **67**, 458–463
174. Owen, E. C. (1947) Nutrition and symbiosis. *Nature (Lond.)* **160**, 78–81
175. White, J. C. D., Smith, J. A. B. & Lea, C. H. (1947) The effect of a high pre-heating temperature with and without ethyl gallate on the storage life of whole milk powder spray-dried on a Gray-Jensen drier. *J. Dairy Res.* **15**, 127–139
176. Higginbottom, C. (1948) Dye reduction tests in the bacteriological examination of dried milks. *J. Dairy Res.* **15**, 280–284
177. Higginbottom, C. (1948) The bacteriological quality of British spray-dried milk powders. *J. Dairy Res.* **15**, 277–279
178. Higginbottom, C. (1948) Bacterial growth in reconstituted spray-dried milk. *J. Dairy Res.* **15**, 285–291
179. Malpress, F. H. & Owen, E. C. (1947) Observation on the excretion of diethylstilboestrol by the ruminant. *J. Endocrinol.* **5**, lxii-lxiii
180. Devi, P., Pontecorvo, G. & Higginbottom, C. (1947) X-ray induced mutations in dried bacteria. *Nature (Lond.)* **160**, 503
181. Henry, K. M., Kon, S. J., Lea, C. H. & White, J. C. D. (1948) Deterioration on storage of dried skim milk. *J. Dairy Res.* **15**, 292–363
182. Owen, E. C. (1947) The role of micro-organisms in the nutrition of farm animals. *Proc. Nutr. Soc.* **5**, 186–199
183. Lea, C. H. & White, J. C. D. (1947) Observations on the amino-aldehyde reaction in the deterioration of stored foods, with particular reference to milk powder. *Proc. 11th Int. Congr. Pure Appl. Chem.*
184. Wright, N. C. (1947) Research and the farmer. The dairy industry. *Scott. Agric.* **26**, 220–224
185. Watts, P. S., McLeod, D. H. & Wheater, D. W. F. (1948) Observations on the temperature compensated resazurin test under winter conditions. *Dairy Ind.* **13**, 31–37
186. Watts, P. S. (1947) The dosage of penicillin in *Streptococcus agalactiae* mastitis and some factors affecting the results. *Vet. Rec.* **59**, 619–621
187. Fraser, H. & Fowler, A. B. (1947) Re-seeding and management of marginal land in south Ayrshire. *Scott. Agric.* **27**, 108–115
188. Waite, R. (1947) The effect of metals in milk and milk products. *J. Soc. Dairy Technol.* **1**, 8 11
189. Watts, P. S. (1948) Metal plug for permanent rumen fistula in bovines. *Vet. Rec.* **60**, 4

Institute reprint no.

190. Holmes, W. (1947) General methods of fodder conservation. *Agric. Prog.* **22,** 79–86
191. Holmes, W. (1948) Protein from dried green crops. *Farming* (*Norwich*) **2,** 170–173
192. Findlay, J. D. & Yang, S. H. (1948) Capillary distribution in cow skin. *Nature* (*Lond.*) **161,** 1012–1013
193. Watts, P. S. & McLeod, D. H. (1948) The concentration of penicillin in the milk after intramammary infusion. *J. Comp. Pathol. Ther.* **58,** 232–243
194. Owen, E. C. (1948) The effect of thyroxine on the metabolism of lactating cows. I. General results and nitrogen metabolism. *Biochem. J.* **43,** 235–243
195. Owen, E. C. (1948) The effect of thyroxine on the metabolism of lactating cows. II. Calcium and phosphorus metabolism. *Biochem. J.* **43,** 243–247
196. Aziz-Ud-Din, I. M. (1948) The visual lesions in certain pathological conditions of the genital tract in cows. *Vet. Rec.* **60,** 345–347
197. Smith, J. A. B. (1948) Nutritive value of fats, with special reference to butter and margarine. *Br. J. Nutr.* **2,** 190–201
198. Smith, J. A. B. (1948) The Hannah Dairy Research Institute. *Farming* (*Norwich*) **2,** 249–252
199. Higginbottom, C. (1948) Platform rejection tests in south-west Scotland. *Dairy Ind.* **13,** 1187–1196
200. Blaxter, K. L. (1948) Do this and get a lot more milk. *Dairy Farm.* (*Ipswich*) **21**(10)
201. Findlay, J. D. (1948) Experiments on the ventilation of cattle byres. *J. Agric. Sci.* **38,** 411–424
202. Holmes, W. (1948) The intensive production of herbage for crop-drying. Part I. A study of the productivity of two annual crops and two leys and of their responses in yield and chemical composition to applications of nitrogenous manure. *J. Agric. Sci.* **38,** 425–436
203. Higginbottom, C. (1947) The microflora of dried milks. *Proc. 4th Int. Congr. Microbiol.*, 516–517
204. Holmes, W. (1949) Doubling our grassland yields. *Scott. Agric.* **28,** 205–212
205. Waite, R. & Sastry, K. N. S. (1949) The carotene content of dried grass. *J. Agric. Sci.* **39,** 174–182
206. Murray Smillie, D., Hail, W., Hainsworth, J., Wyllie, A., Higginbottom, C. & Wheater, D. W. F. (1949) Hypochlorite sterilization trials in south-west Scotland. *Dairy Ind.* **14,** 479–485
207. Waite, R. & Sastry, K. N. S. (1949) The composition of timothy (*Phleum pratense*) and some other grasses during seasonal growth. *Emp. J. Exp. Agric.* **17,** 179–187
208. Devi, D. & Higginbottom, C. (1949) Quantitative studies on mutants of *Bacterium aerogenes* requiring nicotinic acid and histidine. *Biochem. J.* **45,** 391–396
209. Waite, R. (1949) The relation between moisture content and moulding in cured hay. *Ann. Appl. Biol.* **36,** 496–503
210. McNaught, M. L., Smith, J. A. B., Henry, K. M. & Kon, S. K. (1950) The utilization of non-protein nitrogen in the bovine rumen. 5. The isolation and nutritive value of a preparation of dried rumen bacteria. *Biochem. J.* **46,** 32–36
211. McNaught, M. L., Owen, E. C. & Smith, J. A. B. (1950) The utilization of non-protein nitrogen in the bovine rumen. 6. The effect of metals on the activity of the rumen bacteria. *Biochem. J.* **46,** 36–43
212. Holmes, W. & Waite, R. (1949) The electric fence. Its use for grazing dairy cattle. *Scott. Agric.* **29,** 135–140
213. Holmes, W. (1950) Ten years without coupons. *Farm. Stock-Breed.* 17 Jan., p. 174
214. Waite, R. & White, J. C. D. (1949) The composition of the soluble and insoluble portions of reconstituted milk powders. *J. Dairy Res.* **16,** 379–389

Institute reprint no.

215. Watts, P. S. & Robertson, J. (1950) The influence of the attested herd scheme on the local incidence and degree of tuberculosis in slaughtered cows and calves. *Vet. Rec.* **62**, 127–131
216. Higginbottom, C. & Wheater, D. W. F. (1950) Changes in milk temperature from farm to creamery. *Scott. Agric.* **29**, 186–188
217. Findlay, J. D., Goodall, A. M. & Yang, S. H. (1950) The number of sweat glands in the helix of the cow's ear and the milk yield. *J. Dairy Res.* **17**, 23–36
218. Holmes, W. (1949) The intensive production of herbage for crop-drying. Part II. A study of the effect of massive dressings of nitrogenous fertilizer and of the time of their application on the yield, chemical and botanical composition of two grass leys. *J. Agric. Sci.* **39**, 128–141
219. Smith, J. A. B. (1950) Dairying problems. *Scott. Agric.* **30**, 43–47
220. Smith, J. A. B. (1950) The food value of milk and milk products. *J. Soc. Dairy Technol.* **3**, 254–263
221. Findlay, J. D. & Yang, S. H. (1950) The sweat glands in Ayrshire cattle. *J. Agric. Sci.* **40**, 126–133
222. Sikka, L. A. (1950) A study of lactation as affected by heredity and environment. *J. Dairy Res.* **17**, 231–252
223. Watts, P. S. (1950) Studies on vitamins A and C in bovines. Vitamin C in the liver, kidney and plasma of cows, calves and foetuses. *J. Comp. Pathol. Ther.* **60**, 283–293
224. Holmes, W., Waite, R., Fergusson, D. L. & Campbell, J. I. (1950) Studies in grazing management. I. A comparison of the production obtained from close-folding and rotational grazing of dairy cows. *J. Agric. Sci.* **40**, 381–391
225. Waite, R., Holmes, W., Campbell, J. I. & Ferguson, D. L. (1950) Studies in grazing management. II. The amount and chemical composition of herbage eaten by dairy cattle under close-folding and rotational method of grazing. *J. Agric. Sci.* **40**, 392–402
226. Holmes, W. (1951) Here's a year's record of a famous herd fed only on grass. *Dairy Farm.* (*Ipswich*), Jan.
227. Blaxter, K. L. (1950) Iodinated protein and lactation. *Agriculture* (*Lond.*) **57**, 413–417
228. Watts, P. S. (1950) Mastitis in Ayrshire from 1938–1950. *Vet. Rec.* **63**, 32–39
229. Beakley, W. R. (1950) The design of thermistor thermometers with linear calibration. *J. Sci. Instrum.* **28**, 176–179
230. Holmes, W. (1950) The intensive production of herbage for crop-drying. III. The effect of continued application of nitrogenous fertilizers to grassland. *J. Agric. Sci.* **41**, 64–69
231. Mahadevan, P. (1950) The effect of environment and heredity on lactation. I. Milk yield. *J. Agric. Sci.* **41**, 80–88
232. Mahadevan, P. (1950) The effect of environment and heredity on lactation. II. Persistency of lactation. *J. Agric. Sci.* **41**, 89–93
233. Mahadevan, P. (1950) The effect of environment and heredity on lactation. III. Butterfat percentage. *J. Agric. Sci.* **41**, 94–97
234. Holmes, W. (1950) The intensive production of herbage for crop-drying. IV. The effect of massive applications of nitrogen with and without phosphate and potash on the yield of grassland herbage. *J. Agric. Sci.* **41**, 70–79
235. Blaxter, K. L. & Wood, W. A. (1950) The nutrition of the young Ayrshire calf. 1. The endogenous nitrogen and basal energy metabolism of the calf. *Br. J. Nutr.* **5**, 11–25
236. Blaxter, K. L. & Howells, A. (1950) The nutrition of the young Ayrshire calf. 2.

Institute reprint no.

A spirometer for the determination of the respiratory exchange of the calf. *Br. J. Nutr.* **5**, 25–29

237. Blaxter, K. L. & Wood, W. A. (1950) The nutrition of the young Ayrshire calf. 3. The metabolism of the calf during starvation and subsequent realimentation. *Br. J. Nutr.* **5**, 29–55
238. Blaxter, K. L. & Wood, W. A. (1950) The nutrition of the young Ayrshire calf. 4. Some factors affecting the biological value of protein determined by nitrogen-balance methods. *Br. J. Nutr.* **5**, 55–67
239. Chanda, R., Owen, E. C. & Cramond, B. (1950) The composition of human milk, with special reference to the relation between phosphatase and to the partition of certain vitamins. *Br. J. Nutr.* **5**, 228–242
240. Devi, P., Pontecorvo, G. & Higginbottom, C. (1951) Mutations affecting the nutritional requirements of *Aerobacter aerogenes* induced by irradiation of dried cells. *J. Gen. Microbiol.* **5**, 781–787
241. Higginbottom, C. (1951) The can immersion and surface-cooling of milk on the farm. *Dairy Ind.* **16**, 241–249
242. McNaught, M. L. (1951) The utilization of non-protein nitrogen in the bovine rumen. 7. A qualitative and quantitative study of the breakdown of carbohydrate which accompanies protein formation in bovine rumen contents during *in vitro* incubations. *Biochem. J.* **49**, 325–332
243. Waite, R., Macdonald, W. B. & Holmes, W. (1951) Studies in grazing management. III. The behaviour of dairy cows grazing under the close-folding and rotational systems of management. *J. Agric. Sci.* **41**, 163–173
244. Chanda, R., Clapham, H. M., McNaught, M. L. & Owen, E. C. (1951) The use of chromium sesquioxide to measure the digestibility of carotene by goats and cows. *J. Agric. Sci.* **41**, 179–186
245. Blaxter, K. L. (1950) The protein and energy nutrition of the young calf. *Agric. Prog.* **25**, 1–9
246. Brown, F. & Blaxter, K. L. (1951) Separation of the tocopherols by paper chromatography. *Chem. Ind.* (*Lond.*) **70**, 633–634
247. Blaxter, K. L. (1951) Conversion factors for vegetable and animal foods for human consumption. *Br. J. Nutr.* **5**, 250–255
248. Chanda, R. & Owen, E. C. (1951) The effect of thyroxine and thiouracil on the composition of milk. 1. The partition of phosphorus in cow milk in relation to phosphatase and other constituents. *Biochem. J.* **50**, 100–109
249. Chanda, R., Clapham, H. M., McNaught, M. L. & Owen, E. C. (1951) The digestibility of carotene by the cow and the goat as affected by thyroxine and thiouracil. *Biochem. J.* **50**, 95–99
250. Blaxter, K. L. (1951) The importance of fat-soluble vitamins for calves. *J. R. Agric. Soc. Eng.* **112**, 25–35
251. Holmes, W. (1951) The feeding value of grass and grassland products. *Proc. Br. Soc. Anim. Prod.*, 90–112
252. Blaxter, K. L. (1952) The utilization of the minerals, vitamins and other constituents of grass. *Br. J. Nutr.* **6**, 110–117
253. Blaxter, K. L. & Wood, W. A. (1952) The nutrition of the young Ayrshire calf. 5. The nutritive value of cow's whole milk. *Br. J. Nutr.* **6**, 1–12
254. Blaxter, K. L. (1952) The nutrition of the young Ayrshire calf. 6. The utilization of the energy of whole milk. *Br. J. Nutr.* **6**, 13–20
255. Blaxter, K. L. & Wood, W. A. (1952) The nutrition of the young Ayrshire calf. 7. The biological value of gelatin and of casein when given as the sole source of protein. *Br. J. Nutr.* **6**, 56–71

Institute reprint no.

256. Yang, S. H. (1952) Histochemical studies of bovine sweat glands. *J. Agric. Sci.* **42,** 155–158
257. Goodall, A. M. & Yang, S. H. (1952) Myoepithelial cells in bovine sweat glands. *J. Agric. Sci.* **42,** 159–161
258. Blaxter, K. L., Watts, P. S. & Wood, W. A. (1952) The nutrition of the young Ayrshire calf. 8. Muscular dystrophy in the growing calf. *Br. J. Nutr.* **6,** 125–144
259. Blaxter, K. L. & Wood, W. A. (1952) The nutrition of the young Ayrshire calf. 9. The composition of the tissues of normal and dystrophic calves. *Br. J. Nutr.* **6,** 144–163
260. MacDonald, A. M., Blaxter, K. L., Watts, P. S. & Wood, W. A. (1952) The nutrition of the young Ayrshire calf. 10. Histopathology of muscular dystrophy and its relation to muscle chemistry. *Br. J. Nutr.* **6,** 164–169
261. Blaxter, K. L., Wood, W. A. & MacDonald, A. M. (1953) The nutrition of the young Ayrshire calf. 11. The toxicity of cod-liver oil. *Br. J. Nutr.* **7,** 34–50
262. Blaxter, K. L., Brown, F. & MacDonald, A. M. (1953) The nutrition of the young Ayrshire calf. 12. Factors affecting the tocopherol reserves, muscle composition and muscle histology of 4-day-old calves. *Br. J. Nutr.* **7,** 105–123
263. Blaxter, K. L., Brown, F. & MacDonald, A. M. (1953) The nutrition of the young Ayrshire calf. 13. The toxicity of the usaturated acids of cod-liver oil. *Br. J. Nutr.* **7,** 287–298
264. Brown, F. (1952) The estimation of vitamin E. 1. Separation of tocopherol mixtures occurring in natural products by paper chromatography. *Biochem. J.* **51,** 237–239
265. White, J. C. D. (1953) Composition of whales' milk. *Nature (Lond.)* **171,** 612
266. Chanda, R. & Owen, E. C. (1952) The effect of thyroxine and thiouracil on the composition of milk. 2. The carotene and vitamin A content. *Biochem. J.* **51,** 404–417
267. Chanda, R., McNaught, M. L. & Owen, E. C. (1952) The effect of thyroxine and thiouracil on the composition of milk. 3. The concentration of the major constituents and of some of the water-soluble vitamins in cow's milk. *Biochem. J.* **51,** 543–552
268. Holmes, W., Waite, R., Fergusson, D. L. & MacLusky, D. S. (1952) Studies in grazing management. IV. A comparison of close-folding and rotational grazing of dairy cows on intensively fertilized pasture. *J. Agric. Sci.* **42,** 304–313
269. Waite, R., Holmes, W. & Boyd, J. (1952) Studies in grazing management. V. Further determinations of the nutrient intake of dairy cows under close-folding and rotational methods of grazing. *J. Agric. Sci.* **42,** 314–321
270. Blaxter, K. L. & Rook, J. A. F. (1952) The heat of combustion of the tissues of cattle in relation to their chemical composition. *Br. J. Nutr.* **7,** 83–91
271. Findlay, J. D. (1953) The effect of climate on cattle and their productivity. *N.A.A.S. Q. Rev.* **19,** 285–290
272. Chanda, R. & Owen, E. C. (1952) The partition of phosphorus and aneurin in relation to phosphatase in the colostrum and milk of the cow. *J. Agric. Sci.* **42,** 403–412
273. Yang, S. H. (1952) A method of assessing cutaneous pigmentation in bovine skin. *J. Agric. Sci.* **42,** 465–467
274. Higginbottom, C. (1952) Chilled water surface coolers for cooling milk on the farm. *Dairy Ind.* **17,** 327–332
275. Holmes, W. (1952) Self-sufficiency on the dairy farm – some economic aspects. *Scott. Agric.* **32,** 83–89

Institute reprint no.

276. Chanda, R. (1953) The partition of carotenoids and vitamin A in the colostrum and milk of the cow and the goat. *J. Agric. Sci.* **43**, 54–71
277. White, J. C. D., Waite, R., Hawley, H. B., Clark, J. G. & Henry, K. M. (1952) The effect of preheating milk to 230°F. (110°C.) on the keeping quality of spray-dried whole milk. *J. Dairy Res.* **19**, 339–347
278. Brown, F. (1952) The estimation of vitamin E. 2. Quantitative analysis of tocopherol mixtures by paper chromatography. *Biochem. J.* **52**, 523–526
279. Owen, E. C. (1952) Bone as a mineral reserve. *Br. J. Nutr.* **6**, 415–423
280. Higginbottom, C. (1953) Cooling milk on the farm. *Scott. Agric.* **32**, 192–195
281. Waite, R., Fensom, A. & Lovett, S. (1953) The amino-acid composition of extracted grass-protein. *J. Sci. Food Agric.* **4**, 28–33
282. Higginbottom, C. (1952) The effect of storage at different relative humidities on the survival of micro-organisms in milk powder and in pure cultures dried in milk. *J. Dairy Res.* **20**, 65–75
283. Chanda, R. (1953) The effect of mastitis on the carotenoids, vitamin A and phosphorus compounds of milk. *Biochem. J.* **54**, 68–77
284. Brown, F. (1953) Occurrence of δ-tocopherol in seaweed. *Chem. Ind.* (*Lond.*) 174
285. Blaxter, K. L. & Rook, J. A. F. (1953) Indirect determination of water retention in cattle as a means of estimating energy exchange. *Nature* (*Lond.*) **171**, 609–610
286. Alexander, H. & Higginbottom, C. (1953) Bacteriological studies on pasteurized milk. *J. Dairy Res.* **20**, 156–176
287. Holmes, W. & MacLusky, D. S. (1954) The intensive production of herbage for crop-drying. Part V. The effect of continued massive applications of nitrogen with and without phosphate and potash on the yield of grassland herbage. *J. Agric. Sci.* **45**, 129–140
288. Goodall, A. M. & Yang, S. H. (1954) The vascular supply of the skin of Ayrshire calves and embryos. *J. Agric. Sci.* **44**, 1–4
289. Higginbottom, C. (1953) Roll tubes and agar strip tubes for the bacterial colony count of milk. *Proc. Soc. Appl. Bacteriol.* **16**, 16–23
290. Higginbottom, C. (1952) A note on the use of dehydrated milk agar medium for bacterial counts on raw and dried milk. *Proc. Soc. Appl. Bacteriol.* **15**, 112–115
291. Waite, R. (1953) The water-soluble carbohydrates of grasses. I. Changes occurring during the normal life-cycle. *J. Sci. Food Agric.* **4**, 197–204
292. Brown, F. (1953) The tocopherol content of farm feeding-stuffs. *J. Sci. Food Agric.* **4**, 161–165
293. Waite, R. (1953) The water-soluble carbohydrates of grasses. II. Grasses cut at grazing height several times during the grazing season. *J. Sci. Food Agric.* **4**, 257–261
294. Brown, F. (1953) Occurrence of vitamin E in cod and other fish-liver oils. *Nature* (*Lond.*) **171**, 790–791
295. Holmes, W. (1953) Self-sufficiency on the dairy farm. Planning production. *Scott. Agric.* **32**, 211–216
296. Stewart, A. B. & Holmes, W. (1953) Manuring of grassland. I. Some effects of heavy dressings of nitrogen on the mineral composition of grassland herbage. *J. Sci. Food Agric.* **4**, 401–408
297. Blaxter, K. L., Brown, F., Wood, W. A. & MacDonald, A. M. (1953) The nutrition of the young Ayrshire calf. 14. Some effects of natural and synthetic anti-oxidants on the incidence of muscular dystrophy induced by cod-liver oil. *Br. J. Nutr.* **7**, 337–349
298. Blaxter, K. L. (1953) Diet and muscular disease in calves. *Proc. Br. Soc. Anim. Prod.*, 94–101

Institute reprint no.

299. McNaught, M. L., Owen, E. C., Henry, K. M. & Kon, S. K. (1954) The utilization of non-protein nitrogen in the bovine rumen. 8. The nutritive value of the proteins of preparations of dried rumen bacteria, rumen protozoa and brewer's yeast for rats. *Biochem. J.* **56**, 151–156

300. Blaxter, K. L. & Sharman, G. A. M. (1953) Prevention and cure of enzootic muscular dystrophy in beef cattle. *Nature (Lond.)* **172**, 1006–1007

301. Findlay, J. D. (1953) Farm animals and high temperatures. *Br. Agric. Bull.* **6**, 212–216

302. Blaxter, K. L. (1953) Muscular dystrophy. *Vet. Rec.* **65**, 835–838

303. Blaxter, K. L. & Wood, W. A. (1953) Some observations on the biochemical and physiological events associated with diarrhoea in calves. *Vet. Rec.* **65**, 889–892

304. Higginbottom, C. & Wheater, D. W. F. (1954) The incidence of *Streptococcus bovis* in cattle. *J. Agric. Sci.* **44**, 434–442

305. Chanda, R., Clapham, H. M. & Owen, E. C. (1954) Dietary carotene and the degree of esterification of vitamin A in the milk and blood of cows. *Biochem. J.* **56**, 453–458

306. Blaxter, K. L., Graham, N. McC. & Rook, J. A. F. (1954) Apparatus for the determination of the energy exchange of calves and of sheep. *J. Agric. Sci.* **45**, 10–18

307. Holmes, W. (1954) Modern methods of grassland management in feeding the dairy herd. *Proc. Nutr. Soc.* **13**, 19–22

308. Beakley, W. R. (1954) A method of modifying the thermal e.m.f./temperature characteristics of constantan thermocouple wire. *J. Sci. Instrum.* **31**, 259–260

309. Beakley, W. R. (1954) Maintaining optimum cycling intervals in a two-step controller. *Electron. Eng.* **26**, 99

310. Blaxter, K. L., Rook, J. A. F. & MacDonald, A. M. (1954) Experimental magnesium deficiency in calves. I. Clinical and pathological observations. *J. Comp. Pathol. Ther.* **64**, 157–175

311. Blaxter, K. L. & Rook, J. A. F. (1954) Experimental magnesium deficiency in calves. 2. The metabolism of calcium, magnesium and nitrogen and magnesium requirements. *J. Comp. Pathol. Ther.* **64**, 176–186

312. Holmes, W. (1954) High milk yields per acre from grassland. *J. Br. Grassl. Soc.* **9**, 17–27

313. Beakley, W. R. (1954) A stable d.c. source of low voltage with low internal resistance. *J. Sci. Instrum.* **31**, 219–220

314. McNaught, M. L., Smith, J. A. B. & Black, W. A. P. (1954) The utilization of carbohydrates of seaweed by rumen microflora *in vitro*. *J. Sci. Food Agric.* **5**, 350–352

315. Owen, E. C. (1954) The carotene, carotenoid and chlorophyll contents of some Scottish seaweeds. *J. Sci. Food Agric.* **5**, 449–453

316. Findlay, J. D. (1954) The climatic physiology of farm animals. *Meteorol. Monogr.* **2**, 19–29

317. Bianca, W. (1955) The effect of thermal stress on the acid-base balance of the Ayrshire calf. *J. Agric. Sci.* **45**, 428–430

318. McLean, J. A. (1954) A method for constructing direct reading thermistor thermometers. *J. Sci. Instrum.* **31**, 455–457

319. Blaxter, K. L. (1954) Problems associated with feeding of whole milk. *Proc. Br. Vet. Assoc. Conf.*, pp. 38–43

320. Beakley, W. R. & Findlay, J. D. (1955) The effect of environmental temperature and humidity on the skin temperature of Ayrshire calves. *J. Agric. Sci.* **45**, 353–364

Institute reprint no.

321. Beakley, W. R. & Findlay, J. D. (1955) The effect of environmental temperature and humidity on the ear temperatures of Ayrshire calves. *J. Agric. Sci.* **45,** 373–379
322. Beakley, W. R. & Findlay, J. D. (1955) The effect of environmental temperature and humidity on the temperature of the skin of the scrotum in Ayrshire calves. *J. Agric. Sci.* **45,** 365–372
323. Goodall, A. M. (1955) Arterio-venous anastomoses in the skin of the head and ears of the calf. *J. Anat.* **89,** 100–105
324. Beakley, W. R. & Findlay, J. D. (1955) The effect of environmental temperature and humidity on the rectal temperature of calves. *J. Agric. Sci.* **45,** 339–352
325. Beakley, W. R. & Findlay, J. D. (1955) The effect of environmental temperature and humidity on the frequency of the heart beat of Ayrshire calves. *J. Agric. Sci.* **45,** 461–468
326. Beakley, W. R. & Findlay, J. D. (1955) The effect of environmental temperature and humidity on the respiration rate of Ayrshire calves. *J. Agric. Sci.* **45,** 452–460
327. Higginbottom, C. (1955) The effect of menadione (2-methyl-1:4 naphthoquinone) on the keeping quality of milk. *J. Dairy Res.* **22,** 48–59
328. Blaxter, K. L. & Sharman, G. A. M. (1955) Hypomagnesaemic tetany in beef cattle. *Vet. Rec.* **67,** 108–115
329. Chanda, R., Clapham, H. M. & Owen, E. C. (1955) Blood and milk vitamin A alcohol and ester as affected by treatment of the cow with thyrotrophin in the presence and absence of dietary carotene. *Voeding* **16,** 778–781
330. Blaxter, K. L. & Rook, J. A. F. (1955) Energy and carbohydrate metabolism in magnesium-deficient calves. *Br. J. Nutr.* **9,** 121–132
331. Chanda, R., Clapham, H. M. & Owen, E. C. (1955) The effects of deprivation of carotene and anterior pituitary hormone on the partition of vitamin A in the blood and milk of the cow. *Biochem. J.* **60,** 391–399
332. Blaxter, K. L. & Graham, N. McC. (1955) Plane of nutrition and starch equivalents. *J. Agric. Sci.* **46,** 292–306
333. Holmes, W. & McLusky, D. S. (1955) The intensive production of herbage for crop-drying. Part VI. A study of the effect of intensive nitrogen fertilizer treatment on species and strains of grass, grown alone and with white clover. *J. Agric. Sci.* **46,** 267–286
334. Blaxter, K. L. & Graham, N. McC. (1955) Methods of assessing the energy values of foods for ruminant animals. *Proc. Nutr. Soc.* **14,** 131–139
335. Bligh, J. (1955) Comparison of rectal and deep body temperature in the calf. *Nature* (*Lond.*) **176,** 402–403
336. Crocker, W. P., Jenkins, D. I., Provan, A. L., Macdonald, F. J., Rowland, S. J. & White, J. C. D. (1955) A comparison of the Gerber and Röse Gottlieb methods for the determination of fat in milk. *J. Dairy Res.* **22,** 336–339
337. Bianca, W. (1955) A method of serial blood sampling and of injection for experimental farm animals. *Vet. Rec.* **67,** 791–792
338. McLusky, D. S. (1955) The quantities of herbage eaten by grazing dairy cows. *Proc. Br. Soc. Anim. Prod.*, 45–51
339. Waite, R. (1956) Variations in milk quality. *Scott. Agric.* **36,** 41–43
340. Manson, W. (1956) Acid-soluble nucleotides of lactating mammary gland. *Biochem. Biophys. Acta* **19,** 398–399
341. Holmes, W., Waite, R., McLusky, D. S. & Watson, J. N. (1956) Winter feeding of dairy cows. I. The influence of level and source of protein and of the level of energy in the feed on milk yield and composition. *J. Dairy Res.* **23,** 1–12

Institute reprint no.

342. Holmes, W. (1956) Winter feeding of dairy cows. II. Dried grass produced in spring and in autumn compared with other concentrates and with spring grazing. *J. Dairy Res.* **23**, 13–20
343. Waite, R., White, J. C. D. & Robertson, A. (1956) Variations in the chemical composition of milk with particular reference to the solids-not-fat. I. The effect of stage of lactation, season of year and age of cow. *J. Dairy Res.* **23**, 65–81
344. Robertson, A., Waite, R. & White, J. C. D. (1956) Variations in the chemical composition of milk with particular reference to the solids-not-fat. II. The effect of heredity. *J. Dairy Res.* **23**, 82–91
345. Blaxter, K. L. & Graham, N. McC. (1956) The effect of the grinding and cubing process on the utilization of the energy of dried grass. *J. Agric. Sci.* **47**, 207–217
346. Blaxter, K. L. & McGill, R. F. (1956) Magnesium metabolism in cattle. *Vet. Rev. Annot.* **2**, 35–55
347. Nisbet, W. (1956) An instrument for measuring respiratory rates and volumes. *J. Sci. Instrum.* **33**, 154–158
348. Waite, R. & Abbot, J. (1956) The measurement of milk solids-not-fat by density methods. *Proc. 14th Int. Dairy Congr.* **3**, pp. 699–707
349. White, J. C. D. & Davies, D. T. (1956) The relation between the rate of heat coagulation of milk and its protein and mineral composition. *Proc. 14th Int. Dairy Congr.* **1**, pp. 288–298
350. Taylor, M. M. (1956) Sterilized milk as a growth medium for spore-forming bacilli. *Proc. 14th Int. Dairy Congr.* **1**, pp. 487–494
351. Blaxter, K. L., Graham, N. McC. & Wainman, F. W. (1956) Some observations on the digestibility of food by sheep, and on related problems. *Br. J. Nutr.* **10**, 69–91
352. Garton, G. A., Duncan, W. R. H., Blaxter, K. L., McGill, R. F., Sharman, G. A. M. & Hutcheson, M. K. (1956) Muscular dystrophy of beef cattle and unsaturated fats. *Nature (Lond.)* **177**, 792–793
353. Blackburn, P. S. (1956) Antibiotic treatment of mastitis and its effect on the cell content of the milk. *J. Dairy Res.* **23**, 225–228
354. Reid, D. & Holmes, W. (1956) Winter feeding of dairy cows. III. The influence on milk yield of high and low protein concentrates, each fed at two levels in addition to dried grass and grass silage. *J. Dairy Res.* **23**, 159–168
355. Armstrong, D. G. & Blaxter, K. L. (1956) Heat increments of feeding in ruminants. *Nature (Lond.)* **177**, 1183–1184
356. Marr, T. G., Sharman, G. A. M. & Blaxter, K. L. (1956) A note on the occurrence of muscular dystrophy in hoggs in the north of Scotland. *Vet. Rec.* **68**, 408–410
357. Blackburn, P. S. (1956) The periductal lobules of the bovine mammary gland. *Br. Vet. J.* **112**, 366–369
358. Edwards, D. C. & Darroch, R. A. (1956) The synthesis of lysine by the lactating goat. *Br. J. Nutr.* **10**, 286–292
359. Blaxter, K. L. & Rook, J. A. F. (1956) The indirect determination of energy retention in farm animals. I. The development of method. *J. Agric. Sci.* **48**, 194–209
360. Blaxter, K. L. & Rook, J. A. F. (1956) The indirect determination of energy retention in farm animals. II. Studies on the validity of the method. *J. Agric. Sci.* **48**, 210–219
361. Modi, V. V. & Owen, E. C. (1956) Riboflavin in milk. *Nature (Lond.)* **178**, 1120
362. Blaxter, K. L. (1956) Starch equivalents, ration standards and milk production. *Proc. Br. Soc. Anim. Prod.*, 3–31

Institute reprint no.

363. Nisbet, A. M. (1956) Arterio-venous anastomoses in the teat of a cow. *Nature (Lond.)* **178**, 1477
364. Holmes, W., Reid, D., MacLusky, D. S., Waite, R. & Watson, J. N. Winter feeding of dairy cows. IV. The influence of four levels of concentrate feeding in addition to a basal ration of grass products on the production obtained from milking cows. *J. Dairy Res.* **24**, 1–10
365. MacLusky, D. S. (1956) The role of grass in feeding the dairy herd. *J. R. Agric. Soc. Engl.* **117**, 33–42
366. Blaxter, K. L. (1957) The effects of defective nutrition during pregnancy in farm livestock. *Proc. Nutr. Soc.* **16**, 52–58
367. Findlay, J. D. (1957) The respiratory activity of calves subjected to thermal stress. *J. Physiol. (Lond.)* **136**, 300–309
368. Bligh, J. (1957) The initiation of thermal polypnoea in the calf. *J. Physiol. (Lond.)* **136**, 413–419
369. Bligh, J. (1957) The relationship between the temperature in the rectum and of the blood in the bicarotid trunk of the calf during exposure to heat stress. *J. Physiol. (Lond.)* **136**, 393–403
370. Bligh, J. (1957) A comparison of the temperature of the blood in the pulmonary artery and in the bicarotid trunk of the calf during thermal polypnoea. *J. Physiol. (Lond.)* **136**, 404–412
371. Manson, W. & Modi, V. V. (1957) Flavin adenine dinucleotide (FAD) metabolism and lactation. *Biochim. Biophys. Acta* **24**, 423–425
372. Owen, E. C. & Darroch, R. A. (1957) The effect of thyrotropin on milk yield and heart rate of the goat. *J. Dairy Res.* **24**, 157–161
373. Waite, R. (1957) The water-soluble carbohydrates of grasses. III. First and second year growth. *J. Sci. Food Agric.* **8**, 422–428
374. Armstrong, D. G., Blaxter, K. L. & Graham, N. McC. (1957) Utilisation of the end products of ruminant digestion. *Proc. Br. Soc. Anim. Prod.*, 3–15
375. Blaxter, K. L. (1957) Myopathic conditions in animals. *Proc. 75th Annu. Congr. Br. Vet. Assoc.*, pp. 3–8
376. Blaxter, K. L., Sharman, G. A. M. & Macdonald, A. M. (1957) Iron-deficiency anaemia in calves. *Br. J. Nutr.* **11**, 234–246
377. Armstrong, D. G. & Blaxter, K. L. (1957) The heat increment of steam-volatile fatty acids in fasting sheep. *Br. J. Nutr.* **11**, 247–272
378. Lovatt Evans, C., Nisbet, A. & Ross, K. A. (1957) A histological study of the sweat glands of normal and dry-coated horses. *J. Comp. Pathol. Ther.* **67**, 397–405
379. Nisbet, W. (1957) An electromechanical timer. *J. Sci. Instrum.* **34**, 286
380. Bianca, W. (1957) The effect of repeated short exposures to heat on the volume and hydration of the blood of the calf. *Br. Vet. J.* **113**, 227–241
381. Waite, R. & Blackburn, P. S. (1957) The chemical composition and the cell count of milk. *J. Dairy Res.* **24**, 328–339
382. Armstrong, D. G., Blaxter, K. L. & Graham, N. McC. (1957) The heat increments of mixtures of steam-volatile fatty acids in fasting sheep. *Br. J. Nutr.* **11**, 392–408
383. Armstrong, D. G. & Blaxter, K. L. (1957) The utilization of acetic, propionic and butyric acids by fattening sheep. *Br. J. Nutr.* **11**, 413–425
384. Waite, R. (1958) The water-soluble carbohydrates of grasses. IV. The effect of different levels of fertilizer treatment. *J. Sci. Food Agric.* **9**, 39–43
385. Hofman, T. (1958) Inhomogeneity of α-casein from goat milk. *Nature (Lond.)* **181**, 633–634
386. Hofman, T. (1958) Studies on casein. I. Purification and properties of phosphoprotein phosphatase. *Biochem. J.* **69**, 135–139

Institute reprint no.

387. Hofman, T. (1958) Studies on casein. 2. The action of phosphatases on caseins and low-molecular-weight phosphates. *Biochem. J.* **69**, 139–145

388. Armstrong, D. G., Blaxter, K. L., Graham, N. McC. & Wainman, F. W. (1958) The utilization of the energy of two mixtures of steam-volatile fatty acids by fattening sheep. *Br. J. Nutr.* **12**, 177–188

389. Campling, R. C., MacLusky, D. S. & Holmes, W. (1958) Studies in grazing management. VI. The influence of free- and strip- grazing and of nitrogenous fertilizers on production from dairy cows. *J. Agric. Sci.* **51**, 62–69

390. Campling, R. C. & Holmes, W. (1958) Winter feeding of dairy cows. V. A comparison of the relative value for milk production of artificially dried cocksfoot and a timothy/meadow-fescue mixture grown both in spring and in autumn. *J. Dairy Res.* **25**, 147–153

391. White, J. C. D. & Davies, D. T. (1958) The relation between the chemical composition of milk and the stability of the caseinate complex. I. General introduction, description of samples, methods and chemical composition of samples. *J. Dairy Res.* **25**, 236–255

392. Davies, D. T. & White, J. C. D. (1958) The relation between the chemical composition of milk and the stability of the caseinate complex. II. Coagulation by ethanol. *J. Dairy Res.* **25**, 256–266

393. White, J. C. D. & Davies, D. T. (1958) The relation between the chemical composition of milk and the stability of the caseinate complex. III. Coagulation by rennet. *J. Dairy Res.* **25**, 267–280

394. White, J. C. D. & Davies, D. T. (1958) The relation between the chemical composition of milk and the stability of the caseinate complex. IV. Coagulation by heat. *J. Dairy Res.* **25**, 281–296

395. Crossland, A., Owen, E. C. & Proudfoot, R. (1958) Riboflavin metabolism of cows and goats and rate of biosynthesis of riboflavin by the lactating goat. *Br. J. Nutr.* **12**, 312-329

396. Findlay, J. D. (1958) Physiological reactions of cattle to climatic stress. *Proc. Nutr. Soc.* **17**, 186–190

397. Blaxter, K. L. (1958) Nutrition and climatic stress in farm animals. *Proc. Nutr. Soc.* **17**, 191–197

398. Castle, M. E. (1958) To supplement summer grass? The response in milk yield to the concentrate feeding of grazing cows. *Farm. Stock-Breed.* **72**, no. 3568, (suppl.) 36–37

399. Castle, M. E., MacLusky, D. S., Waite, R. & Watson, J. N. (1958) The effect of level of concentrate feeding and type of roughage on milk production. *J. Dairy Res.* **25**, 365–372

400. Bianca, W. (1958) The relation between respiratory rate and heart rate in the calf subjected to severe heat stress. *J. Agric. Sci.* **51**, 321–324

401. Blackburn, P. S., Castle, M. E. & Strachan, N. H. (1958) The effect of feeding two levels of concentrates on the incidence of ketosis in dairy cows. *Br. Vet. J.* **114**, 323–332

402. Graham, N. McC., Wainman, F. W., Blaxter, K. L. & Armstrong, D. G. (1959) Environmental temperature, energy metabolism and heat regulation in sheep. I. Energy metabolism in closely clipped sheep. *J. Agric. Sci.* **52**, 13–24

403. Blaxter, K. L., Graham, N. McC., Wainman, F. W. & Armstrong, D. G. (1959) Environmental temperature, energy metabolism and heat regulation in sheep. II. The partition of heat losses in closely clipped sheep. *J. Agric. Sci.* **52**, 25–40

404. Blaxter, K. L., Graham, N. McC. & Wainman, F. W. (1959) Environmental

Institute reprint no.

temperature, energy metabolism and heat regulation in sheep. III. The metabolism and thermal exchanges of sheep with fleeces. *J. Agric. Sci.* **52**, 41–49
405. Armstrong, D. G., Blaxter, K. L. Graham, N. McC. & Wainman, F. W. (1959) The effect of environmental conditions on food utilisation by sheep. *Anim. Prod.* **1**, 1–12
406. Bianca, W. (1959) Acclimatization of calves to a hot dry environment. *J. Agric. Sci.* **52**, 296–304
407. Bianca, W. (1959) Acclimatization of calves to a hot humid environment. *J. Agric. Sci.* **52**, 305–312
408. Bianca, W. (1959) The effect of clipping the coat on various reactions of calves to heat. *J. Agric. Sci.* **52**, 380–383
409. Blaxter, K. L. (1958) Dairy cows. *Proc. Conf. Scientific Principles of Feeding Farm Live Stock*, pp. 21–36
410. Castle, M. E., MacLusky, D. S., Morrison, J. & Watson, J. N. (1959) The effect of concentrates of high or low starch equivalent, both fed at two levels, on the milk production of dairy cows. *J. Dairy Res.* **26**, 1–8
411. Castle, M. E. & Watson, J. N. (1959) A comparison between an early weaning and a more conventional system of rearing dairy calves. *Anim. Prod.* **1**, 31–36
412. Davies, D. T. & White, J. C. D. (1959) Determination of heat-induced changes in the protein stability and chemical composition of milk. *Proc. 15th Int. Dairy Congr.* **3**, 1677–1685
413. Higginbottom, C. (1959) The associated growth of the coli-aerogenes group and other bacteria in milk. *Proc. 15th Int. Dairy Congr.* **3**, 1349–1354
414. Mackie, A. M. & Nisbet, A. M. (1959) The histology of the bovine muzzle. *J. Agric. Sci.* **52**, 376–379
415. Owen, E. C., Darroch, R. A. & Proudfoot, R. (1959) Studies of the rate of passage and disappearance in the intestine of the goat of carotene dissolved in fat and mixed with chromium sesquioxide. *Br. J. Nutr.* **13**, 26–37
416. Taylor, M. M. (1959) Factors affecting the growth of spore-forming bacteria in sterilized milk. *Proc. 15th Int. Dairy Congr.* **3**, 1402–1407
417. Waite, R. (1959) Further measurements of milk solids-not-fat by a density method. *Proc. 15th Int. Dairy Congr.* **3**, 1559–1565
418. Waite, R. & Gorrod, A. R. N. (1959) The structural carbohydrates of grasses. *J. Sci. Food Agric.* **10**, 308–317
419. Waite, R. & Gorrod, A. R. N. (1959) The comprehensive analysis of grasses. *J. Sci. Food Agric.* **10**, 317–326
420. Waite, R. & Paterson, J. A. (1959) The composition of bulk milk measured on Scottish milk-recorded farms. *J. Soc. Dairy Technol.* **12**, 117–122
421. Waite, R., Castle, M. E. & Watson, J. N. (1959) The effect on milk composition of feeding spring grass to cows. *J. Dairy Res.* **26**, 173–181
422. Castle, M. E. (1959) Bulk milk collection. *Ayrsh. Cattle Soc. J.* **31**, 19–20, 22
423. Blackburn, P. S., Castle, M. E., Drysdale, A. D. & Strachan, N. H. (1959) The effect of feeding hay or silage on the incidence of ketosis in dairy cows. *Br. Vet. J.* **115**, 145–155
424. Castle, M. E. & Walker, R. F. R. (1959) The outdoor rearing of Ayrshire calves on pasture with and without supplementary feeding. *J. Br. Grassl. Soc.* **14**, 88–93
425. Castle, M. E. (1959) Grass tetany and high farming. *Farm. News* **111**(30), 27
426. Blackburn, P. S. (1959) Botryomycosis of the udder. *Br. Vet. J.* **115**, 311–316
427. Reid, D. (1959) Studies on the cutting management of grass-clover swards. I. The

Institute reprint no.

effect of varying the closeness of cutting on the yields from an established grass-clover sward. *J. Agric. Sci.* **53**, 299–312
428. Blaxter, K. L., Cowlishaw, B. & Rook, J. A. F. (1960) Potassium and hypomagnesaemic tetany in calves. *Anim. Prod.* **2**, 1–10
429. Modi, V. V., Owen, E. C. & Darroch, R. A. (1959) Partition of riboflavin in cow's milk. *J. Dairy Res.* **26**, 277–280
430. Findlay, J. D., McLean, J. A. & Bennet, R. D. (1959) A climatic laboratory for farm animals. *Heat. Ventil. Eng.* **33**, 95–100
431. Blackburn, P. S. & Castle, M. E. (1959) The use of the oestroscope and its value in detecting oestrus in dairy cattle. *Br. Vet. J.* **115**, 399–407
432. MacLusky, D. S. (1959) Drinking habits of grazing cows. *Agriculture (Lond.)* **66**, 383–386
433. Graham, N. McC., Blaxter, K. L. & Armstrong, D. G. (1958) Analytical and other techniques used in respiration calorimetry and their errors. *Proc. 1st Int. Symp. Energy Metabolism*, pp. 157–163
434. Wainman, F. W. & Blaxter, K. L. (1958) Closed-circuit respiration apparatus for the cow and steer. *Proc. 1st Int. Symp. Energy Metabolism*, pp. 80–84
435. Wainman, F. W. & Blaxter, K. L. (1958) Developments in the design of closed-circuit respiration apparatus for animals weighing up to 70 kg. *Proc. 1st Int. Symp. Energy Metabolism*, pp. 85–92
436. Reid, D. & MacLusky, D. S. (1960) Studies on the cutting management of grass-clover swards. Part II. The effects of close cutting with either a gang mower or a reciprocating-knife mower on the yields from an established grass-clover sward. *J. Agric. Sci.* **54**, 158–165
437. Blaxter, K. L. (1960) The utilization of the energy of grassland products. *Proc. 8th Int. Grassl. Congr.*, pp. 479–484
438. Castle, M. E. & Reid, D. (1960) Irrigation of grassland in south-west Scotland and its influence on the utilization of fertilizer nitrogen. *Proc. 8th Int. Grassl. Congr.*, pp. 146–150
439. Castle, M. E. (1960) Milk production at the Hannah. *Farm. Rev.* no. 14, 18–20
440. Blackburn, P. S. (1960) Sub-clinical mastitis. *Dairy Farm.* **7**(4), 30–31
441. Higginbottom, C. & Taylor, M. M. (1960) The growth of aerobic spore-forming bacilli in sterilized milk. *J. Dairy Res.* **27**, 221–234
442. Higginbottom, C. & Taylor, M. M. (1960) Certain non-biological properties of sterilized milk in relation to the treatment of the milk and the temperature of heating. *J. Dairy Res.* **27**, 235–244
443. Higginbottom, C. & Taylor, M. M. (1960) The oxidation-reduction potential of sterilized milk. *J. Dairy Res.* **27**, 245–257
444. Higginbottom, C. & Taylor, M. M. (1960) The alcohol test applied to sterilized milk. *J. Dairy Res.* **27**, 259–265
445. Davies, D. T. & White, J. C. D. (1960) The use of ultrafiltration and dialysis in isolating the aqueous phase of milk and in determining the partition of milk constituents between the aqueous and disperse phases. *J. Dairy Res.* **27**, 171–190
446. MacLusky, D. S. (1960) Some estimates of the areas of pasture fouled by the excreta of dairy cows. *J. Br. Grassl. Soc.* **15**, 181–188
447. Armstrong, D. G. (1960) Calorimetric determination of the net energy value of dried S.23 ryegrass at four stages of growth. *Proc. 8th Int. Grassl. Congr.*, pp. 485–489
448. Castle, M. E. & Reid, D. (1960) Effects of fertiliser treatments and cutting managements on yield and quality of grass and clover swards, in *Soc. Chem. Ind. Monogr.* no. 9, 55–68

Institute reprint no.

449. Findlay, J. D. & Jenkinson, D. McE. (1960) The morphology of bovine sweat glands and the effect of heat on the sweat glands of the Ayrshire calf. *J. Agric. Sci.* **55**, 247–249
450. Manson, W. (1960) The application of the Sakaguchi reaction to the determination of DNP-arginine. *Biochim. Biophys. Acta* **44**, 182–184
451. Castle, M. E. & Holmes, W. (1960) The intensive production of herbage for crop-drying. VII. The effect of further continued massive applications of nitrogen with and without phosphate and potash on the yield of grassland herbage. *J. Agric. Sci.* **55**, 251–260
452. Reith, J. W. S., Inkson, R. H. E., Stewart, A. B., Holmes, W., MacLusky, D. S., Reid, D., Heddle, R. G., Clouston, D. & Copeman, G. J. F. (1961) The effects of fertilizers on herbage production. Part I. The effect of nitrogen phosphate and potash on yield. *J. Agric. Sci.* **56**, 17–29
453. Castle, M. E., Drysdale, A. D. & Watson, J. N. (1960) The effect of feed supplements on the yield and composition of milk from cows grazing good pasture. *J. Diary Res.* **27**, 419–426
454. Armstrong, D. G., Blaxter, K. L., Clapperton, J. L., Graham, N. McC. & Wainman, F. W. (1960) Heat production and heat emission of two breeds of sheep. *J. Agric. Sci.* **55**, 395–401
455. Castle, M. E. (1960) Making the most of bulk feeds. *Dairy Farm.* **7**(11), 28–29
456. Castle, M. E. & Reid, D. (1960) A comparison of lucerne varieties in south-west Scotland. *J. Br. Grassl. Soc.* **15**, 281–286
457. Blaxter, K. L. & Wainman, F. W. (1961) Environmental temperature and the energy metabolism and heat emission of steers. *J. Agric. Sci.* **56**, 81–90
458. Blaxter, K. L. (1960) Energy utilization in the ruminant. *Proc. Univ. Nottingham 7th Easter School Agric. Sci.*, pp. 183–197
459. Castle, M. E. & Reid, D. (1960) Grassland irrigation. *Scott. Agric.* **42**, 143–145
460. Findlay, J. D. & Ingram, D. L. (1961) Brain temperature as a factor in the control of thermal polypnoea in the ox (*Bos taurus*). *J. Physiol.* (*Lond.*) **155**, 72–85
461. Blaxter, K. L., Wainman, F. W. & Wilson, R. S. (1961) The regulation of food intake by sheep. *Anim. Prod.* **3**, 51–61
462. Castle, M. E., Drysdale, A. D. & Waite, R. (1961) The effect of root feeding on the intake and production of dairy cows. *J. Dairy Res.* **28**, 67–74
463. Waite, R. (1961) A note on a method of overcoming the effect of udder disease or injury in experiments involving milk yield and composition. *J. Dairy Res.* **28**, 75–79
464. Reid, D. (1961) Factors influencing the role of clover in grass-clover leys fertilized with nitrogen at different rates. I. The effects of the varicty of companion grass on the yields of total herbage and of clover. *J. Agric. Sci.* **56**, 143–153
465. Reid, D. (1961) Factors influencing the role of clover in grass-clover leys fertilized with nitrogen at different rates. II. The effects of the variety of white clover on the yields of total herbage and of clover. *J. Agric. Sci.* **56**, 155–160
466. Blackburn, P. S., Castle, M. E., Drysdale, A. D. & Strachan, N. H. (1961) The effect of feeding a high or a low mineral concentrate on the incidence of ketosis in dairy cows. *Br. Vet. J.* **117**, 158–166
467. Armstrong, D. G., Blaxter, K. L. & Graham, N. McC. (1961) The heat increment in fasting sheep of acetic acid partially neutralized with sodium hydroxide. *Br. J. Nutr.* **15**, 169–175
468. Ingram, D. L., McLean, J. A. & Whittow, G. C. (1961) Increase of evaporative loss of water from the skin of the ox in response to local heating of the hypothalamus. *Nature* (*Lond.*) **191**, 81–82

Institute reprint no.

469. Nisbet, W. (1961) Switch with floating pre-set differential. *J. Sci. Instrum.* **38**, 378–379
470. Reid, D. (1961) Factors influencing the role of clover in grass-clover leys fertilized with nitrogen at different rates. III. The effects of the variety of companion grass and of the variety of clover on the botanical composition of the sward. *J. Agric. Sci.* **57**, 231–236
471. Abbot, J., Waite, R. & Hearne, J. F. (1961) Gas-packing milk powder with a mixture of nitrogen and hydrogen in the presence of palladium catalyst. *J. Dairy Res.* **28**, 285–292
472. Castle, M. E. & Watson, J. N. (1961) The effect of level of concentrate feeding before and after calving on the production of dairy cows. *J. Dairy Res.* **28**, 231–243
473. Manson, W. (1962) Recent views on the structure of casein. *Proc. 16th Int. Dairy Congr.* **B**, 513–521
474. Higginbottom, C. (1962) Bacteriological studies on raw milk before and after the adoption of bulk milk cooling on the farm. *Proc. 16th Int. Dairy Congr.* **A**, 426–432
475. Abbot, J. & Waite, R. (1962) Physico-chemical changes in 'instant' skimmed milk powder during storage. *Proc. 16th Int. Dairy Congr.* **B**, 943–951
476. Manson, W. (1961) The N-terminal amino acid residues of α_s-casein. *Arch. Biochem. Biophys.* **95**, 336–342
477. Ingram, D. L. & Whittow, G. C. (1961) Measurement of changes in the body temperature of the ox (*Bos taurus*). *Br. Vet. J.* **117**, 479–484
478. Blaxter, K. L. & Wainman, F. W. (1961) The utilization of food by sheep and cattle. *J. Agric. Sci.* **57**, 419–425
479. Blackburn, P. S., Castle, M. E., Drysdale, A. D., & Strachan N. H. (1961) Potassium chlorate as a prophylactic treatment for subclinical ketosis. *Br. Vet. J.* **117**, 491–496
480. Manson, W. (1962) Electrophoresis of casein in a column stabilised by a density gradient. *J. Electroanal. Chem.* **3**, 203–208
481. Whittow, G. C. (1962) The significance of the extremities of the ox (*Bos taurus*) in thermoregulation. *J. Agric. Sci.* **58**, 109–120
482. Castle, M. E. & Watson, J. N. (1961) The production and use of cocksfoot foggage for winter grazing in south-west Scotland. *J. Br. Grassl. Soc.* **16**, 247–252
483. Bianca, W., Findlay, J. D. & Mabon, R. M. (1962) Respiration and acid-base status in calves – normal values. *Res. Vet. Sci.* **3**, 34–37
484. Bianca, W. & Findlay, J. D. (1962) The effect of thermally-induced hyperpnoea on the acid-base status of the blood of calves. *Res. Vet. Sci.* **3**, 38–49
485. Findlay, J. D. (1961) Cattle and climate. *Agric. Prog.* **36**, 5–19
486. Abbot, J. & Waite, R. (1962) The effect of antioxidants on the keeping quality of whole milk powder. I. Flavones, gallates, butylhydroxyanisole and nordihydroguaiaretic acid. *J. Dairy Res.* **29**, 55–61
487. Castle, M. E. & Drysdale, A. D. (1962) Liquid manure as a grassland fertiliser. I. The response to liquid manure and to dry fertiliser. *J. Agric. Sci.* **58**, 165–171
488. Castle, M. E., Drysdale, A. D. & Watson, J. N. (1962) A comparison between ryegrass and cocksfoot hay for milk production. *J. Dairy Res.* **29**, 199–206
489. Drysdale, A. D. (1962) Don't waste liquid manure. *Dairy Farm.* **9**(3), 55, 57
490. Gonzalez-Jimenez, E. & Blaxter, K. L. (1962) The metabolism and thermal regulation of calves in the first month of life. *Br. J. Nutr.* **16**, 199–212
491. Armstrong, D. G. & Blaxter, K. L. (1961) The utilization of the energy of carbohydrate by ruminants. *Proc. 2nd Symp. Energy Metabolism*, pp. 187–197

Institute reprint no.

492. Martin, A. K. & Blaxter, K. L. (1961) The utilization of the energy of protein by ruminants. *Proc. 2nd Symp. Energy Metabolism*, pp. 200–209
493. Blaxter, K. L. (1961) The utilization of the energy of food by ruminants. *Proc. 2nd Symp. Energy Metabolism*, pp. 211–224
494. Barlow, R. M., Robertson, J. M., Owen, E. C. & Proudfoot, R. (1962) A condition in the goat resembling swayback in lambs. *Vet. Rec.* **74**, 737–739
495. Bianca, W. (1962) Relative importance of dry- and wet-bulb temperatures in causing heat stress in cattle. *Nature* (*Lond.*) **195**, 251–252
496. Blaxter, K. L. & Martin, A. K. (1962) The utilization of protein as a source of energy in fattening sheep. *Br. J. Nutr.* **16**, 397–407
497. Ingram, D. L. & Whittow, G. C. (1962) The effects of variations in respiratory activity and in the skin temperatures of the ears on the temperature of the blood in the external jugular vein of the ox (*Bos taurus*). *J. Physiol.* (*Lond.*) **163**, 211–221
498. Ingram, D. L. & Whittow, G. C. (1962) The effect of heating the hypothalamus on respiration in the ox (*Bos taurus*). *J. Physiol.* (*Lond.*) **163**, 200–210
499. Blackburn, P. S. & Castle, M. E. (1962) The use of the consistometer in the diagnosis of pregnancy in dairy cattle. *Br. Vet. J.* **118**, 321–326
500. Blaxter, K. L. (1962) Muscular dystrophy in farm animals: its cause and prevention. *Proc. Nutr. Soc.* **21**, 211–216
501. Blaxter, K. L. & Wilson, R. S. (1962) The voluntary intake of roughages by steers. *Anim. Prod.* **4**, 351–358
502. Davies, D. T. & White, J. C. D. (1962) The determination of calcium and magnesium in milk and milk diffusate. *J. Dairy Res.* **29**, 285–296
503. Manson, W. (1962) The effect upon casein of aqueous solutions of urea. *Biochem. Biophys. Acta* **63**, 515–517
504. Bianca, W. (1962) Tolerance to severe heat and the behaviour of respiratory minute volume in cattle. *Nature* (*Lond.*) **195**, 1208–1209
505. Reid, D. (1962) Studies on the cutting management of grass-clover swards. III. – The effects of prolonged close and lax cutting on herbage yields and quality. *J. Agric. Sci.* **59**, 359–368
506. Blaxter, K. L. (1962) The fasting metabolism of adult wether sheep. *Br. J. Nutr.* **16**, 615–626
507. Jenkinson, D. McE. and Peacock, P. R. (1963) The effect of several carcinogenic agents on the skin of the Ayrshire calf. *Res. Vet. Sci.* **4**, 114–120
508. Bianca, W. (1963) Rectal temperature and respiratory rate as indicators of heat tolerance in cattle. *J. Agric. Sci.* **60**, 113–120
509. Blaxter, K. L. (1962) Progress in assessing the energy value of feeding-stuffs for ruminants. *J. R. Agric. Soc. Eng.* **123**, 7–21
510. Waite, R. & Blackburn, P. S. (1963) The relationship between milk yield composition and tissue damage in a case of subclinical mastitis. *J. Dairy Res.* **30**, 23–33
511. Blaxter, K. L. (1963) The effect of selenium administration on the growth and health of sheep on Scottish farms. *Br. J. Nutr.* **17**, 105–115
512. Blaxter, K. L. & Wilson, R. S. (1963) The assessment of a crop husbandry technique in terms of animal production. *Anim. Prod.* **5**, 27–42
513. Castle, M. E. & Reid, D. (1963) Nitrogen and herbage production. *J. Br. Grassl. Soc.* **18**, 1–6
514. Castle, M. E., Drysdale, A. D., Waite, R. & Watson, J. N. (1963) The effect of the replacement of concentrates by roots on the intake and production of dairy cows. *J. Dairy Res.* **30**, 199–207

Institute reprint no.

515. Waite, R., Abbot, J. & Blackburn, P. S. (1963) The use of quarter samples in the assessment of the effects of feeding treatments on milk composition. *J. Dairy Res.* **30**, 209–215

516. McLean, J. A. (1963) Measurement of cutaneous moisture vaporization from cattle by ventilated capsules. *J. Physiol.* (*Lond.*) **167**, 417–426

517. McLean, J. A. (1963) The partition of insensible losses of body weight and heat from cattle under various climatic conditions. *J. Physiol.* (*Lond.*) **167**, 427–447

518. White, J. C. D. and Davies, D. T. (1963) The determination of citric acid in milk and milk sera. *J. Dairy Res.* **30**, 171–189

519. Wainman, F. W. & Paterson, D. (1963) A note on the collection of urine from male cattle and sheep. *J. Agric. Sci.* **61**, 253–254

520. Blaxter, K. L., Joyce, J. P. & Wainman, F. W. (1963) Effect of air velocity on the heat losses of sheep and cattle. *Nature* (*Lond.*) **198**, 1115–1116

521. Ingram, D. L. & Whittow, G. C. (1963) Changes of arterial blood pressure and heart rate in the ox (*Bos taurus*) with changes of body temperature. *J. Physiol.* (*Lond.*) **168**, 736–746

522. McLean, J. A. (1963) The regional distribution of cutaneous moisture vaporization in the Ayrshire calf. *J. Agric. Sci.* **61**, 275–280

523. Waite, R. (1963) Botanical and chemical changes in maturing grass and their effect on its digestibility. *Agric. Prog.* **38**, 50–56

524. Ingram, D. L., McLean, J. A. & Whittow, G. C. (1963) The effect on heating the hypothalamus and the skin on the rate of moisture vaporization from the skin of the ox (*Bos taurus*). *J. Physiol.* (*Lond.*) **169**, 394–403

525. Blaxter, K. L. (1963) Ruminants as experimental animals. *Lab. Anim. Cent. Collect. Pap.* **12**, 21–28

526. Drysdale, A. D. (1963) Liquid manure as a grassland fertilizer. II. The response to winter applications. *J. Agric. Sci.* **61**, 353–360

527. Blaxter, K. L. & Joyce, J. P. (1963) The accuracy and ease with which measurements of respiratory metabolism can be made with tracheostomized sheep. *Br. J. Nutr.* **17**, 523–537

528. Findlay, J. D. & Jenkinson, D. McE. (1964) Sweat gland function in the Ayrshire calf. *Res. Vet. Sci.* **5**, 109–115

529. Bianca, W. (1964) Thermoregulatory responses of the dehydrated ox to drinking cold and warm water in a warm environment. *Res. Vet. Sci.* **5**, 75–80

530. Castle, M. E. (1963) Feeding for milk production. *Inst. Corn Agric. Merch. J.* **11**(4), 37–39

531. Joyce, J. P. & Blaxter, K. L. (1964) The effect of air movement, air temperature and infrared radiation on the energy requirements of sheep. *Br. J. Nutr.* **18**, 5–27

532. Clapperton, J. L. (1964) The effect of walking upon the utilization of food by sheep. *Br. J. Nutr.* **18**, 39–46

533. Clapperton, J. L. (1964) The energy metabolism of sheep walking on the level and on gradients. *Br. J. Nutr.* **18**, 47–54

534. Nay, T. & Jenkinson, D. McE. (1964) Skin structure and milk production. *J. Dairy Res.* **31**, 53–58

535. Blaxter, K. L. & Wainman, F. W. (1964) The effect of increased air movement on the heat production and emission of steers. *J. Agric. Sci.* **62**, 207–214

536. Castle, M. E. (1964) Concentrates, grass and milk. *Ayrsh. Cattle Soc. J.* **36**, 96, 98 101

537. Reith, J. W. S., Inkson, R. H. E., Holmes, W., Maclusky, D. S., Reid, D., Heddle, R. G. & Copeman, G. J. F. (1964) The effects of fertilizers on herbage produc-

Institute reprint no.

tion. II. The effect of nitrogen, phosphorus and potassium on botanical and chemical composition. *J. Agric. Sci.* **63**, 209–219

538. Robertshaw, D. (1964) Methohexital sodium anaesthesia in calves. *Vet. Rec.* **76**, 357–359

539. Waite, R., Johnston, Margaret J. & Armstrong, D. G. (1964) The evaluation of artificially dried grass as a source of energy for sheep. I. The effect of stage of maturity on the apparent digestibility of rye-grass, cocksfoot and timothy. *J. Agric. Sci.* **62**, 301–398

540. Armstrong, D. G. (1964) Evaluation of artificially dried grass as a source of energy for sheep. II. The energy value of cocksfoot, timothy and two strains of rye-grass at varying stages of maturity. *J. Agric. Sci.* **62**, 399–416

541. Armstrong, D. G., Blaxter, K. L. & Waite, R. (1964) The evaluation of artificially dried grass as a source of energy for sheep. III. The prediction of nutritive value from chemical and biological measurements. *J. Agric. Sci.* **62**, 417–424

542. Waite, R. (1964) Some recent developments in the manufacture of milk powders. *J. Soc. Dairy Technol.* **17**, 122–131

543. Blaxter, K. L. & Wainman, F. W. (1964) The utilization of the energy of different rations by sheep and cattle for maintenance and for fattening. *J. Agric. Sci.* **63**, 113–128

544. Bateman, J. V. & Blaxter, K. L. (1964) The utilization of the energy of artificially dried lucerne. *J. Agric. Sci.* **63**, 129–131

545. Blaxter, K. L. (1964) The effect of outdoor climate in Scotland on sheep and cattle. *Vet. Rec.* **76**, 1445–1451

546. Castle, M. E. & Watson, J. N. (1964) Grassland management and milk production at the Hannah Institute 1958–62. *J. Br. Grassl. Soc.* **19**, 343–348

547. Joyce, J. P. & Blaxter, K. L. (1964) Respiration in sheep in cold environments. *Res. Vet. Sci.* **5**, 506–516

548. Higginbottom, C., Jones, M. & Taylor, M. (1964) The influence of a change in farm dairy practice on the bacterial flora of fresh and stored raw milk. *J. Appl. Bacteriol.* **27**, 385–391

549. Clapperton, J. L., Joyce, J. P. & Blaxter, K. L. (1965) Estimates of the contribution of solar radiation to the thermal exchanges of sheep at a latitude of 55° north. *J. Agric. Sci.* **64**, 37–49

550. Castle, M. E., Reid, D. & Heddle, R. G. (1965) The effect of varying the date of application of fertilizer nitrogen on the yield and seasonal productivity of grassland. *J. Agric. Sci.* **64**, 177–184

551. Castle, M. E., Drysdale, A. D. & Watson, J. N. (1964) The feeding of supplementary concentrates to dairy cows grazing good pasture. *J. Br. Grassl. Soc.* **19**, 381–386

552. Ekern, A., Blaxter, K. L. & Sawers, D. (1965) The effect of artificial drying and of freezing on the energy value of pasture herbage. *Proc. 3rd Symp. Energy Metabolism*, pp. 217–224

553. Armstrong, D. G. & Blaxter, K. L. (1965) Effects of acetic and propionic acids on energy retention and milk secretion in goats. *Proc. 3rd Symp. Energy Metabolism*, pp. 59–72

554. Martin, A. K. & Blaxter, K. L. (1965) The energy cost of urea synthesis in sheep. *Proc. 3rd Symp. Energy Metabolism*, pp. 83–91

555. Joyce, J. P. & Blaxter, K. L. (1965) The effect of wind on heat losses of sheep. *Proc. 3rd Symp. Energy Metabolism*, pp. 355–367

556. Bianca, W., Findlay, J. D. & McLean, J. A. (1965) Responses of steers to water restriction. *Res. Vet. Sci.* **6**, 38–55

Institute reprint no.

557. Reid, D. & Castle, M. E. (1965) The response of grass-clover and pure grass leys to irrigation and fertilizer nitrogen treatment. I. Irrigation effects. *J. Agric. Sci.* **64**, 185–194
558. Bianca, W. (1965) Sweating in dehydrated steers. *Res. Vet. Sci.* **6**, 33–37
559. Manson, W. (1965) The separation of major components of the casein of bovine milk by electrophoresis in a density gradient. *Biochem. J.* **94**, 452–457
560. Blaxter, K. L. (1965) The efficiency of feed conversion by livestock. *J. R. Agric. Soc. Eng.* **125**, 87–99
561. Waite, R. & McPhillips, J. (1965) The gravimetric measurement of milk total solids in large numbers. *J. Soc. Dairy Technol.* **18**, 27–31
562. Johnston, Margaret J. & Waite, R. (1965) Studies in the lignification of grasses. I. Perennial rye-grass (S24) and cocksfoot (S37). *J. Agric. Sci.* **64**, 211–219
563. Whittow, G. C. & Robertshaw, D. (1965) The use of an exteriorized carotid artery for cardiovascular studies in the ox (*Bos taurus*). *Am. J. Vet. Res.* **26**, 183–186
564. Drysdale, A. D. (1965) What is effluent worth? *Dairy Farm.* **12**(2), 57–59
565. Waite, R. (1965) The chemical composition of grasses in relation to agronomical practice. *Proc. Nutr. Soc.* **24**, 38–46
566. Blackburn, P. S. (1965) A solution for use in assessing the cell count of cow's milk. *Br. Vet. J.* **121**, 154–158
567. Findlay, J. D. & Robertshaw, D. (1965) The role of the sympatho-adrenal system in the control of sweating in the ox (*Bos taurus*). *J. Physiol.* (*Lond.*) **179**, 285–297
568. Abbot, J. & Waite, R. (1965) The effect of antioxidants on the keeping quality of whole milk powder. II. Tocopherols. *J. Dairy Res.* **21**, 143–146
569. Owen, E. C., Proudfoot, R. Robertson, J. M., Barlow, R. M., Butler, E. J. & Smith, B. S. W. (1965) Pathological and biochemical studies of an outbreak of swayback in goats. *J. Comp. Pathol. Ther.* **75**, 241–251
570. Reid, D. & Castle, M. E. (1965) The response of grass-clover and pure-grass leys to irrigation and fertilizer nitrogen treatment. II. Clover and fertilizer nitrogen effects. *J. Agric. Sci.* **65**, 109–119
571. Whittow, G. C. (1965) The effect of hyperthermia on the systemic and pulmonary circulation of the ox (*Bos taurus*). *Q. J. Exp. Physiol. Cogn. Med. Sci.* **50**, 300–311
572. Ekern, A., Blaxter, K. L. & Sawers, D. (1965) The effect of artificial drying on the energy value of grass. *Br. J. Nutr.* **19**, 417–434
573. Blaxter, K. L. (1965) Animal production in adverse environments. *J. Univ. Newcastle upon Tyne Agric. Soc.* **19**, 10–15
574. Czerkawski, J. W. & Blaxter, K. L. (1965) Octadecenoic acids in sheep rumen. *Biochem. J.* **96**, 25c–27c
575. Blaxter, K. L. & Clapperton, J. L. (1965) Prediction of the amount of methane produced by ruminants. *Br. J. Nutr.* **19**, 511–522
576. Vercoe, J. E. & Blaxter, K. L. (1965) The metabolism of formic acid in sheep. *Br. J. Nutr.* **19**, 523–530
577. Drysdale, A. D. (1965) Liquid manure as a grassland fertilizer. III. The effect of liquid manure on the yield and botanical composition of pasture, and its interaction with nitrogen, phosphate and potash fertilizers. *J. Agric. Sci.* **65**, 333–340
578. Whittow, G. C. (1963) Cardiovascular responses of the ox (*Bos taurus*) to hyperthermia. *Proc. 3rd Int. Cong. Biometeorol.*, pp. 460–463
579. Bianca, W. (1966) Heat tolerance in dehydrated steers. *J. Agric. Sci.* **66**, 57–60
580. Reid, D. (1966) Studies on the cutting management of grass–clover swards. IV. The effects of close and lax cutting on the yield of herbage from swards cut at different frequencies. *J. Agric. Sci.* **66**, 101–106

Institute reprint no.

581. Jenkinson, D. McE., Sengupta, B. P. & Blackburn, P. S. (1966) The distribution of nerves, monoamine oxidase and cholinesterase in the skin of cattle. *J. Anat.* **100**, 593–613
582. Davies, D. T. & White, J. C. D. (1966) The stability of milk protein to heat. I. Subjective measurement of heat stability of milk. *J. Dairy Res.* **33**, 67–81
583. Davies, D. T. & White, J. C. D. (1966) The stability of milk protein to heat. II. Effect on heat stability of ageing milk at different temperatures. *J. Dairy Res.* **33**, 83–91
584. White, J. C. D. & Davies, D. T. (1966) The stability of milk protein to heat. III. Objective measurement of heat stability of milk. *J. Dairy Res.* **33**, 93–102
585. Blaxter, K. L., Wainman, F. W. & Davidson, J. L. (1966) The voluntary intake of food by sheep and cattle in relation to their energy requirements for maintenance. *Anim. Prod.* **8**, 75–83
586. Blaxter, K. L. & Wainman, F. W. (1966) The fasting metabolism of cattle. *Br. J. Nutr.* **20**, 103–111
587. Blaxter, K. L., Clapperton, J. L. & Wainman, F. W. (1966) Utilization of the energy and protein of the same diet by cattle of different ages. *J. Agric. Sci.* **67**, 67–75
588. Martin, A. K. (1966) Some errors in the determination of nitrogen retention of sheep by nitrogen balance studies. *Br. J. Nutr.* **20**, 325–337
589. Blaxter, K. L., Clapperton, J. L. & Wainman, F. W. (1966) The extent of differences between six British breeds of sheep in their metabolism, feed intake and utilization, and resistance to climatic stress. *Br. J. Nutr.* **20**, 283–294
590. Czerkawski, J. W., Blaxter, K. L. & Wainman, F. W. (1966) The metabolism of oleic, linoleic and linolenic acids by sheep with reference to their effects on methane production. *Br. J. Nutr.* **20**, 349–362
591. Castle, M. E., Drysdale, A. D. & Watson, J. N. (1966) The effect of feeding dried sugar-beet pulp on the intake and production of dairy cows. *J. Dairy Res.* **33**, 123–128
592. Blackburn, P. S. (1966) The variation in the cell count of cow's milk throughout lactation and from one lactation to the next. *J. Dairy Res.* **33**, 193–198
593. Owen, E. C. (1965) Minerals and the ruminant with special reference to the discovery of swayback in the goat. *Food Cosmet. Toxicol.* **3**, 701–706
594. Czerkawski, J. W., Blaxter, K. L. & Wainman, F. W. (1966) The effect of functional groups other than carboxyl on the metabolism of C_{18} and C_{12} alkyl compounds by sheep. *Br. J. Nutr.* **20**, 495–508
595. Czerkawski, J. W., Blaxter, K. L. & Wainman, F. W. (1966) The effect of linseed oil and of linseed oil fatty acids incorporated in the diet on the metabolism of sheep. *Br. J. Nutr.* **20**, 485–494
596. Joyce, J. P., Blaxter, K. L. & Park, C. (1966) The effect of natural outdoor environments on the energy requirements of sheep. *Res. Vet. Sci.* **7**, 342–359
597. Blaxter, K. L., Clapperton, J. L. & Martin, A. K. (1966) The heat of combustion of the urine of sheep and cattle in relation to its chemical composition and to diet. *Br. J. Nutr.* **20**, 449–460
598. Findlay, J. D. & Whittow, G. C. (1966) The role of arterial oxygen tension in the respiratory response to localized heating of the hypothalamus and to hyperthermia. *J. Physiol.* (*Lond.*) **186**, 333–346
599. Waite, R. & Taylor, M. M. (1966) The chemical composition and microbial flora of milk stored in the presence of potassium dichromate. *Proc. 17th Int. Dairy Congr.*, section B: 2, pp. 87–93
600. Taylor, M. M. (1966) Factors influencing the incidence of psychrophilic organisms

Institute reprint no.

in milk held at low temperatures. *Proc. 17th Int. Dairy Congr.*, section A: 2, pp. 329–332

601. Abbot, J. & Waite, R. (1966) Characterization of the ease of reconstitution of milk powders; nomenclature and tests. *Proc. 17th Int. Dairy Congr.*, section E: 3, pp. 203–207
602. Reid, D. (1966) The response of herbage yields and quality to a wide range of nitrogen application rates. *Proc. 10th Int. Grassl. Congr.*, section 1, pp. 209–213
603. Webster, A. J. F. (1966) The establishment of thermal equilibrium in sheep exposed to cold environments. *Res. Vet. Sci.* **7**, 454–465
604. Webster, A. J. F. & Blaxter, K. L. (1966) The thermal regulation of two breeds of sheep exposed to air temperatures below freezing point. *Res. Vet. Sci.* **7**, 466–479
605. Drysdale, A. D. & Strachan, N. H. (1966) Liquid manure as a grassland fertilizer. IV. The effect of liquid manure on the mineral content of grass and clover. *J. Agric. Sci.* **67**, 337–343
606. Castle, M. E. & Drysdale, A. D. (1966) Liquid manure as a grassland fertilizer. V. The response to mixtures of liquid manure (urine) and dung. *J. Agric. Sci.* **67**, 397–404
607. Czerkawski, J. W. (1966) The effect on digestion in the rumen of a gradual increase in the content of fatty acids in the diet of sheep. *Br. J. Nutr.* **20**, 833–842
608. Bianca, W. (1966) A simple method for estimating the isotonicity of blood. *Br. Vet. J.* **122**, 530–534
609. Robertshaw, D. & Whittow, G. C. (1966) The effect of hyperthermia and localized heating of the anterior hypothalamus on the sympatho-adrenal system of the ox (*Bos taurus*). *J. Physiol.* (*Lond.*) **187**, 351–360
610. Martin, A. K. (1966) Metabolism of benzoic acid by sheep. *J. Sci. Food Agric.* **17**, 496–500
611. Jenkinson, D. McE. & Blackburn, P. S. (1967) The distribution of nerves, monoamine oxidase and cholinesterase in the skin of the sheep and goat. *J. Anat.* **101**, 333–341
612. Bianca, W. (1966) The impact of the atmospheric environment on the integument of domestic animals. *Proc. 4th Int. Biometeorol. Congr.*, pp. 83–102
613. Owen, E. C. (1965) Metabolites of riboflavine and related lyochromes in milk and urine. *Proc. 1st Int. Congr. Food Sci. Technol.* vol. III, pp. 669–678
614. Blaxter, K. L. & Czerkawski, J. W. (1966) Modifications of the methane production of the sheep by supplementation of its diet. *J. Sci. Food Agric.* **17**, 417–421
615. Abd El-Salam, M. H. & Manson, W. (1967) The C-terminal sequence of amino acid residues of κ-casein isolated from buffalo's milk. *J. Dairy Res.* **34**, 85–88
616. Findlay, J. D. & Robertshaw, D. (1967) The mechanism of body temperature changes induced by intraventricular injections of adrenaline, noradrenaline and 5-hydroxytryptamine in the ox (*Bos taurus*). *J. Physiol.* (*Lond.*) **189**, 329–336
617. Owen, E. C. (1967) Nitrogen balances. *Proc. Nutr. Soc.* **26**, 116–124
618. Reid, D. (1967) Studies on the cutting management of grass-clover swards. V. The effect of changes in the closeness of cutting at different times in the season on the yield and quality of herbage from a perennial ryegrass–white clover sward. *J. Agric. Sci.* **68**, 249–254
619. Czerkawski, J. W. (1967) The determination of lignin. *Br. J. Nutr.* **21**, 325–332
620. Taylor, M. M. (1967) A hygiene test for milk. *Dairy Ind.* **32**, 278–283
621. Noble, R. C. & Moore, J. H. (1967) The partition of lipids between the yolk and yolk sac membrane during the development of the chick embryo. *Can. J. Biochem.* **45**, 949–958

Institute reprint no.

622. Noble, R. C. & Moore, J. H. (1967) The transport of phospholipids from the yolk to the yolk-sac membrane during the development of the chick embryo. *Can. J. Biochem.* **45**, 1125–1133
623. Czerkawski, J. W. (1967) Effect of storage on the fatty acids of dried ryegrass. *Br. J. Nutr.* **21**, 599–608
624. Castle, M. E., Drysdale, A. D. & Watson, J. N. (1967) A note on the effect of Scotch grain distillers' solubles on the milk production of dairy cows. *Anim. Prod.* **9**, 409–412
625. Hales, J. R. S., Findlay, J. D. & Mabon, R. M. (1967) Tissue hypoxia in oxen exposed to severe heat. *Respir. Physiol.* **3**, 43–46
626. Hart, L. I., Owen, E. C. & Proudfoot, R. (1967) The influence of dietary molybdenum on the xanthine oxidase activity of the milk of ruminants. *Br. J. Nutr.* **21**, 617–630
627. Webster, A. J. F. (1967) Continuous measurement of heart rate as an indicator of the energy expenditure of sheep. *Br. J. Nutr.* **21**, 769–785
628. Moore, J. H., Noble, R. C. & O'Donoghue, P. N. (1967) Effects of intravenous administration of adrenocorticotrophin on the adrenal lipids of the rabbit. *Can. J. Biochem.* **45**, 1335–1341
629. Blackburn, P. S. & Castle, M. E. (1968) Cow's milk of abnormally low composition. *Br. Vet. J.* **124**, 27–35
630. Manson, W. (1967) Reaction of basic dyes with α_s- and β-casein. *Arch. Biochem. Biophys.* **122**, 370–377
631. Czerkawski, J. W. (1967) Incubation inside the bovine rumen. *Br. J. Nutr.* **21**, 865–878
632. Webster, A. J. F. & Park, C. (1967) The effect of jute coats on the heat losses of two breeds of sheep exposed to different environments. *Anim. Prod.* **9**, 483–490
633. Jenkinson, D. McE., Blackburn, P. S. & Proudfoot, R. (1967) Seasonal changes in the skin glands of the goat. *Br. Vet. J.* **123**, 541–549
634. Jenkinson, D. McE. & Blackburn, P. S. (1967) The distribution of nerves, monoamine oxidase and cholinesterase in the skin of the pig. *Res. Vet. Sci.* **8**, 306–312
635. Whittow, G. C. & Findlay, J. D. (1968) Oxygen cost of thermal panting. *Am. J. Physiol.* **214**, 94–99
636. Manson, W. (1966) The reaction of casein with the basic dye safranine. *Proc. 17th Int. Dairy Congr.*, section B: 2, pp. 193–199
637. Owen, E. C. & West, D. W. (1968) Bacterial degradation of riboflavin. *J. Chem. Soc. Sect. C Org. Chem.*, no. 1, 34–36
638. Findlay, J. D. & Thompson, G. E. (1968) The effect of intraventricular injections of noradrenaline, 5-hydroxytryptamine, acetylcholine and tranylcypromine on the ox (*Bos taurus*) at different environmental temperatures. *J. Physiol.* (*Lond.*) **194**, 809–816
639. Reid, D. (1968) Studies on the cutting management of grass-clover swards. VI. The effects of different closeness and frequency of cutting treatments on the yield and quality of herbage from a cocksfoot–white clover sward. *J. Agric. Sci.* **70**, 59–64
640. Blackburn, P. S. (1968) The cell count of cow's milk and the micro-organisms cultured from the milk. *J. Dairy Res.* **35**, 59–65
641. Hales, J. R. S., Findlay, J. D. & Robertshaw, D. (1968) Evaporative heat loss mechanisms of the newborn calf, *Bos taurus*. *Br. Vet. J.* **124**, 83–88
642. Jenkinson, D. McE. & Blackburn, P. S. (1968) The distribution of nerves, monoamine oxidase and cholinesterase in the skin of the horse. *Res. Vet. Sci.* **9**, 165–169

Institute reprint no.

643. Jenkinson, D. McE., Noble, R. C. & Thompson, G. E. (1968) Adipose tissue and heat production in the new-born ox (*Bos taurus*). *J. Physiol.* (*Lond.*) **195**, 639–646
644. Ulyatt, M. J., Blaxter, K. L. & McDonald, I. (1967) The relations between the apparent digestibility of roughages in the rumen and lower gut of sheep, the volume of fluid in the rumen and voluntary feed intake. *Anim. Prod.* **9**, 463–470
645. Moore, J. H. (1968) Dietary fat and milk fat secretion in the cow. *Proc. Nutr. Soc.* **27**, 66–70
646. Andersson, B., Ekman, L., Hökfelt, B., Jobin, M., Olsson, K. & Robertshaw, D. (1967) Studies of the importance of the thyroid and the sympathetic system in the defence to cold of the goat. *Acta Physiol. Scand.* **69**, 111–118
647. Jenkinson, D. McE. & Blackburn P. S. (1968) The distribution of nerves, monoamine oxidase and cholinesterase in the skin of poultry. *Res. Vet. Sci.* **9**, 429–434
648. Owen, E. C. & Proudfoot, R. (1968) The effect of tungstate ingestion on xanthine oxidase in milk and liver. *Br. J. Nutr.* **22**, 331–340
649. Waite, R., Castle, M. E., Watson, J. N. & Drysdale, A. D. (1968) Biuret and urea in concentrates for milking cows. *J. Dairy Res.* **35**, 191–202
650. Waite, R. & Wilson, A. G. (1968) The composition of rumen fluid from cows fed biuret and urea. *J. Dairy Res.* **35**, 203–212
651. Owen, E. C. & West, D. W. (1968) Photolytic decomposition of 7,8-dimethyl-10-formylmethylisoalloxazine. *Chem. Ind.* (*Lond.*) 881–882
652. Castle, M. E. & Reid, D. (1968) The effects of single compared with split applications of fertilizer nitrogen on the yield and seasonal production of a pure grass sward. *J. Agric. Sci.* **70**, 383–389
653. Hales, J. R. S. & Findlay, J. D. (1968) Respiration of the ox: normal values and the effects of exposure to hot environments. *Respir. Physiol.* **4**, 333–352
654. Hales, J. R. S. & Findlay, J. D. (1968) The oxygen cost of thermally-induced and CO_2-induced hyperventilation in the ox. *Respir. Physiol.* **4**, 353–362
655. Castle, M. E., Drysdale, A. D. & Watson, J. N. (1968) The effect of stocking rate and supplementary concentrate feeding on milk production. *J. Br. Grassl. Soc.* **23**, 137–143
656. Christie, W. W. (1968) Chromatography of the isomeric methylene-interrupted methyl *cis*, *cis*-octadecadienoates. 2. Gas liquid chromatography. *J. Chromatogr.* **37**, 27–32
657. Robertshaw, D. (1968) The pattern and control of sweating in the sheep and the goat. *J. Physiol.* (*Lond.*) **198**, 531–539
658. Jenkinson, D. McE. & Nay, T. (1968) Sweat gland and hair follicle measurements as indicators of skin type in cattle. *Aust. J. Biol. Sci.* **21**, 1001–1011
659. Whittow, G. C. (1968) Cardiovascular response to localized heating of the anterior hypothalamus. *J. Physiol.* (*Lond.*) **198**, 541–548
660. Waite, R. (1968) The composition of the milk of young and of old cows. *J. Dairy Res.* **35**, 407–422
661. Czerkawski, J. W. & Clapperton, J. L. (1968) Analysis of gases produced by metabolism of micro-organisms. *Lab. Pract.* **17**, 994–996, 1012
662. Yaguchi, M., Davies, D. T. & Kim, Y. K. (1968) Preparation of κ-casein by gel filtration. *J. Dairy Sci.* **51**, 473–477
663. Rose, D., Davies, D. T. & Yaguchi, M. (1969) Quantitative determination of the major components of casein mixtures by column chromatography on DEAE-cellulose. *J. Dairy Sci.* **52**, 8–11
664. Bianca, W. (1968) Effects of water deprivation on the water content of cattle skin. *Int. J. Biometeorol.* **12**, 153–157

Institute reprint no.

665. Orr, D. A. & Taylor, M. M. (1968) The disinfection of teat skin. *Br. Vet. J.* **124**, 481–491

666. Owen, E. C. (1968) Simple apparatus for elution of small amounts of material from paper or preparative layer chromatograms. *Lab. Pract.* **17**, 1137

667. Jenkinson, D. McE. & Blackburn, P. S. (1968) The distribution of nerves, monoamine oxidase and cholinesterase in the skin of the cat and dog. *Res. Vet. Sci.* **9**, 521–528

668. Moore, J. H., Noble, R. C. & Steele, W. (1968) Factors affecting the polyunsaturated fatty acid content of the plasma lipids of sheep. *Br. J. Nutr.* **22**, 681–688

669. Czerkawski, J. W. & Breckenridge, G. (1969) The fermentation of sugar-beet pulp and sucrose in an artificial rumen and the effect of linseed oil fatty acids on the fermentation. *Br. J. Nutr.* **23**, 51–66

670. Czerkawski, J. W. & Breckenridge, G. (1969) The effect of oxygen on fermentation of sucrose by rumen micro-organisms *in vitro*. *Br. J. Nutr.* **23**, 67–80

671. Moore, J. H. (1969) The effect of diets containing different proportions of butterfat and maize oil on plasma lipids and aortic atherosis in rabbits. *Br. J. Nutr.* **23**, 125–134

672. Moore, J. H., Noble, R. C. & Steele, W. (1969) The incorporation of linolenic and linoleic acids into the plasma lipids of sheep given intra-abomasal infusions of linseed oil, maize oil or linoleic acid. *Br. J. Nutr.* **23**, 141–152

673. Thompson, G. E. & Jenkinson, D. McE. (1969) Nonshivering thermogenesis in the newborn lamb. *Can. J. Physiol. Pharmacol.* **47**, 249–253

674. Thompson, G. E., Robertshaw, D. & Findlay, J. D. (1969) Noradrenergic innervation of the arrectores pilorum muscles of the ox (*Bos taurus*). *Can. J. Physiol. Pharmacol.* **47**, 310–311

675. Martin, A. K. (1969) Urinary excretion of aromatic acids by sheep given diets containing different amounts of protein and roughage. *Br. J. Nutr.* **23**, 389–399

676. Moore, J. H., Christie, W. W., Braude, R. & Mitchell, K. G. (1969) The effect of dietary copper on the fatty acid composition and physical properties of pig adipose tissues. *Br. J. Nutr.* **23**, 281–287

677. Christie, W. W. & Moore, J. H. (1969) A semimicro method for the stereospecific analysis of triglycerides. *Biochim. Biophys. Acta* **176**, 445–452

678. Czerkawski, J. W. (1969) Metabolism of long chain fatty acids in the rumen. *Proc. 4th Symp. Energy Metabolism*, pp. 103–108

679. Annan, W. D. & Manson, W. (1969) A fractionation of the α_s-casein complex of bovine milk. *J. Dairy Res.* **36**, 259–268

680. Findlay, J. D. & Hales, J. R. S. (1969) Hypothalamic temperature and the regulation of respiration of the ox exposed to severe heat. *J. Physiol.* (*Lond.*) **203**, 651–663

681. Czerkawski, J. W. & Breckenridge, G. (1969) Distribution of polyethylene glycol in suspensions of food particles, especially sugar-beet pulp and dried grass pellets. *Br. J. Nutr.* **23**, 559–565

682. Martin, A. K. (1969) The urinary excretion of aromatic acids by starved sheep. *Br. J. Nutr.* **23**, 715–725

683. Christie, W. W. (1969) The glyceride structure of *Sapium sebiferum* seed oil. *Biochim. Biophys. Acta* **187**, 1–5

684. Moore, J. H. & Noble, R. C. (1969) Dietary linoleic acid and the plasma phospholipids of the rabbit. *Br. J. Nutr.* **23**, 611–618

685. Noble, R. C., Steele, W. & Moore, J. H. (1969) The incorporation of linoleic acid

Institute reprint no.

into the plasma lipids of sheep given intraruminal infusions of maize oil or free linoleic acid. *Br. J. Nutr.* **23**, 709–714

686. Jenkinson, D. McE. & Maloiy, G. M. O. (1969) The distribution of nerves, monoamine oxidase and cholinesterase in the skin of the red deer (*Cervus elaphus*). *Res. Vet. Sci.* **10**, 448–452

687. Steele, W. (1969) The effects of dietary palmitic and stearic acids on milk yield and composition in the cow. *J. Dairy Res.* **36**, 369–373

688. Noble, R. C., Steele, W. & Moore, J. H. (1969) The effects of dietary palmitic and stearic acids on milk fat composition in the cow. *J. Dairy Res.* **36**, 375–381

689. Moore, J. H., Steele, W. & Noble, R. C. (1969) The relationships between dietary fatty acids, plasma lipid composition and milk fat secretion in the cow. *J. Dairy Res.* **36**, 383–392

690. Noble, R. C., Steele, W. & Moore, J. H. (1969) Dietary fatty acids and the plasma phospholipids of the lactating cow. *J. Dairy Res.* **36**, 393–398

691. Christie, W. W. & Moore, J. H. (1969) The effect of dietary copper on the structure and physical properties of adipose tissue triglycerides in pigs. *Lipids* **4**, 345–349

692. Czerkawski, J. W. & Breckenridge, G. (1969) Fermentation of various soluble carbohydrates by rumen micro-organisms with particular reference to methane production. *Br. J. Nutr.* **23**, 925–937

693. Clapperton, J. L. & Czerkawski, J. W. (1969) Methane production and soluble carbohydrates in the rumen of sheep in relation to the time of feeding and the effects of short-term intraruminal infusions of unsaturated fatty acids. *Br. J. Nutr.* **23**, 813–826

694. Castle, M. E. & Watson, J. N. (1969) The effect of level of protein in silage on the intake and production of dairy cows. *J. Br. Grassl. Soc.* **24**, 187–192

695. Moore, J. H., Noble, R. C., Steele, W. and Czerkawski, J. W. (1969) Differences in the metabolism of esterified and unesterified linoleic acid by rumen micro-organisms. *Br. J. Nutr.* **23**, 869–878

696. West, D. W. & Owen, E. C. (1969) The urinary excretion of metabolites of riboflavine by man. *Br. J. Nutr.* **23**, 889–898

697. Thompson, G. E. & Clough, D. P. (1970) Temperature regulation in the new-born ox (*Bos taurus*). *Biol. Neonate* **15**, 19–25

698. Blackburn, P. S. (1969) The incidence of infection of the udder in a herd of cows over a period of years. *Br. Vet. J.* **125**, 504–510

699. Robertshaw, D. & Taylor, C. R. (1969) Sweat gland function of the donkey (*Equus asinus*). *J. Physiol. (Lond.)* **205**, 79–89

700. Noble, R. C., Thompson, G. E. & Moore, J. H. (1969) The effects of the intravenous administration of noradrenalin on the composition of the plasma lipids of sheep. *Res. Vet. Sci.* **10**, 555–558

701. Hales, J. R. S. (1969) Changes in respiratory activity and body temperature of the severely heat-stressed ox and sheep. *Comp. Biochem. Physiol.* **31**, 975–985

702. Mabon, R. M. (1969) Erythrocyte glutathione and growth in the calf. *Br. Vet. J.* **125**, 591–595

703. Bianca, W. & Hales, J. R. S. (1970) Sweating, panting and body temperatures of newborn and one-year-old calves at high environmental temperatures. *Br. Vet. J.* **126**, 45–53

704. Jenkinson, D. McE. (1970) The distribution of nerves, monoamine oxidase and cholinesterase in the skin of the guinea-pig, hamster, mouse, rabbit and rat. *Res. Vet. Sci.* **11**, 60–70

705. Reid, D. (1970) The effects of a wide range of nitrogen application rates on the

Institute reprint no.

yields from a perennial ryegrass sward with and without white clover. *J. Agric. Sci.* **74**, 227–240

706. McLean, J. A. (1969) The environmental needs of farm animals and their heat output. *J. Inst. Heat. Ventil. Eng.* **37**, 201–203

707. Owen, E. C. & West, D. W. (1970) Metabolites of riboflavine in milk, urine and tissues of animals in relation to alimentary symbiotic bacteria. *Br. J. Nutr.* **24**, 45–60

708. Owen, E. C., West, D. W. & Coates, M. E. (1970) Metabolism of riboflavine in germ-free and conventional rabbits. *Br. J. Nutr.* **24**, 259–267

709. Bianca, W. (1970) Effects of dehydration, rehydration and overhydration on the blood and urine of oxen. *Br. Vet. J.* **126**, 121–133

710. Waite, R. (1970) The structural carbohydrates and the *in vitro* digestibility of a ryegrass and a cocksfoot at two levels of nitrogenous fertilizer. *J. Agric. Sci.* **74**, 457–462

711. Castle, M. E. & Watson, J. N. (1970) Silage and milk production, a comparison between grass silages made with and without formic acid. *J. Br. Grassl. Soc.* **25**, 65–71

712. Phillips, J. D. & Taylor, M. M. (1970) The isolation of coryneform organisms from cow fore-milk samples. *Br. Vet. J.* **126**, 210–215

713. Mabon, R. M. (1970) Two possible catecholamine metabolites in the secretion from ox skin. *Res. Vet. Sci.* **11**, 287–288

714. Taylor, M. M., Phillips, J. D., Blackburn, P. S. & Johnston, W. G. (1970) The total somatic cell count related to the bacterial growth from cow fore-milk samples. *Br. Vet. J.* **126**, 254–259

715. Noble, R. C., Steele, W. & Moore, J. H. (1970) The composition of ewe's milk fat during early and late lactation. *J. Dairy Res.* **37**, 297–301

716. Christie, W. W. & Moore, J. H. (1970) A comparison of the structures of triglycerides from various pig tissues. *Biochim. Biophys. Acta* **210**, 46–56

717. Owen, E. C. & Pringle, J. (1970) The inhibitory effect of tungstate on the assay of riboflavin by *Lactobacillus casei* (*ATCC* 7469) and its reversal by molybdate. *Lab. Pract.* **19**, 591–592

718. Day, N., Jenkinson, D. McE. & Walker-Love, J. (1971) Freeze-branding of young beef animals. *Anim. Prod.* **13**, 93–99

719. Reid, D. & Castle, M. E. (1970) A comparison of the effects of anhydrous ammonia and a solid ammonium nitrate fertilizer on herbage production from a pure perennial ryegrass sward. *J. Agric. Sci.* **75**, 347–353

720. Thompson, G. E. & Clough, D. P. (1970) The effect of intravenous infusion of dopamine on the unesterified fatty acids of sheep plasma. *Res. Vet. Sci.* **11**, 428–430

721. Jenkinson, D. McE., Thompson, G. E., Kenny, J. D. R. & Pearson, J. M. (1970) Histochemical studies on mast cells in cattle skin. *Histochem. J.* **2**, 419–424

722. Czerkawski, J. W. & Breckenridge, G. (1970) Small scale apparatus for studying rumen fermentation in vitro. *Lab. Pract.* **19**, 717–719, 728

723. Martin, A. K. (1970) The urinary aromatic acids excreted by sheep given S24 perennial ryegrass cut at six stages of maturity. *Br. J. Nutr.* **24**, 943–959

724. Christie, W. W., & Moore, J. H. (1970) The structure of egg yolk triglycerides. *Biochim. Biophys. Acta* **218**, 83–88

725. Christie, W. W., Noble, R. C. & Moore, J. H. (1970) Determination of lipid classes by a gas-chromatographic procedure. *Analyst* **95**, 940–944

726. Martin, A. K. (1970) Effect of stage of maturity of perennial ryegrass on its

Institute reprint no.

content of some organic acids and phenolic compounds. *J. Sci. Food Agric.* **21**, 496–501

727. Reid, D. & Castle, M. E. (1970) The effects of the date of applying anhydrous ammonia or a solid nitrogen fertilizer on the spring growth from a pure perennial ryegrass sward. *J. Agric. Sci.* **75**, 523–532
728. Christie, W. W. & Moore, J. H. (1970) The variation of triglyceride structure with fatty acid composition in pig adipose tissue. *Lipids* **5**, 921–928
729. Jenkinson, D. McE, & Robertshaw, D. (1971) Studies on the nature of sweat gland 'fatigue' in the goat. *J. Physiol.* (*Lond.*) **212**, 455–465
730. Castle, M. E. & Watson, J. N. (1970) Silage and milk production, a comparison between wilted and unwilted grass silages made with and without formic acid. *J. Br. Grassl. Soc.* **25**, 278–284
731. Whittow, G. C. (1970) Respiratory variations of blood pressure in the ox (*Bos taurus*) during panting. *Br. Vet. J.* **126**, 652–655
732. Noble, R. C., Steele, W. & Moore, J. H. (1971) Diet and the fatty acids in the plasma of lambs during the first eight days after birth. *Lipids* **6**, 26–34
733. Noble, R. C., Steele, W. & Moore, J. H. (1970) Diet and the diacyl phospholipids in the livers of lambs. *Biochim. Biophys. Acta* **218**, 359–362
734. Noble, R. C., Steele, W. & Moore, J. H. (1971) The plasma lipids of the ewe during pregnancy and lactation. *Res. Vet. Sci.* **12**, 47–53
735. Steele, W., Noble, R. C. & Moore, J. H. (1971) The effects of 2 methods of incorporating soybean oil into the diet on milk yield and composition in the cow. *J. Dairy Res.* **38**, 43–48
736. Steele, W., Noble, R. C. & Moore, J. H. (1971) The effects of dietary soybean oil on milk-fat composition in the cow. *J. Dairy Res.* **38**, 49–56
737. Steele, W., Noble, R. C. & Moore, J. H. (1971) The relationship between plasma lipid composition and milk fat secretion in cows given diets containing soybean oil. *J. Dairy Res.* **38**, 57–64
738. Thompson, G. E., Clough, D. P., Scobie, A. M. W. & Kenny, J. D. R. (1971) The localisation of noradrenaline and dopamine in brown adipose tissue of the newborn lamb. *Biol. Neonate* **17**, 394–398
739. Tümmler, R., Steinfelder, K., Owen, E. C. & West, D. W. (1971) Negative ion mass spectrometry of the new riboflavin metabolite: 7,8-Dimethyl-10-formylmethyl-isoalloxazine ethyl-hemi-alcoholate (flavin-10-acetaldehyde ethyl-hemi-alcoholate). *Org. Mass Spectrom.* **5**, 41-52
740. Thompson, G. E. & Clough, D. P. (1971) The effect of intraventricular noradrenaline on plasma unesterified fatty acid concentration in the ox. *Life Sci.* **10**, 415–420
741. Jenkinson, D. McE. (1971) Myoepithelial cells of the sweat glands of domestic animals. *Res. Vet. Sci.* **12**, 152–155
742. McLean, J. A. (1971) A gradient layer calorimeter for large animals. *J. Inst. Heat. Ventil. Eng.* **39**, 1–8
743. Castle, M. E. & Henderson, R. (1971) A note on the relationship between the rate of machine-milking and milk production of Ayrshire heifers. *Anim. Prod.* **13**, 353–356
744. Clough, D. P., Kenny, J. D. R., Scobie, A. M. W. & Thompson, G. E. (1971) Dopamine and mast cells in the adipose tissue of the ox and the effect of noradrenaline and dopamine infusions on the plasma unesterified fatty acids. *Res. Vet. Sci.* **12**, 228–233
745. Christie, W. W. & Moore, J. H. (1971) Structures of triglycerides isolated from various sheep tissues. *J. Sci. Food Agric.* **22**, 120–124

Institute reprint no.

746. Noble, R. C., Steele, W. & Moore, J. H. (1971) Fatty acid composition of liver lipids of young lambs. *Br. J. Nutr.* **26**, 97–105

747. Czerkawski, J. W. & Breckenridge, G. H. McG. (1971) Determination of concentration of hydrogen and some other gases dissolved in biological fluids. *Lab. Pract.* **20**, 403–405, 413

748. Manson, W. & Annan, W. D. (1971) The structure of a phosphopeptide derived from β-casein. *Arch. Biochem. Biophys.* **145**, 16–26

749. Castle, M. E. & Watson, J. N. (1971) A comparison between a diploid and a tetraploid ryegrass for milk production. *J. Agric. Sci.* **77**, 69–76

750. Taylor, M. M. (1971) The water agar test for rapid estimation of psychrotrophs. *Dairy Ind.* **36**, 400–402

751. Taylor, M. M. (1971) The bacteriological control of pasteurising plant using the water agar test. *Dairy Ind.* **36**, 445–447

752. Whittow, G. C. (1971) Cardio-acceleration in the ox (*Bos taurus*) during hyperthermia. *Res. Vet. Sci.* **12**, 495–496

753. Clapperton, J. L. & Czerkawski, J. W. (1971) The effect of tertiary branched-chain carboxylic acids on the energy metabolism of sheep. *Br. J. Nutr.* **26**, 459–468

754. Abbot, J. (1971) The effect of antioxidants on the keeping quality of spray-dried whole milk powder. *J. Soc. Dairy Technol.* **24**, 182–183

755. Czerkawski, J. W. & Breckenridge, G. (1972) Fermentation of various glycolytic intermediates and other compounds by rumen micro-organisms, with particular reference to methane production. *Br. J. Nutr.* **27**, 131–146

756. Castle, M. E. & Watson, J. N. (1971) Silage and milk production. A comparison between a diploid and a tetraploid ryegrass. *J. Br. Grassl. Soc.* **26**, 265–270

757. Christie, W. W., Moore, J. H., Lorimer, A. R. & Lawrie, T. D. V. (1971) The structures of triglycerides from atherosclerotic plaques and other human tissues. *Lipids* **6**, 854–856

758. Jenkinson, D. McE. (1972) The skin structure of British deer. *Res. Vet. Sci.* **13**, 70–73

759. Christie, W. W. & Moore, J. H. (1972) The lipid components of the plasma, liver and ovarian follicle in the domestic chicken (*Gallus gallus*). *Comp. Biochem. Physiol.* **41B**, 287-295

760. Noble, R. C., Christie, W. W. & Moore, J. H. (1971) Diet and the lipid composition of adipose tissue in the young lamb. *J. Sci. Food Agric.* **22**, 616–619

761. Christie, W. W. & Moore, J. H. (1972) The structures of adipose tissue and heart muscle triglycerides in the domestic chicken (*Gallus gallus*). *J. Sci. Food Agric.* **23**, 73–77

762. Noble, R. C., Steele, W. & Moore, J. H. (1972) Postnatal changes in the phospholipid composition of livers from young lambs. *Lipids* **6**, 926–929

763. Christie, W. W. & Moore, J. H. (1972) The lipid composition and triglyceride structure of eggs from several avian species. *Comp. Biochem. Physiol.* **41B**, 297–306

764. Christie, W. W. (1972) Quantitative esterification of lipids on thin-layer adsorbents. *Analyst* **97**, 221–223

765. Jenkinson, D. McE. & Nay, T. (1972) The sweat glands and hair follicles of European cattle. *Aust. J. Biol. Sci.* **25**, 585–595

766. Castle, M. E. (1972) A comparative study of the feeding value of dried sugar-beet pulp for milk production. *J. Agric. Sci.* **78**, 371–377

767. McLean, J. A. & Calvert, D. T. (1972) Influence of air humidity on the partition of heat exchanges of cattle. *J. Agric. Sci.* **78**, 303–307

Institute reprint no.

768. Calvert, D. T., Clough, D. P., Findlay, J. D. & Thompson, G. E. (1972) Hypothalamic cooling, heat production and plasma lipids in the ox. *Life Sci.* **11**, 223–229
769. McLean, J. A. (1972) On the calculation of heat production from open-circuit calorimetric measurements. *Br. J. Nutr.* **27**, 597–600
770. Noble, R. C., Steele, W. & Moore, J. H. (1972) The metabolism of linoleic acid by the young lamb. *Br. J. Nutr.* **27**, 503–508
771. Reid, D. (1972) The effects of the long-term application of a wide range of nitrogen rates on the yields from perennial ryegrass swards with and without white clover. *J. Agric. Sci.* **79**, 291–301
772. Clapperton, J. L. & Czerkawski, J. W. (1972) Metabolism of propane-1:2-diol infused into the rumen of sheep. *Br. J. Nutr.* **27**, 553–560
773. Thompson, G. E. & Clough, D. P. (1972) The effect of cold exposure on plasma lipids of the new-born and adult ox. *Q. J. Exp. Physiol. Cogn. Med. Sci.* **57**, 192–198
774. Rollinson, D. H. L., Injidi, M. H. & Jenkinson, D. McE. (1972) The distribution of nerves, monoamine oxidase and cholinesterase in the skin of the camel (*Camelus dromedarius*). *Res. Vet. Sci.* **13**, 304–305
775. Taylor, M. M. & Phillips, J. D. (1972) A comparison between the bacterial growth from milk on two different media and the somatic cell count. *Br. Vet. J.* **128**, 242–248
776. Taylor, M. M. (1972) *Eubacterium fissicatena* sp. nov. isolated from the alimentary tract of the goat. *J. Gen. Microbiol.* **71**, 457–463
777. Jessen, C., McLean, J. A., Calvert, D. T. & Findlay, J. D. (1972) Balanced and unbalanced temperature signals generated in spinal cord of the ox. *Am. J. Physiol.* **222**, 1343–1347
778. Castle, M. E., MacDaid, E. & Watson, J. N. (1972) Some factors affecting milk production from grassland at the Hannah Institute, 1951–70. *J. Br. Grassl. Soc.* **27**, 87–92
779. Clough, D. P. & Thompson, G. E. (1972) Effect of intraventricular noradrenaline and acetylcholine on plasma unesterified fatty acid concentration in sheep. *Neuroendocrinology* **9**, 365–374
780. Manson, W. & Carolan, T. (1972) The alkali-induced elimination of phosphate from β-casein. *J. Dairy Res.* **39**, 189–194
781. Waite, R. & Smith, G. M. (1972) Measurement of the protein content of milk from mastitic quarters by the Amido Black method. *J. Dairy Res.* **39**, 195–201
782. Castle, M. E. (1972) Acid additives for silage making. *Scott. Agric.* **51**, 302–304
783. Morrison, I. M. (1972) A semi-micro method for the determination of lignin and its use in predicting the digestibility of forage crops. *J. Sci. Food Agric.* **23**, 455–463
784. Noble, R. C., O'Kelly, J. C. & Moore, J. H. (1972) Observations on the lecithin: cholesterol acyltransferase system in bovine plasma. *Biochim. Biophys. Acta* **207**, 519–528
785. Castle, M. E. & MacDaid, E. (1972) The decomposition of cattle dung and its effect on pasture. *J. Br. Grassl. Soc.* **27**, 133–137
786. Czerkawski, J. W. (1972) Fate of metabolic hydrogen in the rumen. *Proc. Nutr. Soc.* **31**, 141–146
787. Thomas, P. C. & Clapperton, J. L. (1972) Significance to the host of changes in fermentation activity. *Proc. Nutr. Soc.* **31**, 165–170
788. Christie, W. W., Jenkinson, D. McE. & Moore, J. H. (1972) Variation in lipid composition through the skin and subcutaneous adipose tissue of pigs. *J. Sci. Food Agric.* **23**, 1125–1129

Institute reprint no.

789. Findlay, J. D. (1972) Climatologic data needed to specify climatic stress. *World Rev. Anim. Prod.* **8**, 38–44

790. Castle, M. E. (1972) A study of the intake of drinking water by dairy cows at grass. *J. Br. Grassl. Soc.* **27**, 207–210

791. Jenkinson, D. McE. & Nay, T. (1973) The sweat glands and hair follicles of Asian, African and South American cattle. *Aust. J. Biol. Sci.* **26**, 259–275

792. West, D. W. & Owen, E. C. (1973) Degradation of riboflavin by alimentary bacteria of the ruminant and man: production of 7,8-dimethyl-10-carboxymethylisoalloxazine. *Br. J. Nutr.* **29**, 33–41

793. West, D. W. & Owen, E. C. (1973) Metabolism of riboflavin: reduction of 7,8-dimethyl-10-formylmethylisoalloxazine by an enzyme in liver. *Br. J. Nutr.* **29**, 43–50

794. Czerkawski, J. W., Harfoot, C. G. & Breckenridge, G. (1972) The relationship between methane production and concentrations of hydrogen in the aqueous and gaseous phases during rumen fermentation *in vitro*. *J. Appl. Bacteriol.* **35**, 537–551

795. Castle, M. E. (1973) A simple grazing system for dairy cows. *Scott. Agric.* **51**, 467–470

796. Morrison, I. M. (1972) Improvements in the acetyl bromide technique to determine lignin and digestibility and its application to legumes. *J. Sci. Food Agric.* **23**, 1463–1469

797. Czerkawski, J. W. & Breckenridge, G. (1973) Dissimilation of 1,2-propanediol by rumen micro-organisms. *Br. J. Nutr.* **29**, 317–330

798. Noble, R. C. & Czerkawski, J. W. (1973) A gas-chromatographic method for the determination of low concentrations of acrylic acid in mixtures of C_2 to C_5 fatty acids in biological materials. *Analyst* **98**, 122–125

799. Morrison, I. M. (1973) The isolation of plant cell wall preparations with low nitrogen contents. *J. Agric. Sci.* **80**, 407–410

800. Castle, M. E. & Watson, J. N. (1973) A comparison between a paddock system and a 'Wye College' system of grazing for milk production. *J. Br. Grassl. Soc.* **28**, 7–11

801. Hunter, M. L., Christie, W. W. & Moore, J. H. (1973) The structures of the principal glycerolipids of pig liver. *Lipids* **8**, 65–70

802. Castle, M. E. (1972) Beet pulp for milk production. *Br. Sugar Beet Rev.* **40**, 269–272

803. Noble, R. C., O'Kelly, J. C. & Moore, J. H. (1973) Observations on changes in lipid composition and lecithin-cholesterol-acyl transferase reaction of bovine plasma induced by heat exposure. *Lipids* **8**, 216–223

804. Jenkinson, D. McE. & Mabon, R. M. (1973) The effect of temperature and humidity on skin surface pH and the ionic composition of skin secretions in Ayrshire cattle. *Br. Vet. J.* **129**, 282–295

805. Moore, J. H. (1974) Lipid biochemistry – from forage to milk, in *Industrial Aspects of Biochemistry* (Spencer, B., ed.), pp. 835–863, North-Holland, Amsterdam

806. Castle, M. E. & Watson, J. N. (1973) Silage and milk production. A comparison between wilted grass silage made with and without formic acid. *J. Br. Grassl. Soc.* **28**, 73–80

807. Harfoot, C. G., Noble, R. C. & Moore, J. H. (1973) Factors influencing the extent of biohydrogenation of linoleic acid by rumen micro-organisms *in vitro*. *J. Sci. Food Agric.* **24**, 961–970

808. Manson, W. (1973) The lability of the phosphate groups of β-casein towards alkali. *Neth. Milk Dairy J.* **27**, 181–187

Institute reprint no.

809. Martin, A. K. (1973) Urinary aromatic acid excretion by fed and fasted sheep in relation to protein metabolism in the rumen. *Br. J. Nutr.* **30**, 251–267

810. Christie, W. W. (1973) The structures of bile phosphatidylcholines. *Biochim. Biophys. Acta* **316**, 204–211

811. Clough, D. P., Darling, K. F., Findlay, J. D. & Thompson, G. E. (1973) Cold sensitivity in the spinal cord of sheep. *Pfluegers Arch. Eur. J. Physiol.* **342**, 137–144

812. Castle, M. E. & Watson, J. N. (1973) The relationship between the DM content of herbage for silage making and effluent production. *J. Br. Grassl. Soc.* **28**, 135–138

813. Czerkawski, J. W. (1973) Effect of linseed oil fatty acids and linseed oil on rumen fermentation in sheep. *J. Agric. Sci.* **81**, 517–531

814. Christie, W. W. & Hunter, M. L. (1973) Stereospecific analysis of triglycerides that contain isotopically-labelled fatty acids. *Biochim. Biophys. Acta* **316**, 282–287

815. Jenkinson, D. McE., Mabon, R. M. & Manson, W. (1974) Sweat proteins. *Br. J. Dermatol.* **90**, 175–181

816. Christie, W. W., Hunter, M. L. & Harfoot, C. G. (1973) The biosynthetic preparation of 1-^{14}C-*trans*-11-octadecenoic acid. *J. Labelled Compd.* **IX**, 483–488

817. Christie, W. W. & Moore, J. H. (1974) The phospholipids of pig adipose tissue. *Comp. Biochem. Physiol.* **47B**, 409–415

818. Morrison, I. M. (1973) Isolation and analysis of lignin-carbohydrate complexes from *Lolium multiflorum. Phytochemistry* (*Oxf.*) **12**, 2979–2984

819. Castle, M. E. & Watson, J. N. (1973) The intake of drinking water by grazing dairy cows. The effect of water availability. *J. Br. Grassl. Soc.* **28**, 203–207

820. Holt, C., Dalgleish, D. G. & Parker, T. G. (1973) Particle size distributions in skim milk. *Biochim. Biophys. Acta* **328**, 428–432

821. Noble, R. C., Moore, J. H. & Harfoot, C. G. (1974) Observations on the pattern of biohydrogenation of esterified and unesterified linoleic acid in the rumen. *Br. J. Nutr.* **31**, 99–108

822. Czerkawski, J. W. (1973) Manipulation of rumen fermentation. *Process Biochem.* **8** (10), 25–30

823. Dalgleish, D. G. (1973) A study of the interaction of calcium ions with bovine α_{s_1}-casein using fluorescence spectroscopy. *Eur. J. Biochem.* **40**, 375–380

824. Czerkawski, J. W. (1974) Methods for determining 2-6-diaminopimelic acid and 2-aminoethylphosphonic acid in gut contents. *J. Sci. Food Agric.* **25**, 45–55

825. Jenkinson, D. McE., McLean, J. A. & Walton, G. S. (1974) The potential use of iontophoresis in the treatment of skin disorders. *Vet. Rec.* **94**, 8–12

826. McLean, J. A. (1974) Improved air-sampling system for open-circuit calorimetry. *Lab. Pract.* **23**, 14–15

827. Clapperton, J. L., Thomas, P. C., Stokes, M. R. & Henderson, J. (1974) A compact machine for feeding sheep up to 12 times daily. *Br. J. Nutr.* **31**, 271–272

828. Bell, A. W., Thompson, G. E. & Findlay, J. D. (1974) The contribution of the shivering hind leg to the metabolic response to cold of the young ox (*Bos taurus*). *Pfluegers Arch. Eur. J. Physiol.* **346**, 341–350

829. Morrison, I. M. (1974) Structural investigations on the lignin-carbohydrate complexes of *Lolium perenne. Biochem. J.* **139**, 197–204

830. Morrison, I. M. (1974) Lignin-carbohydrate complexes from *Lolium perenne. Phytochemistry* (*Oxf.*) **13**, 1161–1165

831. Christie, W. W., Moore, J. H. & Gottenbos, J. J. (1974) Effect of dietary saturated fatty acids and linoleic acid upon the structures of triglycerides in rabbit tissues. *Lipids* **9**, 201–207

Institute reprint no.

832. Davies, D. T. (1974) The quantitative partition of the albumin fraction of milk serum proteins by gel chromatography. *J. Dairy Res.* **41**, 217–228

833. Clegg, K. M., Lim, C. L. & Manson, W. (1974) The structure of a bitter peptide derived from casein by digestion with papain. *J. Dairy Res.* **41**, 283–287

834. Gardner, J. W. & Thompson, G. E. (1974) A method for the determination of volatile fatty acids in the blood plasma of ruminant animals. *Analyst* **99**, 326–329

835. Jenkinson, D. McE., Mabon, R. M. & Manson, W. (1974) The effect of temperature and humidity on the losses of nitrogenous substances from the skin of Ayrshire calves. *Res. Vet. Sci.* **17**, 75–80

836. Castle, M. E. & Watson, J. N. (1974) Red clover silage for milk production. *J. Br. Grassl. Soc.* **29**, 101–108

837. Clapperton, J. L. (1974) The effect of trichloroacetamide, chloroform and linseed oil given into the rumen of sheep on some of the end-products of rumen digestion. *Br. J. Nutr.* **32**, 155–161

838. Darling, K. F., Findlay, J. D. & Thompson, G. E. (1974) Effect of intraventricular acetylcholine and eserine on the metabolism of sheep. *Pfluegers Arch. Eur. J. Physiol.* **349**, 235–245

839. Morrison, I. M. (1974) Changes in the hemicellulosic polysaccharides of ryegrass with increasing maturity. *Carbohyd. Res.* **36**, 45–51

840. Noble, R. C. & Moore, J. H. (1974) Heat exposure and the fatty acid composition of the plasma of the young lamb. *Res. Vet. Sci.* **17**, 204–209

841. Bell, A. W. & Thompson, G. E. (1974) Effects of cold exposure and feeding on net exchange of plasma free fatty acids and glycerol across the hind leg of the young ox. *Res. Vet. Sci.* **17**, 265–267

842. Bell, A. W., Gardner, J. W. & Thompson, G. E. (1974) The effects of acute cold exposure and feeding on volatile fatty acid metabolism in the hind leg of the young ox. *Br. J. Nutr.* **32**, 471–477

843. Hadley, N. F. & Christie, W. W. (1974) The lipid composition and triglyceride structure of eggs and fat bodies of the lizard *Sceloporus jarrovi*. *Comp. Biochem. Physiol.* **48B**, 275–284

844. Noble, R. C., Crouchman, M. L. & Moore, J. H. (1974) The presence of linoleic acid in the skin surface lipids of the ox. *Res. Vet. Sci.* **17**, 372–376

845. Bell, A. W. & Thompson, G. E. (1974) The effects of acute cold exposure and feeding on the circulation of the young ox (*Bos taurus*), with special reference to the hind leg. *Res. Vet. Sci.* **17**, 384–389

846. Sweetsur, A. W. M. & White, J. C. D. (1974) Studies on the heat stability of milk protein. I. Interconversion of type A and type B milk heat-stability curves. *J. Dairy Res.* **41**, 349–358

847. Sweetsur, A. W. M. (1974) An ion-selective method for the determination of chloride in milk. *Analyst* **99**, 690–692

848. Castle, M. E. (1974) Beet pulp for the dairy cow. *Br. Sugar Beet Rev.* **42**, 146–151

849. Morrison, I. M. (1975) Delignification and hemicellulose extraction of cell walls of *Lolium perenne* and *Trifolium pratense*. *Phytochemistry (Oxf.)* **14**, 505–508

850. Smith, M. E. & Jenkinson, D. McE. (1975) The effect of age, sex and season on sebum output of Ayrshire calves. *J. Agric. Sci.* **84**, 57–60

851. Reid, D. & Strachan, N. H. (1974) The effects of a wide range of nitrogen rates on some chemical constituents of the herbage from perennial ryegrass swards with and without white clover. *J. Agric. Sci.* **83**, 393–401

852. Christie, W. W. (1974) Biosynthesis of triglycerides in freshly secreted milk from goats. *Lipids* **9**, 876–882

Institute reprint no.

853. Mabon, R. M. (1974) Deoxyribonucleic acid (DNA) in cattle skin washings. *Br. J. Dermatol.* **91**, 271–274

854. Jenkinson, D. McE. & Nay, T. (1975) The sweat glands and hair follicles of different species of *Bovidae*. *Aust. J. Biol. Sci.* **28**, 55–68

855. Harfoot, C. G., Crouchman, M. L., Noble R. C. & Moore, J. H. (1974) Competition between food particles and rumen bacteria in the uptake of long-chain fatty acids and triglycerides. *J. Appl. Bacteriol.* **37**, 633–641

856. Calvert, D. T. (1975) Core temperature and heat loss in the ox. *Bibl. Radiol.* **6**, 224–230

857. Holt, C., Parker, T. G. & Dalgleish, D. G. (1975) The thermochemistry of reactions between α_{s_1}-casein and calcium chloride. *Biochim. Biophys. Acta* **379**, 638–644

858. Bell, A. W., Gardner, J. W., Manson, W. & Thompson, G. E. (1975) Acute cold exposure and the metabolism of blood glucose, lactate and pyruvate, and plasma amino acids in the hind leg of the fed and fasted young ox. *Br. J. Nutr.* **33**, 207–217

859. Noble, R. C., Crouchman, M. L. & Moore, J. H. (1975) Plasma cholesterol ester formation in the neonatal lamb. *Biol. Neonate* **26**, 117–121

860. Noble, R. C. & Moore, J. H. (1975) The essential fatty acid status of the new-born ruminant – a case for improvement. *IRCS Med. Sci.* **3**, 1–4

861. Sweetsur, A. W. M. & White, J. C. D. (1975) Studies on the heat stability of milk protein. II. Effect of exposing milk to light. *J. Dairy Res.* **42**, 57–71

862. Sweetsur, A. W. M. & White, J. C. D. (1975) Studies on the heat stability of milk protein. III. Effect of heat-induced acidity in milk. *J. Dairy Res.* **42**, 73–88

863. Thompson, G. E., Gardner, J. W. & Bell, A. W. (1975) The oxygen consumption, fatty acid and glycerol uptake of the liver in fed and fasted sheep during cold exposure. *Q. J. Exp. Physiol. Cogn. Med. Sci.* **60**, 107–121

864. Noble, R. C., Crouchman, M. L., Jenkinson, D. McE. & Moore, J. H. (1975) Relationship between lipids in plasma and skin secretions of neonatal calf with particular reference to linoleic acid. *Lipids* **10**, 128–133

865. Castle, M. E. & Watson, J. N. (1975) Further comparisons between a rigid rotational 'Wye College' system and other systems of grazing for milk production. *J. Br. Grassl. Soc.* **30**, 1–6

866. Castle, M. E. & Thomas, T. P. (1975) The water intake of British Friesian cows on rations containing various forages. *Anim. Prod.* **20**, 181–189

867. Vernon, R. G. (1975) Effect of dietary safflower oil upon lipogenesis in neonatal lamb. *Lipids* **10**, 284–289

868. Hodgson, J. C. & Thomas, P. C. (1975) A relationship between the molar proportion of propionic acid and the clearance rate of the liquid phase in the rumen of the sheep. *Br. J. Nutr.* **33**, 447–456

869. Taylor, M. M. (1975) The water agar test: a new test to measure the bacteriological quality of cream. *J. Hyg.* **74**, 345–357

870. Holt, C., Parker, T. G. & Dalgleish, D. G. (1975) Measurement of particle sizes by elastic and quasi-elastic light-scattering. *Biochim. Biophys. Acta* **400**, 283–292

871. Holt, C. (1975) Casein micelle size from elastic and quasi-elastic light scattering measurements. *Biochem. Biophys. Acta* **400**, 293–301

872. Morrison, I. M. (1975) Determination of the degree of polymerisation of oligo- and polysaccharides by gas-liquid chromatography. *J. Chromatogr.* **108**, 361–364

873. Noble, R. C., Lorimer, A. R., Moore, J. H. & Lawrie, T. D. V. (1975) The determination of plasma unesterified fatty acid concentrations by titrimetric procedures—a cautionary note. *IRCS Med. Sci.* **3**, 230

Institute reprint no.

874. Noble, R. C. (1975) Use of an internal vial for determining quench correction in scintillation counting of heterogeneous media. *Lipids* **10**, 312–314
875. Sweetsur, A. W. M. & Wilson, A. G. (1975) An ion-selective electrode method for the determination of nitrate in grass and clover. *Analyst* **100**, 485–488
876. Czerkawski, J. W., Christie, W. W., Breckenridge, G. & Hunter, M. L. (1975) Changes in the rumen metabolism of sheep given increasing amounts of linseed oil in their diet. *Br. J. Nutr.* **34**, 25–44.
877. Castle, M. E., Macdaid, E. & Watson, J. N. (1975) The automatic recording of the grazing behaviour of dairy cows. *J. Br. Grassl. Soc.* **30**, 161–163
878. Czerkawski, J. W. & Breckenridge, G. (1975) Separation and determination of mass and radioactivity of fermentation gases. *Anal. Biochem.* **67**, 476–484
879. Castle, M. E. (1975) Silage and milk production. *Agric. Prog.* **50**, 53–60
880. Taylor, M. M. (1975) Bacteriological tests as indices for the development of off-flavours in cream. *J. Hyg.* **75**, 113–119
881. Smith, M. E., Noble, R. C. & Jenkinson, D. McE. (1975) The effect of environment on sebum output and composition in cattle. *Res. Vet. Sci.* **19**, 253–258
882. Noble, R. C. & Shand, J. H. (1975) A simple method for the differential counting of ^{14}C and ^{32}P using a polyethylene minivial. *Anal. Biochem.* **68**, 345–347
883. Taylor, M. M. (1975) The development of taints in pasteurized cream. *J. Soc. Dairy Technol.* **28**, 215–219
884. Jenkinson, D. McE., Mason, I. L. and Nay, T. (1975) Inheritance of some sweat gland and hair follicle characteristics in cattle. *Aust. J. Biol. Sci.* **28**, 417–424
885. Castle, M. E. & Watson, J. N. (1975) Silage and milk production. A comparison between barley and dried grass as supplements to silage of high digestibility. *J. Br. Grassl. Soc.* **30**, 217–222
886. Smith, M. E. & Jenkinson, D. McE. (1975) The mode of secretion of the sebaceous glands of cattle. *Br. Vet. J.* **131**, 610–618
887. Christie, W. W. & Vernon, R. G. (1975) The positional distribution of fatty acids incorporated into triacylglycerols by rat adipose tissue slices *in vitro*. *Biochem. Biophys. Res. Commun.* **66**, 243–249
888. Christie, W. W. (1975) Structural analyses of triglycerides containing isotopically-labelled fatty acids. *J. Chromatogr. Sci.* **13**, 411–415
889. Calvert, D. T. & Findlay, J. D. (1975) Localization of the effective thermosensitive site in the preoptic region of the ox. *J. Appl. Physiol.* **39**, 702–706
890. Christie, W. W., Moore, J. H., Noble, R. C. & Vernon, R. G. (1975) Changes with diet in the composition of phosphatidyl choline of sheep bile. *Lipids* **10**, 645–648
891. Christie, W. W., Hunter, M. L., Moore, J. H., Noble, R. C. & Vernon, R. G. (1975) Effect of diet on the composition of cholesteryl esters of sheep adrenals. *Lipids* **10**, 649–651
892. Czerkawski, J. W. & Breckenridge, G. (1975) New inhibitors of methane production by rumen micro-organisms. Development and testing of inhibitors *in vitro*. *Br. J. Nutr.* **34**, 429–446
893. Czerkawski, J. W. & Breckenridge, G. (1975) New inhibitors of methane production by rumen micro-organisms. Experiments with animals and other practical possibilities. *Br. J. Nutr.* **34**, 447–457
894. Kelly, N. C. & Thomas, P. C. (1975) The effect of intravenous infusions of methionine on the voluntary intake of grass silage by sheep. *J. Br. Grassl. Soc.* **30**, 237–239
895. Taylor, M. M. (1975) An investigation of the bacteriological quality of bakery cream. *J. Food Technol.* **10**, 637–643

Institute reprint no.

896. Harfoot, C. G., Noble, R. C. & Moore, J. H. (1975) The role of plant particles, bacteria and cell-free supernatant fractions of rumen contents in the hydrolysis of trilinolein and the subsequent hydrogenation of linoleic acid. *Antonie Van Leeuwenhoek J. Microbiol. Serol.* **41**, 533–542

897. Christie, W. W. & Wooding, F. B. P. (1975) The site of triglyceride biosynthesis in milk. *Experientia* **31**, 1445–1447

898. Manson, W. (1975) Nutritional aspects of milk protein. *J. Soc. Dairy Technol.* **28**, 203–209

899. Noble, R. C., Crouchman, M. L. & Moore, J. H. (1975) Synthesis of cholesterol esters in the plasma and liver of sheep. *Lipids* **10**, 790–799

900. Bell, A. W., Clarke, P. L. & Thompson, G. E. (1975) Changes in the metabolism of the shivering hind leg of the young ox during several days of continuous cold exposure. *Q. J. Exp. Physiol. Cogn. Med. Sci.* **60**, 267–284

901. Sweetsur, A. W. M. & Phillips, J. D. (1976) Effect of storage on the electronic cell count of milk. *J. Dairy Res.* **43**, 53–62

902. Thomas, P. C. & Kelly, M. E. (1976) The effect of frequency of feeding on milk secretion in the Ayrshire cow. *J. Dairy Res.* **43**, 1–7

903. Thompson, G. E. & Bell, A. W. (1976) Heat production in the newborn ox during noradrenaline infusion. *Biol. Neonate* **28**, 375–381

904. Manson, W., Annan, W. D. & Barnes, G. K. (1976) α_{s_0}-casein: its preparation and characterization. *J. Dairy Res.* **43**, 133–136

905. Morrison, I. M. (1976) Xylanase (hemicellulase) activity in cell-free rumen fluid. *Carbohydr. Res.* **47**, 129–139

906. Chamberlain, D. G., Thomas, P. C. & Wilson, A. G. (1976) Efficiency of bacterial protein synthesis in the rumen of sheep receiving a diet of sugar beet pulp and barley. *J. Sci. Food Agric.* **27**, 231–238

907. Czerkawski, J. W. (1976) The artificial rumen. *Lab. Pract.* **25**, 15–20

908. Bell, A. W., Hilditch, T. E., Horton, P. W. & Thompson, G. E. (1976) The distribution of blood flow between individual muscles and non-muscular tissues in the hind limb of the young ox (*Bos taurus*): values at thermoneutrality and during exposure to cold. *J. Physiol.* (*Lond.*) **257**, 229–243

909. Castle, M. E. (1976) A simple disc instrument for estimating herbage yield. *J. Br. Grassl. Soc.* **31**, 37–40

910. McLean, J. A. & Watts, P. R. (1976) Analytical refinements in animal calorimetry. *J. Appl. Physiol.* **40**, 827–831

911. Thomas, P. C., Kelly, N. C. & Wait, M. K. (1976) The effect of physical form of a silage on its voluntary consumption and digestibility by sheep. *J. Br. Grassl. Soc.* **31**, 19–22

912. Banks, W., Clapperton, J. L. & Ferrie, M. E. (1976) The physical properties of milk fats of different chemical compositions. *J. Soc. Dairy Technol.* **29**, 86–90

913. Christie, W. W., Vernon, R. G. & Wooding, F. B. P. (1976) Lipid metabolism in cytoplasmic droplets from freshly secreted milk. *Biochem. Soc. Trans.* **4**, 242–243

914. West, D. W. & Dalgleish, D. G. (1976) A kinetic analysis of the dephosphorylation, by bovine spleen phosphoprotein phosphatase (EC3.1.3.16) of a phosphopeptide derived from β-casein. *Biochim. Biophys. Acta* **438**, 169–175

915. Noble, R. C., Mabon, R. M. & Jenkinson, D. McE. (1976) The effects of exposure to warm environments on bovine plasma lipid composition. *Res. Vet. Sci.* **21**, 90–93

916. Jenkinson, D. McE., Lloyd, D. H. & Mabon, R. M. (1976) Location of the source of the soluble skin surface proteins of cattle by immunofluorescence. *Res. Vet. Sci.* **21**, 124–126

Institute reprint no.

917. Banks, W., Clapperton, J. L., Ferrie, M. E. & Wilson, A. G. (1976) Effect of feeding fat to dairy cows receiving a fat-deficient basal diet. I. Milk yield and composition. *J. Dairy Res.* **43**, 213–218
918. Banks, W., Clapperton, J. L. & Ferrie, M. E. (1976) Effect of feeding fat to dairy cows receiving a fat-deficient basal diet. II. Fatty acid composition of the milk fat. *J. Dairy Res.* **43**, 219–227
919. Banks, W., Mazumder, N. K. and Spooner, R. L. (1976) The action pattern of bovine pancreatic alpha-amylase. *Int. J. Biochem.* **7**, 107–110
920. Czerkawski, J. W. (1976) Chemical composition of microbial matter in the rumen. *J. Sci. Food Agric.* **27**, 621–632
921. Jenkinson, D. McE. (1976) The skin surface and body defence. *ARC Res. Rev.* **2**, 5–8
922. Morrison, I. M. (1976) New laboratory methods for predicting the nutritive value of forage crops. *World Rev. Anim. Prod.* **12**(3), 75–82
923. Williams, A. G. & Harfoot, C. G. (1976) Factors affecting the uptake and metabolism of soluble carbohydrates by the rumen ciliate *Dasytricha ruminantium* isolated from ovine rumen contents by filtration. *J. Gen. Microbiol.* **96**, 125–136
924. Smith, M. E. & Ahmed, S. U. (1976) The lipid composition of cattle sebaceous glands: a comparison with skin surface lipid. *Res. Vet. Sci.* **21**, 250–252
925. Shand, J. H. & Noble, R. C. (1976) The differential counting of ^{125}I and ^{3}H by scintillation counting using a polyethylene miniature vial. *IRCS Med. Sci.* **4**, 340
926. Clarke, P. L., Thomson, E. M. & Thompson, G. E. (1976) Acute cold exposure and milk secretion in the goat. *J. Physiol. (Lond.)* **263**, 176P
927. Vernon, R. G. (1976) Effect of dietary fats on ovine adipose tissue metabolism. *Lipids* **11**, 662–669
928. Sweetsur, A. W. M. (1976) The stability of instantized skimmed milk powder to hot coffee. *J. Soc. Dairy Technol.* **29**, 157–160
929. Thomas, P. C., Chamberlain, D. G. & Alwash, A. M. (1976) Digestion in the stomach and intestines of sheep receiving diets of red clover silage with various supplements. *J. Br. Grassl. Soc.* **31**, 123–128
930. Noble, R. C., Shand, J. H. & West, I. G. (1976) A radio-chromatographic scanning procedure for ^{32}P-labelled compounds. *Lab. Pract.* **25**, 681–682
931. Christie, W. W., Hunter, M. L. & Vernon, R. G. (1976) Triacylglycerol biosynthesis in rat adipose-tissue homogenates. *Biochem. J.* **159**, 571–577
932. Shand, J. H. & Noble, R. C. (1976) A simple method for the differential measurement of ^{125}I and ^{3}H by liquid-scintillation spectrometry. *Clin. Sci. Mol. Med.* **51**, 511–513
933. Lloyd, D. H., Mabon, R. M. & Jenkinson, D. McE. (1977) The antigenic constituents of cattle skin washings. *J. Comp. Pathol.* **87**, 75–82
934. Czerkawski, J. W. (1976) The use of pivalic acid as a reference substance in measurements of production of volatile fatty acids by rumen micro-organisms *in vitro*. *Br. J. Nutr.* **36**, 311–315
935. Hodgson, J. C., Thomas, P. C. & Wilson, A. G. (1976) The influence of the level of feeding on fermentation in the rumen of sheep receiving a diet of ground barley, ground hay and flaked maize. *J. Agric. Sci.* **87**, 297–302
936. Watts, P. R. (1976) A system for automatic collection and removal of urine from male farm animals. *Br. J. Nutr.* **36**, 295–297
937. Noble, R. C., Evans, A. J. & Crouchman, M. L. (1977) Lecithin-cholesterol-acyltransferase activity in the plasma of immature, oestrogenised and laying fowl (*Gallus domesticus*). *Res. Vet. Sci.* **22**, 35–39
938. Thomas, P. C. (1975) Diet and milk secretion in the ruminant. *World Rev. Anim. Prod.* **11**(4), 33–44

Institute reprint no.

939. Thomas, P. C. (1976) Diet and milk protein production. *J. Soc. Dairy Technol.* **29**, 31–37
940. West, D. W. & Towers, G. E. (1976) Cellulose acetate electrophoresis of casein proteins. *Anal. Biochem.* **74**, 58–66
941. Holt, C. (1976) The stability of bovine casein micelles. *Proc. Int. Conf. Colloid Surf. Sci.* **1**, 641–644, Akademia Kiado, Budapest
942. Castle, M. E. & Watson, J. N. (1976) Silage and milk production. A comparison between barley and groundnut cake as supplements to silage of high digestibility. *J. Br. Grassl. Soc.* **31**, 191–195
943. Noble, R. C., Shand, J. H. & Moore, J. H. (1977) 'Protected' linoleic acid as an antenatal supplement improves the essential fatty acid status of the new-born ruminant. *IRCS Med. Sci.* **5**, 41
944. Parker, T. G. & Dalgleish, D. G. (1977) The potential application of the theory of branching processes to the association of milk protein. *J. Dairy Res.* **44**, 79–84
945. Parker, T. G. & Dalgleish, D. G. (1977) The theory of branching processes applied to milk proteins. I. Heat-induced coagulation of whole milk. *J. Dairy Res.* **44**, 85–92
946. Muir, D. D. & Sweetsur, A. W. M. (1976) The influence of naturally occurring levels of urea on the heat stability of bulk milk. *J. Dairy Res.* **43**, 495–499
947. Noble, R. C., Shand, J. H. & West, I. G. (1977) A continuous radio-scanning method for chromatograms and electrophoretograms. *Lab. Pract.* **26**, 174–175
948. Griffiths, M. W. & Banks, W. (1976) The maltase activity of bovine serum as an aid to the study of the Amylase I locus. *Anim. Blood Grps Biochem. Genet.* **7**, 201–206
949. Noble, R. C. & Drummond, J. T. (1977) The construction of permanent standards for scintillation counting involving the use of polyethylene mini-vials. *Lab. Pract.* **26**, 23–24
950. Castle, M. E., Retter, W. C., Watson, J. N. & Zewdie, E. (1977) Silage and milk production: a comparison between four rates of groundnut cake supplementation of silage of high digestibility. *J. Br. Grassl. Soc.* **32**, 43–48
951. Henderson, R. J., Christie, W. W. & Moore, J. H. (1977) Some relationships between fatty acid biosynthesis and esterification in rat adipocytes. *Biochem. Soc. Trans.* **5**, 284–285
952. Noble, R. C., Shand, J. H. & Drummond, J. T. (1977) Erythrocyte fragility and the essential fatty acid status of the neonatal lamb. *IRCS Med. Sci.* **5**, 196
953. Clapperton, J. L. (1977) The effect of a methane-suppressing compound, tri-chloroethyl adipate, on rumen fermentation and the growth of sheep. *Anim. Prod.* **24**, 169–181
954. Noble, R. C., Vernon, R. G., Christie, W. W., Moore, J. H. & Evans, A. J. (1977) The effect of dietary fats on the plasma lipid composition of sheep. *Lipids* **12**, 423–433
955. Davies, D. T. & Law, A. J. R. (1977) An improved method for the quantitative fractionation of casein mixtures using ion-exchange chromatography. *J. Dairy Res.* **44**, 312–321
956. Watts, P. R. (1977) A calorimetric analysis of ciracadian rhythms in the distribution of body temperatures in ox calves (*Bos taurus*). *J. Thermal Biol.* **2**, 69–74
957. Watts, P. R., McLean, J. A. & Downie, A. J. (1977) Heat balance in calves (*Bos taurus*) over 24-hr periods. *J. Thermal Biol.* **2**, 61–68
958. Vernon, R. G. (1977) Effect of different fatty acids on lipogenesis in rat and sheep adipose tissue *in vitro*. *Int. J. Biochem.* **8**, 517–523

Institute reprint no.

959. White, J. C. D. & Sweetsur, A. W. M. (1977) Kinetics of the heat-induced aggregation of milk protein. *J. Dairy Res.* **44**, 237 247
960. Banks, W. & Greenwood, C. T. (1977) Mathematical models for the action of alpha amylase on amylose. *Carbohydr. Res.* **57**, 301–315
961. Mackinlay, A. G., West, D. W. & Manson, W. (1977) Specific casein phosphorylation by a casein kinase from lactating bovine mammary gland. *Eur. J. Biochem.* **76**, 233–243
962. Vernon, R. G. (1977) Development of perirenal adipose tissue in the neonatal lamb: effects of dietary safflower oil. *Biol. Neonate* **32**, 15–23
963. Morrison, I. M. (1977) Extraction of hemicellulose from plant cell-walls with water after preliminary treatment with methanolic sodium methoxide. *Carbohydr. Res.* **57**, C4–C6
964. Christie, W. W. (1977) Freshly-secreted milk as a source of mammary gland enzymes. *ARC Res. Rev.* **3**, 15–17
965. Manson, W., Carolan, T. & Annan, W. D. (1977) Bovine α_{s_0}-casein, a phosphorylated homologue of α_{s_1}-casein. *Eur. J. Biochem.* **78**, 411–417
966. Noble, R. C., Drummond, J. T. & Shand, J. H. (1977) The use of minature glass vials in the provision of multi-functional standards for scintillation counting. *Lab. Pract.* **26**, 399
967. Chalmers, J. S., Thomas, P. C. & Kelly, M. E. (1977) The effect of level of feeding and composition of the diet on milk secretion in the Ayrshire cow. *Proc. Nutr. Soc.* **36**, 55A
968. Farhan, S. M. A. & Thomas, P. C. (1977) The effect of intraabomasal infusions of glucose or casein on milk secretion in Saanen goats receiving a low protein diet. *Proc. Nutr. Soc.* **36**, 57A
969. Muir, D. D. & Sweetsur, A. W. M. (1977) Effect of urea on the heat coagulation of the caseinate complex in skim milk. *J. Dairy Res.* **44**, 249–257
970. White, J. C. D. & Sweetsur, A. W. M. (1977) Kinetics of the heat-induced aggregation of milk protein. *J. Dairy Res.* **44**, 237–247
971. Parker, T. G. & Dalgleish, D. G. (1977) The use of light-scattering and turbidity measurements to study the kinetics of extensively aggregating proteins: α_s casein. *Biopolymers*, **16**, 2533–2547
972. Thompson, G. E., Bassett, J. M. & Bell, A. W. (1978) The effects of feeding and acute cold exposure on the visceral release of volatile fatty acids, estimated hepatic uptake of propionate and release of glucose and plasma insulin concentration in sheep. *Br. J. Nutr.* in press
973. Castle, M. E., Retter, W. C. & Watson, J. N. (1977) Silage and milk production. A comparison between additives for silage of high digestibility. *J. Br. Grassl. Soc.* **32**, 157–164
974. Castle, M. E., Retter, W. C. & Metcalfe, J. D. (1977) A note on supplements for dairy cows offered silage of high digestibility. *Anim. Prod.* **25**, 397–400

NUMBERED BRIEF REPORTS

Institute brief report no.

1. Blaxter, K. L., Rook, J. A. F. & McDonald, A. M. (1954) The origin of tetany in magnesium deficiency. *Proc. Nutr. Soc.* **13**, ii
2. Blaxter, K. L. & Graham, N. McC. (1954) Plane of nutrition and energy utilization by sheep. *Proc. Nutr. Soc.* **13**, vii

Institute brief report no.

3. Blaxter, K. L., Graham, N. McC. & Wainman, F. W. (1955) Interrelations between passage of food through the digestive tract and its digestibility. *Proc. Nutr. Soc.* **14**, iv
4. Cowlishaw, B. & Blaxter, K. L. (1955) Triorthocresyl phosphate and vitamin E metabolism in calves. *Proc. Nutr. Soc.* **14**, iii
5. Blaxter, K. L. & Graham, N. McC. (1955) Calorimetric measurements of the nutritive value for sheep of dried grass prepared in different ways. *Proc. Nutr. Soc.* **14**, xv
6. Armstrong, D. G. & Blaxter, K. L. (1956) The heat increment of a mixture of steam-volatile fatty acids and of glucose in the fasting sheep. *Proc. Nutr. Soc.* **15**, i
7. Blaxter, K. L., Graham, N. McC. & Wainman, F. W. (1956) Frequency of feeding and energy utilization by sheep. *Proc. Nutr. Soc.* **15**, ii
8. Findlay, J. D. (1955) The respiratory behaviour of calves exposed to increasing thermal stress. *J. Physiol.* (*Lond.*) **130**, 16P–17P
9. Nisbet, A. M. (1955) Sensory nerve endings in the bovine muzzle (planum nasolabial). *J. Physiol.* (*Lond.*) **130**, 3P–4P
10. Bianca, W. (1955) The effect of repeated short term exposures to high environmental temperatures on the volume and hydration of the blood of the calf. *J. Physiol.* (*Lond.*) **130**, 17P
11. Blaxter, K. L., Hutcheson, M. K., Sharman, G. A. M. & MacDonald, A. M. (1957) Anaemia in calves. *Proc. Nutr. Soc.* **16**, i
12. Blaxter, K. L., Graham, N. McC. & Wainman, F. W. (1957) Energy retention of sheep in relation to environmental temperature. *Proc. Nutr. Soc.* **16**, iii–iv
13. Owen, E. C. & Proudfoot, R. (1958) The use of chromium sesquioxide to determine 'instantaneous' digestibilities. *Proc. Nutr. Soc.* **17**, xiii–xiv
14. Blackburn, P. S., Blaxter, K. L. & Castle, Elizabeth J. (1957) Vitamin D_3 toxicity in calves. *Proc. Nutr. Soc.* **16**, xvi
15. Armstrong, D. G. Blaxter, K. L., Graham, N. McC. & Wainman, F. W. (1957) The energetic efficiency of fat synthesis from mixtures of steam-volatile fatty acids in sheep. *Proc. Nutr. Soc.* **16**, xvii
16. Crossland, A., Owen, E. C. & Proudfoot, R. (1958) Acid-resistant lyochromes, with dithionite-quenchable fluorescence, occuring in ruminant urine and milk. *Biochem. J.* **68**, 14P
17. Armstrong, D. G., Blaxter, K. L., Graham, N. McC. & Wainman, F. W. (1958) The effect of environmental temperature and level of feeding on food utilization by sheep. *Proc. Nutr. Soc.* **17**, xxxii
18. Armstrong, D. G., Blaxter, K. L. & Graham, N. McC. (1958) The heat increment of acetic acid partially neutralized with sodium hydroxide, in fasting sheep. *Proc. Nutr. Soc.* **17**, xxxii–xxxiii
19. Nisbet, W. (1958) A simple electric respiration counter. *J. Physiol.* (*Lond.*) **144**, 4P–5P
20. Crossland, A., Owen, E. C. & Proudfoot, R. (1959) The effect of tungstate on the xanthine oxidase in goat's milk. *Proc. Nutr. Soc.* **18**, i–ii
21. Modi, V. V., Owen, E. C. & Proudfoot, R. (1959) Species differences in the occurrence of xanthine oxidase in milk. *Proc. Nutr. Soc.* **18**, i
22. Owen, E. C. & Proudfoot, R. (1959) Micro-organisms in the rumen of goats eating an artificial diet deficient in riboflavin, lysine and tryptophan. *Proc. Nutr. Soc.* **18**, xix
23. Owen, E. C. & Proudfoot, R. (1959) Effect of reduced intake of greenstuff on β-carotene and vitamin A in goat colostrum. *Proc. Nutr. Soc.* **18**, xxii

Institute brief report no.

24. Montgomery, J., Owen, E. C. & Proudfoot, R. (1959) 'Riboflavin metabolite'. *Proc. Nutr. Soc.* **18**, xxxvii–xxxviii
25. Sharman, G. A. M., Blaxter, K. L. & Wilson, R. S. (1959) Prevention of enzootic muscular dystrophy by selenium administration. *Vet. Rec.* **71**, 536
26. Blackburn, P. S., Castle, M. E. & Drysdale, A. D. (1959) Treatment of ketosis in dairy cows by potassium chlorate. *Vet. Rec.* **71**, 665–666
27. Manson, W. (1959) The N-terminal amino-acids of α-casein. *Nature* (*Lond.*) **184**, 1393–1394
28. Owen, E. C., Ellis, S. E., Voss, R. C., Wilson, A. L. & Robertson, J. (1960) The occurrence of phosphates of calcium on cobalt oxide 'bullets' administered to lambs to prevent pining. *Proc. Nutr. Soc.* **19**, xxi–xxii
29. Blaxter, K. L. & Wainman, F. W. (1960) The effect of environmental temperature on the energy metabolism of steers. *Proc. Nutr. Soc.* **19**, xvi–xvii
30. Clapperton, J. L., Armstrong, D. G. & Blaxter, K. L. (1960) The effect of muscular work on the energy metabolism of sheep. *Proc. Nutr. Soc.* **19**, xvii
31. Owen, E. C. & Hytten, F. E. (1960) Xanthine oxidase in human milk. *Proc. Nutr. Soc.* **19**, xxviii
32. Blaxter, K. L., Wainman, F. W. & Wilson, R. S. (1960) The voluntary food intake of sheep. *Proc. Nutr. Soc.* **19**, xxx–xxxi
33. Armstrong, D. G., Blaxter, K. L. & Graham. N. McC. (1960) Fat synthesis from glucose by sheep. *Proc. Nutr. Soc.* **19**, xxxi–xxxii
34. Blaxter, K. L., McCallum, E. S. R., Wilson, R. S., Sharman, G. A. M. & Donald, L. G. (1961) Prevention of enzootic muscular dystrophy by selenium administration. *Proc. Nutr. Soc.* **20**, vi–vii
35. Martin, A. K. & Blaxter, K. L. (1961) Protein as a source of energy for synthesis of fat in sheep. *Proc. Nutr. Soc.* **20**, vii–viii
36. Clapperton, J. L. (1961) The energy expenditure of sheep in walking on the level and on gradients. *Proc. Nutr. Soc.* **20**, xxxi-xxxii
37. Blaxter, K. L. & Wainman, F. W. (1961) The utilization of the energy of the same food by cattle and sheep. *Proc. Nutr. Soc.* **20**, xxxiii–xxxiv
38. Ingram, D. L. & Whittow, G. C. (1961) The effect of heating the hypothalamus of the ox. *J. Physiol.* (*Lond.*) **159**, 86P–87P
39. Owen, E. C. (1962) Examination of paper chromatograms for metabolites of riboflavin by using the petrological microscope. *Biochem. J.* **82**, 7P–8P
40. Owen, E. C., Montgomery, J. P. & Proudfoot, R. (1962) Properties of a metabolite of riboflavin. *Biochem. J.* **82**, 8P
41. Owen, E. C., Hart, L. I. & Hytten, F. E. (1962) Xanthine oxidase in human milk. *Proc. Nutr. Soc.* **21**, xv–xvi
42. Blaxter, K. L. & Wilson, R. S. (1962) The voluntary food intake of steers and wether sheep. *Proc. Nutr. Soc.* **21**, xxii–xxiii
43. Blaxter, K. L. (1962) The effect of selenium on lamb growth: co-operative experiments on Scottish farms. *Proc. Nutr. Soc.* **21**, xix
44. Owen, E. C. (1962) Metabolites of riboflavin. *Proc. Nutr. Soc.* **21**, xvii–xviii
45. Owen, E. C. & Hytten, F. E. (1962) Riboflavin in human milk. *Proc. Nutr. Soc.* **21**, xviii
46. Waite, R., Johnston, M. J. & Armstrong, D. G. (1962) The digestibility of the carbohydrate constituents of S_{23} rye-grass cut at four stages of growth. *Proc. Nutr. Soc.* **21**, xxi–xxii
47. Owen, E. C. (1962) The characterization of riboflavin metabolite (RM 1) as 9-(2'-hydroxethyl)-6,7-dimethylisoalloxazine. *Biochem. J.* **84**, 96P

Institute brief report no.

48. Clapperton, J. L. (1962) The oxygen consumption of sheep when walking in cold and warm environments. *Proc. Nutr. Soc.* **21**, xxxiii–xxxiv
49. Mabon, R. M. (1963) A method for the rapid determination of alcohol in body fluids. *Analyst* **88**, 320–321
50. Owen, E. C., Darroch, R. A. & Proudfoot, R. (1963) Effect of a carotene-free diet on the metabolic rate of the goat. *Proc. Nutr. Soc.* **22**, xv–xvi
51. Martin, A. K. & Blaxter, K. L. (1963) The specific dynamic effect of protein. *Proc. Nutr. Soc.* **22**, xiv–xv
52. Joyce, J. P. & Blaxter, K. L. (1963) The loss of methane and carbon dioxide by belching in sheep. *Proc. Nutr. Soc.* **22**, xx–xxi
53. Owen, E. C. (1963) An anomaly in the reaction of vitamin A with antimony trichloride. *Biochem. J.* **88**, 64P
54. Armstrong, D. G. (1963) Net energy values for artificially dried ryegrass, cocksfoot and timothy. *Proc. Nutr. Soc.* **22**, xxv–xxvi
55. Waite, R., Johnston, J. & Armstrong, D. G. (1963) The digestibility of the carbohydrate constituents of artificially dried ryegrass, cocksfoot and timothy at various stages of growth. *Proc. Nutr. Soc.* **22**, xxiv–xxv
56. Bianca, W. (1963) Thermoregulatory responses of the dehydrated ox to the ingestion of cold and warm water in a warm environment. *J. Physiol.* (*Lond.*) **169**, 14P–15P
57. Blaxter, K. L. (1964) Dietary factors affecting energy utilization. *Proc. Nutr. Soc.* **23**, 3–11
58. Stein, G., Sturkie, P. D. & Whittow, G. C. (1964) Changes in the cardiac output, blood pressure and heart rate of the chicken during hyperthermia. *J. Physiol.* (*Lond.*) **170**, 61P–62P
59. Blaxter, K. L. (1964) Utilization of the metabolizable energy of grass *Proc. Nutr. Soc.* **23**, 62–71
60. Owen, E. C. & Proudfoot, R. (1964) The effects of a single large oral dose of vitamin A or carotene on vitamin A in the liver of the goat. *Proc. Nutr. Soc.* **23**, xxxiii
61. Findlay, J. D. & Robertshaw, D. (1964) Studies on the control of sweating in cattle. *J. Physiol.* (*Lond.*) **175**, 69P
62. Owen, E. C. & Dzialoszynski, L. (1965) The fate of oral riboflavine. *Proc. Nutr. Soc.* **24**, ix
63. Owen, E. C. & Dzialoszynski, L. (1965) The fate of an oral dose of riboflavine in the dog. *Proc. Nutr. Soc.* **24**, ix–x
64. Clapperton, J. L. & Blaxter, K. L. (1965) Methane production of ruminants given natural feeds. *Proc. Nutr. Soc.* **24**, xvi
65. Vercoe, J. E. & Blaxter, K. L. (1965) Methane production from formic acid. *Proc. Nutr. Soc.* **24**, xvii
66. Czerkawski, J. W., Blaxter, K. L. & Wainman, F. W. (1965) The effects of C_{18} acids on methane production by sheep. *Proc. Nutr. Soc.* **24**, xvii
67. Czerkawski, J. W. (1965) Estimation of lignin. *Proc. Nutr. Soc.* **24**, xvii
68. Owen, E. C. & Proudfoot, R. (1965) Riboflavine and fat in cow milk. *Proc. Nutr. Soc.* **24**, xxii
69. Clapperton, J. L. & Blaxter, K. L. (1965) Absence of long-term adaptation in the energy metabolism of sheep on constant feed. *Proc. Nutr. Soc.* **24**, xxxiii–xxxiv
70. Robertshaw, D. & Whittow, G. C. (1965) The sympatho-adrenal response to hyperthermia and to localized heating of the anterior hypothalamus in the ox (*Bos taurus*). *J. Physiol.* (*Lond.*) **179**, 90P–91P

Institute brief report no.

71. Bianca, W. (1963) Thermoregulatory responses of the dehydrated ox to heat exposure and to rehydration. *Proc. 3rd Int. Biometeorol. Congr.*, p. 485
72. Abbot, J. (1965) A simple method of gas-packing instant powders. *Dairy Ind.* **30**, 876–877
73. Orr, D. A. & Taylor, M. M. (1965) Effect of incubation temperature on the flagellation of gram-negative rods. *Nature* (*Lond.*) **208**, 1017
74. Owen, E. C., Proudfoot, R. & West, D. W. (1966) Riboflavine metabolism in the newborn kid. *Proc. Nutr. Soc.* **25**, xi
75. Owen, E. C., Proudfoot, R. & West D. W. (1966) Metabolism of riboflavine. *Proc. Nutr. Soc.* **25**, x
76. Czerkawski, J. W. (1966) Incubations inside the bovine rumen. *Biochem. J.* **99**, 39P
77. Ulyatt, M. J., Czerkawski, J. W. & Blaxter, K. L. (1966) A technique for quantitative measurement of hydrogenation of long-chain fatty acids in the fore-stomachs of the sheep. *Proc. Nutr. Soc.* **25**, xviii
78. Webster, A. J. F. & Valks, D. (1966) The energy cost of standing in sheep. *Proc. Nutr. Soc.* **25**, xxii
79. Clapperton, J. L., Czerkawski, J. W. & Martin, A. K. (1966) The effect of infusing bile salts into the rumen of sheep. *Proc. Nutr. Soc.* **25**, xxx
80. Findlay, J. D. & Robertshaw, D. (1967) The mechanism of body temperature changes produced by intraventricular injections of monoamines in the ox (*Bos taurus*). *J. Physiol.* (*Lond.*) **188**, 46P–48P
81. Hales, J. R. S. (1968) The partition of respiratory ventilation of the panting ox. *J. Physiol.* (*Lond.*) **188**, 45P–46P
82. Hales, J. R. S. (1968) The oxygen cost of hyperventilation in the ox. *J. Physiol.* (*Lond.*) **194**, 24P–25P
83. Clapperton, J. L. & Czerkawski, J. W. (1967) The effect upon the methane production of sheep of a short infusion of higher fatty acids into the rumen. *Proc. Nutr. Soc.* **26**, xxi–xxiii
84. Czerkawski, J. W., Clapperton, J. L. & Strong, C. (1967) Effect of branched-chain fatty acids on methane production by sheep. *Proc. Nutr. Soc.* **26**, xxiii–xxiv
85. Manson, W. (1967) The reaction of casein with ferrous iron and oxygen. *Proc. Nutr. Soc.* **26**, xvii–xviii
86. West, D. W., Owen, E. C. & Taylor, M. M. (1967) Biogenesis of riboflavine metabolites. *Proc. Nutr. Soc.* **26**, xvii
87. West, D. W. & Owen E. C. (1968) Metabolism of riboflavin in goats with a caecal fistula. *Proc. Nutr. Soc.* **27**, 39A
88. Czerkawski, J. W. & Paterson, D. R. (1968) Abnormal changes in the volume of the rumen contents of sheep given molassed sugar-beet pulp. *Proc. Nutr. Soc.* **27**, 33A
89. Findlay, J. D., Hales, J. R. S., Storie, G. A. T. & Thompson, G. E. (1968) A head clamp for use in large animal surgery. *J. Physiol* (*Lond.*) **198**, 55P–56P
90. Jenkinson, D. McE. & Thompson, G. E. (1968) Brown adipose tissue and the thermogenic effect of noradrenaline in new-born lambs. *J. Physiol.* (*Lond.*) **198**, 88P
91. Findlay, J. D. & Hales, J. R. S. (1968) Hypothalamic temperature and the regulation of respiration of the ox exposed to severe heat. *J. Physiol.* (*Lond.*) **200**, 62P–63P
92. Hobson, P. N., Summers, R., Owen, E. C., Spencer, J. C. & West, D. W. (1969) A study of cultures of rumen anaerobic bacteria in the presence of excess riboflavin (Vitamin B_2). *Proc. Nutr. Soc.* **28**, 53A

Institute brief report no.

93. Owen, E. C. & Dundas, I. (1969) Diminution of xanthine oxidase in goat milk after subcutaneous injection of sodium tungstate ($Na_2WO_4.2H_2O$). *Proc. Nutr. Soc.* **28**, 59A
94. Scott, A. M., Ulyatt, M. J., Kay, R. N. B. & Czerkawski, J. W. (1969) Measurement of the flow of long-chain fatty acids into the duodenum of sheep. *Proc. Nutr. Soc.* **28**, 51A
95. Czerkawski, J. W. & Breckenridge, G. (1969) Fermentation of various soluble carbohydrates by rumen micro-organisms. *Proc. Nutr. Soc.* **28**, 52A
96. Clapperton, J. L. (1969) The effect of adding certain fatty acids on the voluntary food intake of sheep. *Proc. Nutr. Soc.* **28**, 57A
97. Martin, A. K. (1969) Precursors of aromatic acids found in sheep urine. *Proc. Nutr. Soc.* **28**, 65A
98. Hales, J. R. S. & Jessen, C. (1969) Increase of cutaneous moisture loss caused by local heating of the spinal cord in the ox. *J. Physiol.* (*Lond.*) **204**, 40P–42P
99. Thompson, G. E. & Jenkinson, D. McE. (1970) Adipose tissue in the new-born goat. *Res. Vet. Sci.* **11**, 102
100. Clough, D. P. & Thompson, G. E. (1970) The effect of intraventricular noradrenaline on plasma unesterified fatty acids in the ox. *J. Physiol.* (*Lond.*) **208**, 85P–86P
101. Clough, D. P., Kenny, J. D. R., Scobie, A. & Thompson, G. E. (1970) Dopamine in adipose tissue of the ox and the effect of dopamine infusion on plasma unesterified fatty acids. *J. Physiol.* (*Lond.*) **209**, 32P–33P
102. Hales, J. R. S. (1970) Hypothalamic control of heat loss mechanisms in the ox. *Proc. Aust. Physiol. Pharmacol. Soc.* **1**, 43–44
103. Owen, E. C. & Pringle, J. (1970) Inhibition of growth of *Lactobacillus casei* by tungstate and its reversal by molybdate. *Biochem. J.* **119**, 23P–24P
104. Owen, E. C. & Dempster, S. A. R. (1970) Growth of the riboflavin-splitting rumen organism *Eubacterium fissicatena* in the presence of tungstate and molybdate. *Biochem. J.* **119**, 24P
105. McLean, J. A., Hales, J. R. S., Jessen. C. & Calvert, D. T. (1970) Influences of spinal cord temperature on heat exchange of the ox. *Proc. Aust. Physiol. Pharmacol. Soc.* **1**, 32–33
106. Mabon, R. M. & Jenkinson, D. McE. (1971) The excretion of 3-methoxy-4-hydroxy mandelic acid (VMA) by cattle skin. *Res. Vet. Sci.* **12**, 289–292
107. Clough, D. P. & Thompson, G. E. (1971) The effect of intraventricular noradrenaline, and acetylcholine, on plasma unesterified fatty acids in sheep. *J. Physiol.* (*Lond.*) **215**, 52P–53P
108. Kendall, P. P., McLean, J. A. & Smith, J. F. (1971) An integrating gas sampling system for open-circuit calorimetry. *J. Physiol.* (*Lond.*) **216**, 4P–5P
109. Davies, D. T. (1970) Specific staining of sulphydryl-containing milk proteins after gel electrophoresis. *Proc. 18th Int. Dairy Congr.*, vol. 1 E, p. 66
110. Sweetsur, A. W. M. and White, J. C. D. (1970) Kinetics of heat coagulation of milk protein. *Proc. 18th Int. Dairy Congr.*, vol. 1 E, p. 53
111. Waite, R. & Smith, G. M. (1970) Measurement of the protein content of milk from mastitic quarters by the amido black method. *Proc. 18th Int. Dairy Congr.*, vol. 1 E, p. 673
112. Taylor, M. M. (1970) Bacteriological control of production at pasteurizing plants by the water agar test. *Proc. 18th Int. Dairy Congr.*, vol. 1 E, p. 169
113. Phillips, J. D. & Taylor, M. M. (1970) Relationship between bacterial numbers and somatic cell counts. *Proc. 18th Int. Dairy Congr.*, vol. 1 E, p. 616
114. McLean, J. A. (1970) Simultaneous direct and indirect calorimetric measurements on cattle. *Proc. 5th Symp. Energy Metabolism*, pp. 229–232

Institute brief report no.

115. Martin, A. K. (1971) Protein metabolism and phenylacetic acid excretion in the urine of sheep. *Proc. Nutr. Soc.* **30**, 57A
116. Moore, J. H., Christie, W. W., Braude, R. & Mitchell, K. G. (1968) The effect of 250 ppm of copper in the diet of growing pigs on the fatty acid composition of the adipose tissue lipids. *Proc. Nutr. Soc.* **27**, 45A–46A
117. Moore, J. H., Coates, M. E. & Turvey, A. (1969) A comparison of the effects of diets high in maize oil or butterfat on tissue lipids in gnotobiotic and conventional rabbits. *Proc. NATO Adv. Study Inst. The Germ-Free Animal as a Tool in Research*, p. 43
118. Steele, W., Noble, R. C. & Moore, J. H. (1970) Blood lipid composition and milkfat secretion in cows given soyabean diets. *Proc. 18th Int. Dairy Congr.*, vol. 1 E, p. 600
119. Harfoot, C. G. & Czerkawski, J. W. (1971) The inhibition by chloroform of methane production by rumen contents. *Proc. Nutr. Soc.* **30**, 44A
120. Clapperton, J. L., Graham, K. & Czerkawski, J. W. (1971) The effect of infusing propane-1:2-diol into the rumen on the metabolism of sheep. *Proc. Nutr. Soc.* **30**, 46A
121. Czerkawski, J. W. & Lumsden, J. (1972) Invertase activity in the rumen contents of sheep given molassed sugar-beet pulp. *Proc. Nutr. Soc.* **30**, 53A
122. Noble, R. C. & Moore, J. H. (1971) The uptake of linoleic acid by the newborn lamb. *Proc. Nutr. Soc.* **30**, 61A
123. Czerkawski, J. W. (1970) Problems in evaluation of feed components and food constituents, with special reference to methane production. *Proc. 5th Symp. Energy Metabolism of Farm Animals*, pp. 21–24
124. Clapperton, J. L. (1970) Ration changes – their effect upon the energy metabolism of sheep. *Proc. 5th Symp. Energy Metabolism of Farm Animals*, pp. 113–116
125. Abbot, J. (1970) Ascorbyl palmitate as an antioxidant in full-cream milk powder. *Proc. 18th Int. Dairy Congr.*, vol. 1 E, p. 464
126. Owen, E. C. & West, D. W. (1969) Metabolites of riboflavin in animals. *Proc. 8th Int. Congr. Nutrition*, Sect. O, p. 15
127. West, D. W. & Owen, E. C. (1969) Metabolites of riboflavine in man. *Proc. 8th Int. Congr. Nutrition*, Sect. W, pp. 6–7
128. Manson, W. & Annan, W. D. (1970) Structural studies on β-casein. *Proc. 18th Int. Dairy Congr.*, vol. 1 E, p. 33
129. Herrero, Maria del Carmen Alvarez, Clapperton, J. L. & Czerkawski, J. W. (1970) The effect of change of rations on the rate of flow of liquid digesta from the rumen of sheep. *Proc. Nutr. Soc.* **29**, 26A–27A
130. McLean, J. A. (1969) Measurement of heat losses from cattle by direct calorimetry. *Biometeorology* **4**, 25–26
131. Nisbet, W. (1971) Digital display of temperature using copper constantan thermocouples. *J. Physiol.* (*Lond.*) **217**, 2P–3P
132. Hodgson, J. C. & Thomas, P. C. (1972) The chemical composition and dilution rate of rumen fluid in sheep receiving a diet of barley, hay and flaked maize. *Proc. Nutr. Soc.* **31**, 57A
133. Clapperton, J. L. & Czerkawski, J. W. (1972) The energy metabolism of sheep – the effect of infusing small amounts of chloroform into the rumen. *Proc. Nutr. Soc.* **31**, 55A
134. Morrison, I. M. (1972) A semi-micro method for the determination of lignin and its use in predicting the digestibility of forage crops. *J. Sci. Food Agric.* **23**, 791

Institute brief report no.

135. Hodgson, J. C. and Thomas, P. C. (1973) The influence of the level of feeding on the digestion of a diet consisting of barley, hay and flaked maize. *Proc. Nutr. Soc.* **32**, 91A
136. Harfoot, C. G., Noble, R. C. & Moore, J. H. (1973) Food particles as a site for biohydrogenation of unsaturated fatty acids in the rumen. *Biochem. J.* **132**, 829–832
137. Bell, A. W. & Findlay, J. D. (1973) The effect of cold on blood flow and oxygen consumption in the hind leg of the ox. *J. Physiol.* (*Lond.*) **232**, 94P–95P
138. Darling, K. F. & Thompson, G. E. (1973) The effect of injecting acetylcholine and noradrenaline into the third cerebral ventricle, on blood pressure and plasma free fatty acids in sheep. *J. Physiol.* (*Lond.*) **232**, 93P–94P
139. Morrison, I. M. (1973) A note on the evaluation of the nutritive value of forage crops by the acetyl bromide technique. *J. Br. Grassl. Soc.* **28**, 153–154
140. Christie, W. W. (1973) Determination of isotopically labelled lipids by combined thin-layer and gas-liquid chromatography. *Proc. Soc. Anal. Chem.* **10**, 219–220
141. Christie, W. W. (1974) Biosynthesis of lipids in freshly secreted milk from goats. *Biochem. Soc. Trans.* **2**, 141
142. Vernon, R. G. (1974) Effect of linoleic acid in adipose tissue metabolism in the developing lamb. *Biochem. Soc. Trans.* **2**, 142
143. Thomas, P. C. (1973) Microbial protein synthesis. *Proc. Nutr. Soc.* **32**, 85–91
144. Czerkawski, J. W. (1973) An indirect estimation of rate of formation of long-chain fatty acids in the rumen of sheep. *Proc. Nutr. Soc.* **32**, 63A–64A
145. Chamberlain, D. G. & Thomas, P. C. (1974) The influence of the composition of the diet on the digestion of organic matter and nitrogen in sheep receiving diets containing hay, barley and flaked maize. *Proc. Nutr. Soc.* **33**, 59A
146. Czerkawski, J. W. & Breckenridge, G. (1974) Use of tritium compounds in the study of methanogenesis. *Proc. Nutr. Soc.* **33**, 76A
147. Towers, G. E., West, D. W. & Noble, R. C. (1974) A simple and direct method for ^{32}P determination by scintillation counting. *J. Int. Res. Comm.* **2**, 1342
148. Noble, R. C. (1974) The use of an internal vial for count correction of scintillation gels. *J. Int. Res. Comm.* **2**, 1443
149. Noble, R. C. & Shand, J. H. (1975) The use of polyethylene mini vials for the differential counting of ^{14}C and ^{32}P. *IRCS Med. Sci.* **3**, 48
150. Czerkawski, J. W. (1975) Composition of microbial matter in the rumen. *Proc. Nutr. Soc.* **34**, 62A
151. Thompson, G. E. & Darling, K. F. (1975) The hepatic uptake of individual free fatty acids in sheep during noradrenaline infusion. *Res. Vet. Sci.* **18**, 325–327
152. Jenkinson, D. McE. & Mabon, R. M. (1975) The corticosteroid content of cattle skin washings. *Res. Vet. Sci.* **19**, 94–95
153. Calvert, D. T. (1975) A brain thermode unit for use in oxen. *J. Physiol.* (*Lond.*) **251**, 1P
154. Calvert, D. T., Findlay, J. D., Ross, J. & Storie, G. A. T. (1975) A stereotaxic device and head clamp for use in large animal surgery. *J. Physiol.* (*Lond.*) **251**, 1P–2P
155. Vernon, R. G. (1975) Inhibition of rat and sheep adipose tissue lipogenesis by palmitic acid and linoleic acid added *in vitro*. *Biochem. Soc. Trans.* **3**, 717
156. Jenkinson, D. McE. (1976) The skin and hair coat of cattle. *Luing J.* 60–61
157. Noble, R. C., Shand, J. H. & West, I. G. (1976) A technique for a radio-chromatographic scanning procedure. *IRCS Med. Sci.* **4**, 170

OTHER PUBLICATIONS

Morris, S. (1932) Biennial reviews of the progress of dairy science. Section A. The physiology of dairy cattle. III. nutrition. *J. Dairy Res.* **4**, 163–165

Morris, S. (1932) Metabolism in tetany. *Biochem. J.* **26**, 1606–1608

Stirling, J. D. (1932) Biennial reviews of the progress of dairy science. Section A. The physiology of dairy cattle. II. lactation. *J. Dairy Res.* 4, 156–162

Wright, N. C. (1932) Safe milk. *Lancet* **223**, 1024–1025

Blackwood, J. H. & Wright, N. C. (1933) The effect of raw milk on teeth. *Farm. Stock Breed.* **47**, 1289–1290

Blackwood, J. H. & Wright, N. C. (1933) The nutritive properties of milk in relation to pasteurization. *Br. Med. J.* **1**, 792–793

Fowler, A. B. (1933) Milk recording in relation to breeding problems. *Rep. Scott. Milk Rec. Ass.*

Wright, N. C. (1933) Condensed and dried milk. *Rep. Cen. Coun. Milk Rec. Soc.*

Howat, G. R. & Wright, N. C. (1934) Factors affecting the solubility of milk powders. *Proc. 10th Int. Dairy Congr.* **2**, 141

Morris, S. & Wright, N. C. (1934) The biological value of the proteins of certain concentrates for milk production. *Proc. 10th Int. Dairy Congr.* **1**, 79

Blackwood, J. H., Morris, S. & Buchanan Smith, A. D. (1935) Biennial reviews of the progress of dairy science. Section A. The physiology of dairy cattle. *J. Dairy Res.* **6**, 252–258

Edwards, S. J. (1936) Biennial reviews of the progress of dairy science. Section E. The diseases of dairy cattle: I. mastitis; II. contagious abortion; III. tuberculosis. *J. Dairy Res.* **7**, 291–318

Morris, S. (1936) Protein requirements for lactation. *Nutr. Abst. Rev.* **6**, 273–280

Howat, C. R., Jackson, C. J. & Nicols, A. A. (1937) Recent investigations into certain problems connected with the condensing and drying of milk. *Proc. 11th Int. Dairy Congr.* **2**, 241–245

Smith, J. A. B., Morris, S. & Buchanan Smith, A. D. (1937) Biennial reviews of the progress of dairy science. Section A. The physiology of dairy cattle. *J. Dairy Res.* **8**, 105–142

Edwards, S. J. (1938) Biennial reviews of the progress of dairy science. Section E. The diseases of dairy cattle. *J. Dairy Res.* **9**, 356–363

Edwards, S. J. (1938) The diagnosis of streptococcus mastitis by means of a selective liquid medium. *Proc. Soc. Agric. Bacteriol.*, 34–35

Morris, S. (1939) Biennial reviews of the progress of dairy science. Section A. II. nutrition. *J. Dairy Res.* **10**, 153–164

Smith, J. A. B. & Morris, S. (1939) Biennial reviews of the progress of dairy science. Section A. The physiology of dairy cattle. *J. Dairy Res.* **10**, 133–164

Fowler, A. B. (1940) Grass silage. *Ayrsh. Cattle Soc. J.* **12**, 59–61

Fowler, A. B. (1940) Nature and nurture. *Ayrsh. Cattle Soc. J.* **12**, 14–16

Fowler, A. B. (1940) The winter feeding problem. *Ayrsh. Cattle Soc. J.* **12**, 107

Fowler, A. B. (1940) Winter housing of stock. *Ayrsh. Cattle Soc. J.* **12**, 149–150

Watts, P. S. (1940) Biennial reviews of the progress of dairy science. Section E. The diseases of dairy cattle. *J. Dairy Res.* **11**, 316–350

Fowler, A. B. (1941) Cereal crops for silage. *Ayrsh. Cattle. Soc. J.* **13**, 63–64

Fowler, A. B. (1941) Grass silage. *Ayrsh. Cattle Soc. J.* **13**, 3–4

Fowler, A. B. (1941) Milk. *Ayrsh. Cattle Soc. J.* **13**, 183

Fowler, A. B. (1941) The use of home-grown feeding-stuffs. *Ayrsh. Cattle Soc. J.* **13**, 127–128

McNair, J. (1941) Crops for silage and types of silos. *Scott. Farm.* **49**, (suppl.), 2

Owen, E. C. (1941) Biennial reviews of the progress of dairy science. Section A. The physiology of dairy cattle: II. nutrition. *J. Dairy Res.* **12**, 213–223

Owen, E. C., Smith, J. A. B. & Wright, N. C. (1941) Chemical substitutes for dietary protein. *Nature* (*Lond.*) **147**, 710–711

Smith, J. A. B. (1941) Biennial reviews of the progress of dairy science. Section A. The physiology of dairy cattle: I. reproduction and lactation. *J. Dairy Res.* **12**, 78–103

Wright, N. C. (1941) The importance of silage. *Scott. Farm.* **49**, (suppl.), 3

Fowler, A. B. (1942) Feeding for milk. *Ayrsh. Cattle Soc. J.* **14**, 39–42

Fowler, A. B. (1942) Planning for 1943. *Ayrsh. Cattle Soc. J.* **14**, 111

Fowler, A. B. (1942) The spring flush of grass. *Ayrsh. Cattle Soc. J.* **14**, 8

Higginbottom, C. (1943) Some notes on the bacteriology of roller-dried milk and dried whey. *Abstr. Proc. Soc. Agric. Bact.*, 22

Thom. C. L. (1943) Dairy calves. *Ayrsh. Cattle Soc. J.* **15**, 3–4

Smith, J. A. B. (1944) Feeding stuffs from the air. *Farm. Stock. Breed.* **58**, 577

Watts, P. S. (1944) Biennial reviews of the progress of dairy science. Section E. The diseases of dairy cattle. *J. Dairy Res.* **13**, 340–370

Smith, J. A. B. (1945) Biennial reviews of the progress of dairy science. Section A. The physiology of dairy cattle: I. reproduction and lactation. *J. Dairy Res.* **14**, 195–224

Smith, J. A. B. & Lea, C. H. (1945) Milk powder, in *Thorpe's Dictionary of Chemistry* (*new ed.*)

Owen, E. C. (1947) Biennial reviews of the progress of dairy science. Section A. The physiology of dairy cattle: II. nutrition. *J. Dairy Res.* **15**, 142–163

Owen, E. C. (1948) Nutrition and symbiosis. *Discovery* (*Lond.*) **7**, 374–379

Blaxter, K. L. (1949) The physiology of lactation. *Adv. Sci.* **6**, 12–14

Blaxter, K. L., Reineke, E. P., Crampton, E. W. & Petersen, W. E. (1949) The role of thyroidal materials and of synthetic goitrogens in animal production and an appraisal of their practical use. *J. Anim. Sci.* **8**, 307–352

Henry, K. M., Kon, S. K., Lea, C. H. & White, J. C. D. (1949) Protein degradation in stored milk powder. *Proc. 12th Int. Dairy Congr.* **2**, 166–172

Holmes, W. (1949) The proteins and vitamins in dried grass. *Year Book Green Crop Dryers Res. Assoc.*, pp. 117–119

Lea, C. H. & Smith, J. A. B. (1949) Fat oxidation in dried milk and its prevention. *Proc. 12th Int. Dairy Congr.* **2**, 341–347

Waite, R. & White, J. C. D. (1949) The composition of the soluble and insoluble portions of reconstituted milk powder. *Proc. 12th Int. Dairy Congr.* **2**, 259–267

Watts, P. S. (1949) Bovine mastitis. *Veteran* **1**, 158–160

Watts, P. S. (1949) The farmer's part in the eradication of mastitis. *Proc. 12th Int. Dairy Congr.* **1**, 207–213

Watts, P. S. (1949) Reviews of the progress of dairy science. Section E. The diseases of dairy cattle. *J. Dairy Res.* **16**, 242–281

Blaxter, K. L. (1950) Energy feeding standards for dairy cattle. *Nutr. Abstr. Rev.* **20**, 1–21

Findlay, J. D. (1950) Climate and cattle. *Ayrsh. Cattle Soc. J.* **22**, 26–28

Owen, E. C. (1950) The nature of science and other hypotheses. *Bull. Br. Soc. Hist. Sci.* **1**, 4

Smith, J. A. B. (1950) Reviews of the progress of dairy science. Section A. The physiology of dairy cattle: I. reproduction and lactation. *J. Dairy Res.* **17**, 338–366

Chanda, R. & Owen, E. C. (1951) The effect of thyroxine and deprival of carotene on the secretion of carotene and the alcoholic form of vitamin A in cow's milk. *Br. J. Nutr.* **5**, ix

Owen, E. C. (1951) Reviews of the progress of dairy science. Section A. The physiology of dairy cattle: II. nutrition. *J. Dairy Res.* **18**, 113–146

Smith, J. A. B. (1951) The work of the Hannah Dairy Research Institute. *J. Soc. Dairy Technol.* **5**, 12–16

Blackburn, P. S. (1952) Reviews of the progress of dairy science. Section E. Diseases of dairy cattle. *J. Dairy Res.* **19**, 356–376

Blaxter, K. L. (1952) Some effects of thyroxine and iodinated casein on dairy cows and their practical significance, in *Vitamins and Hormones*, Academic Press, New York

Blaxter, K. L. & Brown, F. (1952) Vitamin E in the nutrition of farm animals. *Nutr. Abstr. Rev.* **22**, 1–21

Blaxter, K. L., Brown, F., Wood, W. A. & MacDonald, A. M. (1952) α-Tocopherol and other antioxidants in the prevention of experimental muscular dystrophy in calves. *Proc. 2nd Int. Congr. Biochem.*

Chanda, R. & Owen, E. C. (1952) The effect of dietary carotene on the partition of carotenoids and vitamin A in the blood serum of lactating cows. *Proc. 2nd Int. Congr. Biochem.*

Chanda, R. & Owen, E. C. (1952) L'influence de l'infection des glandes mammaires sur les caroténoides, la vitamine A, et la relation entre la phosphatase et les esters phosphoriques du lait de vache. *Proc. 2nd Int. Congr. Biochem.*

Holmes, W. (1952) The intensive management of pasture for dairy cows. *Proc. 6th Int. Grassl. Congr.*, pp. 377–383

Smith, J. A. B. (1952) The Hannah Dairy Research Institute. *Br. Agric. Bull.* **18**, 336–340

Stewart, J., McCallum, J. W. & Watts, P. S. (1952) Studies on vitamins A and C in bovines. The seasonal variations in the vitamin A reserves of cows. *J. Comp. Pathol. Ther.* **62**, 239–243

Beakley, W. R. (1953) A radiometer for measuring surface temperature. *J. Physiol.* (*Lond.*) **121**, 28P

Beakley, W. R. (1953) A reference junction for thermocouples. *J. Physiol.* (*Lond.*) **121**, 28P

Beakley, W. R. & Findlay, J. D. (1953) A climatic chamber for large animals. *J. Physiol.* (*Lond.*) **121**, 40P

Beakley, W. R. & Findlay, J. D. (1953) The effect of thermal environment on the rectal temperatures of calves. *J. Physiol.* (*Lond.*) **121**, 47P

Beakley, W. R., Bligh, J. & Nisbet, W. (1953) A pneumotachograph for cattle. *J. Physiol.* (*Lond.*) **121**, 40P–41P

Blaxter, K. L. (1953) α-Tocopheryl acetate for musclar dystrophy in calves. *Vet. Rec.* **65**, 496

Blaxter, K. L. & Rook, J. A. F. (1953) The effect of magnesium deficiency on the energy exchange of calves. *J. Physiol.* (*Lond.*) **121**, 48P–49P

Blaxter, K. L., Graham, N. McC. & Rook, J. A. F. (1953) Respiration calorimetry with farm animals. *J. Physiol.* (*Lond.*) **121**, 39P–40P

Chanda, R., Clapham, J. M. & Owen, E. C. (1953) The effect of carotene deprivation on the composition of the blood of the cow. *J. Physiol.* (*Lond.*) **121**, 42P–43P

Findlay, J. D. & Goodall, A. M. (1953) Arteriovenous anastomoses in the perichondrium and skin of the ear of the Ayrshire calf. *J. Physiol.* (*Lond.*) **121**, 46P

Holmes, W. (1953) Feeding value of grass and grassland products for dairy cattle, beef cattle and young stock. *Bull. Dept. Agric. Scotl.*

Rook, J. A. F. (1953) The effect of a high plane of nutrition on the composition of milk. *Proc. Nutr. Soc.* **12**, viii–ix

Smith, J. A. B. (1953) Dried milk. *Farm. News* **105**, 33

Smith, J. A. B. (1953) Reviews of the progress of dairy science. Section A. Physiology of dairy cattle: 1. reproduction and lactation. *J. Dairy Res.* **20**, 224–253

Blackburn, P. S. (1954) Bovine mastitis. *SMMB News Bull.* (Nov.), 12–14

Blackburn, P. S. (1954) Reviews of the progress of dairy science. Section E. Diseases of dairy cattle. *J. Dairy Res.* **21**, 391–407

Blaxter, K. L. (1954) Calf rearing on dairy farms. *Farm. Leader* (26), 3–4, 15

Blaxter, K. L. (1954) Ruminant nutrition, in *Progress in the Physiology of Farm Animals* (Hammond, J., ed.), vol. 1, pp. 3–39, Butterworth, London

Findlay, J. D. (1954) Physiological limitations to the adaptability of cattle to high temperatures. *Proc. XIV Journées d'Étude F.E.Z.*

Findlay, J. D. & Beakley, W. R. (1954) Environmental physiology of farm animals, in *Progress in the Physiology of Farm Animals* (Hammond, J., ed.), vol. 1. pp. 252–298, Butterworth, London

Holmes, W. (1954) The intensive production of high quality grass. *Year Book Ass. Green Crop Dryers*, pp. 51–82

Holmes, W. & McLusky, D. S. (1954) The role of clover in grassland management. *Proc. Eur. Grassl. Congr.*

Owen, E. C. (1954) Reviews of the progress of dairy science. Section A. Physiology of dairy cattle: II. physiology and biochemistry of rumination. *J. Dairy Res.* **21**, 408–450

Waite, R. (1954) Low solids not fat. *SMMB News Bull.* (Feb.), 15–16

Blaxter, K. L. (1955) Enzootic muscular dystrophy in cattle in Scotland. *Proc. 3rd Int. Vitamin E Congr.*

Blaxter, K. L. (1955) The myopathies of cattle, sheep, horses and pigs and the role of vitamin E in their prevention. *Proc. 3rd Int. Vitamin E Congr.*

Blaxter, K. L. (1955) Vitamin E and experimental muscular degeneration in calves. *Proc. 3rd Int. Vitamin E Congr.*

Blaxter, K. L. & McGill, F. (1955) Muscular dystrophy. *Vet. Rev. Annot.* **1**, 91–114

Bligh, J. (1955) A comparison of rectal, bicarotid trunk and pulmonary artery temperatures in the calf under heat stress. *J. Physiol.* (*Lond.*) **130**, 46P

Edwards, D. C. (1955) Reviews of the progress of dairy science. Section A. Physiology of dairy cattle. The biochemistry and microbiology of the rumen. *J. Dairy Res.* **22**, 232–250

Findlay, J. D. (1955) Climate and cattle. *Agric. Rev.* **1**(7), 28–33

Hofman, T. (1955) On the specificity of phosphoprotein-phosphatase. *Proc. 3rd Int. Congr. Biochem.*

Waite, R. (1955) Breeding for S.N.F. *Farm. Weekly* **43**(16), 105–106

Waite, R. (1955) Low solids not fat. *Dairy Shorthorn J.* **24**, 499–501

Blackburn, P. S. (1956) Reviews of the progress of dairy science. Section E. Diseases of dairy cattle. *J. Dairy Res.* **23**, 384–406

Blaxter, K. L. (1956) The magnesium content of bone in hypomagnesaemic disorders of livestock. *Proc. Ciba Foundation Symp. Bone Structure Metabolism*, pp. 117–131

Blaxter, K. L. (1956) The nutritive value of feeds as sources of energy: a review. *J. Dairy Sci.* **39**, 1396–1424

Blaxter, K. L. & MacDonald, A. M. (1956) Myopathy in sheep. *Lancet* **271**, 895

Bligh, J. (1956) The initiation of thermal polypnoea in the bovine animal. *Proc. XXth Int. Physiol. Congr.*, pp. 104–105

Bligh, J. (1957) Localisation of the thermal stimulus to polypnoea. *J. Physiol.* (*Lond.*) **135**, 48P–49P

Davies, D. T. (1956) Ultrafiltration and dialysis in the estimation of some soluble constituents of milk. *Chem. Ind.* (*Lond.*) **33**, 871

Findlay, J. D. (1956) Respiratory and rectal temperature relationships in thermal polypnoea in the bovine. *Proc. XXth Int. Physiol. Congr.*, pp. 288–289

Holmes, W. (1956) Feeding grass. *Dep. Agric. Scotl. Adv. Leaflet* no. 25

Waite, R. (1956) Problems posed by variations in the solids-not-fat content of milk. *Proc. 7th Int. Congr. Anim. Husb.* **5**, 63–72

Waite, R. (1956) Winter feeding and low solids-not-fat. *SMMB News Bull.* (Dec.), 11–14

Blaxter, K. L. (1957) The fat-soluble vitamins. *Annu. Rev. Biochem.* **26**, 275–306

Findlay, J. D. (1957) The physiological effects of climatic stress on the bovine animal. *Proc. Int. J. Soc. Bioclimatology Biometeorology* **3**, 1–8

Taylor, M. M. (1957) The inhibition of small inocula on the surface of nutrient agar. *J. Gen. Microbiol.* **16**, ix

Armstrong, D. G. (1958) An account of some work in progress, or completed at the Hannah Dairy Research Institute. *J. King's Coll. Agric. Soc.* **12**, 5–9

Armstrong, D. G., Blaxter, K. L., Graham, N. McC. & Wainman, F. W. (1958) A summary of work on energy metabolism completed and in progress at the Hannah Dairy Research Institute. *Proc. 1st Int. Symp. Energy Metabolism*, pp. 244–247

Blackburn, P. S. (1958) Reviews of the progress of dairy science. Section E. Diseases of dairy cattle. *J. Dairy Res.* **25**, 486–524

Blaxter, K. L. (1958) The nutrition of dairy cows. *Ayrsh. Cattle Soc. J.* **30**, 10–12, 14–17

Findlay, J. D. (1958) Temperature regulation in cattle at high temperatures. *Tijdschr. Diergeneeskd.* **83**, 992–1000

Owen, E. C. (1958) Le métabolisme comparatif du cobalt. *Rev. Pathol. Gén. Physiol. Clin.* **224**, 243

Smith, J. A. B. (1958) The Hannah Dairy Research Institute. *Scott. Agric.* **38**, 17–21

Armstrong, D. G. (1959) Digestion in the ruminant, in *Veterinary Annual* (Pool, W. A., ed.), pp. 205–208, Wright, Bristol

Bianca, W. (1959) Reaktionen des Rindes auf Kälte and Wärme. *Zuechtungskunde* **31**, 476–489

Blackburn, P. S. (1959) Bovine mastitis. *Farm Rev.* (11), 20–24

Blaxter, K. L. (1959) Feeding standards for livestock. *J. Farmers' Club* (2), 14–26

Blaxter, K. L. (1959) Minerals in relation to disease, in *Veterinary Annual* (Pool, W. A., ed.), pp. 184–193, Wright, Bristol

Findlay, J. D. (1959) Environmental physiology, in *Veterinary Annual* (Pool, W. A., ed.), pp. 286–289, Wright, Bristol

Findlay, J. D. (1959) Some adaptations of farm animals to climatic stress. *Proc. R. Soc. Med.* **52**, 677–680

Owen, E. C. (1959) Biochemical aspects of diseases of livestock with particular reference to enzyme systems. *Vet. Rec.* **71**, 1114–1127

Smith, J. A. B. (1959) Dairy research and the Hannah Dairy Research Institute. *J. Edinburgh Agric. Former Students' Ass.* (34), 20–24

Smith, J. A. B. (1959) Physiological aspects of milk production – recent developments and their application. *Proc. 15th Int. Dairy Congr.* **5**, pp. 293–300

Smith, J. A. B. (1959) Recent dairy research in the United Kingdom. *Dairy Sci. Abstr.* **21**, 219–240

Armstrong, D. G. (1960) Badania nad przemiana energii prowadzone w Hannah Dairy Research Institute. *Postepy Nauk Roln.* **61**, 21–32

Armstrong, D. G. (1960) Digestion in the ruminant, in *Veterinary Annual*, 2nd ed. (Pool, W. A., ed.), pp. 181–186, Wright, Bristol

Bianca, W. (1960) Animals and climate, in *Veterinary Annual*, 2nd ed. (Pool, W. A., ed.), pp. 259–264, Wright, Bristol

Blackburn, P. S. (1960) Reviews of the progress of dairy science. Section E. Diseases of dairy cattle. *J. Dairy Res.* **27**, 435–454

Blaxter, K. L. (1960) Energy metabolism. *State Vet. J.* **15**, 114–120

Blaxter, K. L. (1960) Minerals in relation to disease, in *Veterinary Annual*, 2nd ed. (Pool, W. A., ed.), pp. 168–175, Wright, Bristol

Findlay, J. D. (1960) Reactions of cattle to heat and cold. *Agric. Prog.* **34**, 74–77

Owen, E. C. (1960) Biochemical function of some of the minor elements. *Proc. Nutr. Soc.* **19**, 154–162

Smith, J. A. B. (1960) The Hannah Dairy Research Institute. *Food Trade Rev.* **30**(9), 72–73

Armstrong, D. G. (1961) Digestion in the ruminant, in *Veterinary Annual*, 3rd ed. (Pool, W. A., ed.), pp. 224–230, Wright, Bristol

Armstrong, D. G. (1961) Energy requirements of ruminant animals. *Span* **4**, 104–107

Bianca, W. (1961) Animals and climate, in *Veterinary Annual*, 3rd ed. (Pool, W. A., ed.), pp. 312–318, Wright, Bristol

Bianca, W. (1961) Heat tolerance in cattle – its concept, measurement and dependence on modifying factors. *Int. J. Biometeorol.* **5**, 5–30

Bianca, W. & Blaxter, K. L. (1961) The influence of the environment on animal production and health under housing conditions. *Festschr. 8te Int. Tierzuchtkongr.* pp. 113–147

Blaxter, K. L. (1961) Economics and animal husbandry. *J. Agric. Econ.* **14**, 308–313

Blaxter, K. L. (1961) Efficiency of feed conversion by different classes of livestock in relation to food production. *Fed. Proc.* **20**, 268–274

Blaxter, K. L. (1961) Lactation and the growth of the young, in *Milk: the Mammary Gland and its Secretion* (Kon, S. K. & Cowie, A. T., eds), vol. II, pp. 305–361, Academic Press, London

Blaxter, K. L. (1961) Minerals in relation to disease, in *Veterinary Annual*, 3rd ed. (Pool, W. A., ed.), pp. 230–238, Wright, Bristol

Blaxter, K. L. (1961) The nutritional physiology of livestock under natural conditions *J. R. Soc. Arts* **109**, 764–781

Department of Health for Scotland (1961) *Model Dairy Byelaws*. Report of the Committee appointed by the Secretary of State for Scotland (Smith, J. A. B., chairman), Her Majesty's Stationery Office, Edinburgh

Waite, R. (1961) SNF or protein testing? *Ayrsh. Cattle Soc. J.* **33**, 132–134, 170

Armstrong, D. G. (1962) Energy metabolism studies with farm livestock. *J. Agric. Club. Univ. Reading*

Armstrong, D. G. (1962) Grass as food for livestock. *J. Univ. Coll. Wales Agric. Soc.* **43**, 37–53

Bianca, W. (1962) Animals and climate, in *Veterinary Annual*, 4th ed. (Pool, W. A., ed.), pp. 306–311, Wright, Bristol

Blackburn, P. S. (1962) Reviews of the progress of dairy science. Section E. Diseases of dairy cattle. Mastitis. *J. Dairy Res.* **29**, 329–345

Blaxter, K. L. (1962) *The Energy Metabolism of Ruminants*, Hutchinson, London

Blaxter, K. L. (1962) Utnyttjandet av energien i fodersmältningens slutprodukter *Nord. Jordbrugsforsk.* **2**, 3–10

Blaxter, K. L. (1962) Vitamin E in health and disease of cattle and sheep. *Vitam. Horm.* **20**, 633–643

Castle, M. E. (1962) Cake and the milk yield. *Farm. Weekly* **56**(9), 97–98

Drysdale, A. D. (1962) Grassland dairy farming in Holland. *J. South-West Scotl. Grassl. Soc.* (1), 25–26

Gonzalez-Jimenez, E. & Blaxter, K. L. (1962) The critical temperature of the calf. *Anim. Prod.* **4**, 291 (Abstr.)

Smith, J. A. B. (1962) The Hannah Dairy Research Institute, in *Dictionary of Dairying*, 3rd ed. (Davis, J. G., ed.), Hill, London

Smith, J. A. B. (1962) The physiology and biochemistry of milk secretion with special reference to factors influencing the fat/protein ratio in milk. *Proc. 16th Int. Dairy Congr.* **D**, 7–31

Smith, J. A. B. (1962) The physiology of milk secretion in relation to milk composition, in *Recent Advances in Food Science* (Hawthorn, J. & Leitch, J. M., eds), vol. 1, pp. 113–121, Butterworth, London

Waite, R. (1962) Avoiding the drop in milk solids. *Dairy Farm.* **9**(10), 57–59

Waite, R. (1962) Quality milk. *Farm. Rev.* (22), 5–8

Armstrong, D. G. (1963) Energy requirements of ruminant livestock. *J. King's Coll. Agric. Soc.* **17**, 2–7

Bianca, W. (1963) Animals and climate, in *Veterinary Annual*, 5th ed. (Pool, W. A., ed.), pp. 324–329, Wright, Bristol

Bianca, W. (1963) Report on the sessions of the working group on veterinary biometeorology. *Proc. 3rd Int. Congr. Biometeorol.*

Blaxter, K. L. & Joyce, J. P. (1963) Artificial sheep. *Anim. Prod.* **5**, 216 (Abstr.)

Castle, M. E. (1963) The supplementary feeding of cows grazing pasture. *J. Yorkshire Grassl. Soc.* (5), 26–31

Castle, M. E. & Watson, J. N. (1963) Wilted silage – a progress report. *J. South-West Scotl. Grassl. Soc.* (2), 37–38

Castle, M. E. & Watson, J. N. (1963) Wilted silage. *J. South-West Scotl. Grassl. Soc.* (3), 48–50

Castle, M. E., Drysdale, A. D. & Watson, J. N. (1963) Concentrate feeding and the production of cows on good pasture. *Anim. Prod.* **5**, 216 (Abstr.)

Castle, M. E., Drysdale, A. D. & Watson, J. N. (1963) Summer supplements and milk production. *J. South-West Scotl. Grassl. Soc.* (3), 52–53

Drysdale, A. D. (1963) Liquid manure applications in winter. *Farm. News* **115**(4), 24

Drysdale, A. D. (1963) The value of liquid manure and slurry as fertilizers. *J. South-West Scotl. Grassl. Soc.* (3), 46–47

Findlay, J. D. (1963) Acclimatization to heat in sheep and cattle. *Fed. Proc.* **22**, 688–690

Findlay, J. D. (1963) Veterinary biometeorology. *Int. J. Biometeorol.* **7**, 137–143

Findlay, J. D. & McLean, J. A. (1963) General constructional and physical principles of climatic chambers, in *Medical Biometeorology – Weather, Climate and the Living Organism* (Tromp, S. W., ed.), pp. 617–631, Elsevier, Amsterdam

McLean, J. A. (1963) The measurement of respiratory gaseous exchange in cattle using mask methods. *Proc. 3rd Int. Congr. Biometeorol.*

McLeod, J. W. & Taylor, M. M. (1963) Sterilization of the skin: the time factor in the action of antiseptics and the possibilities of local application of antibiotics. *Scott. Med. J.* **8**, 231–242

Reid, D. (1963) The Henley Manor Farm experiment. *J. South-West Scotl. Grassl. Soc.* (2), 26–28

Reid, D. (1963) How much nitrogen? *Dairy Farm.* **10**(2), 52–64

Robertshaw, D. (1963) Carotid loops. *Am. J. Vet. Res.* **24**, 648

Bianca, W. (1964) Animal housing in hot countries. *Int. J. Biometeorol.* **8**, 99–101

Bianca, W. (1964) Haustier und Klime. *Schweiz. Arch. Tierheilkd.* **106**, 535–559

Blackburn, P. S. (1964) Look out for ketosis. *Dairy Farm.* **11**(4), 30–33

Blaxter, K. L. (1964) Myopathies in animals, in *Disorders of Voluntary Muscle* (Walton, J. N., ed.), pp. 462–488, Churchill, London

Blaxter, K. L. (1964) The nutritional significance of the endogenous secretion of nitrogen in ruminants, in *The Role of the Gastrointestinal Tract in Protein Metabolism* (Munro, H. N., ed.), pp. 143–154, Blackwell, Oxford

Blaxter, K. L. (1964) Problems of cold environments. *Proc. 6th Int. Congr. Nutrition*, pp. 227–242

Blaxter, K. L. (1964) Protein metabolism and requirements in pregnancy and lactation, in *Mammalian Protein Metabolism* (Munro, H. N. & Allison, J. B., eds), vol. 2, pp. 173–223, Academic Press, New York

Drysdale, A. D. (1964) Urea as a grassland fertilizer. *Greensward* (5), 33–34

Pette, J. W. & Smith, J. A. B. (1964) The use of antioxidants in dried whole milk. *Annu. Bull. Int. Dairy Fed.* part IV, 3–9

Reid, D. (1964) The cutting management of grassland. *Greensward* (4), 60–62

Smith, J. A. B. (1964) The International Dairy Federation: what it is and what it does. *J. Soc. Dairy Technol.* **17**, 151–155

Smith, J. A. B. (1964) With IDF behind the iron curtain. *Dairy Ind.* **29**, 758–761

Bianca, W. (1965) Animals and climate, in *Veterinary Annual*, 6th ed. (Pool, W. A., ed.), Wright, Bristol

Bianca, W. (1965) The influence of weather and climate on farm animals, in *A Survey of Human Biometeorology. World Meteorological Organization Technical Note* no. 65, pp. 85–90

Bianca, W. (1965) Report on the sessions of the working group on veterinary biometeorology. *Proc. 4th Int. Congr. Biometeorol.*

Bianca, W. (1965) Reviews of the progress of dairy science. Section A. Physiology. Cattle in a hot environment. *J. Dairy Res.* **32**, 291–345

Blackburn, P. S. (1965) The hair of cattle, horse, dog and cat, in *The Comparative Physiology and Pathology of Skin* (Rook, A. & Walton, G. S., eds), pp. 201–210, Blackwell, Oxford

Blaxter, K. L. (1965) Climatic factors and the productivity of different breeds of livestock, in *The Biological Significance of Climatic Changes in Britain* (Johnson, C. G. & Smith, L. P., eds), Academic Press, London

Blaxter, K. L., Joyce, J. P. & Webster, A. J. F. (1965) Investigation of environmental stress in sheep and cattle. *Proc. 2nd Symp. Shelter Research*

Castle, M. E. (1965) Grassland farming in the Netherlands. *Farm Rev.* (29), 16–19

Castle, M. E. (1965) Some recent grassland experiments and their significance in British grassland farming. *Agric. Prog.* **40**, 35–41

Drysdale, A. D. (1965) Summer milk production without concentrates. *Farm. News* **117**(32), 15

Drysdale, A. D. (1965) What is effluent worth? *Dairy Farm.* **12**(2), 57–59

Findlay, J. D. (1965) Current research in the environmental physiology of cattle. *1st Report FAO Expert Panel on Climatology and Animal Breeding*, pp. 38–44

Jenkinson, D. McE. (1965) The skin of domestic animals, in *The Comparative Physiology and Pathology of Skin* (Rook, A. & Walton, G. S., eds), pp. 591–608, Blackwell, Oxford

Owen, E. C. (1965) Some aspects of the metabolism of vitamin A and carotene. *World Rev. Nutr. Diet.* **5**, 132–208

Reid, D. (1965) How and when to cut. *Dairy Farm.* **12**(3), 53–55

Reid, D. (1965) Judging grassland by herbage production. *Greensward* **7**, 11–12

Whittow, G. C. (1965) Regulation of body temperature, in *Avian Physiology*, 2nd ed. (Sturkie, P. D., ed.), Comstock, Ithaca

Bianca, W. (1966) Animals and climate, in *Veterinary Annual*, pp. 244–251

Bianca, W. (1966) Haematological changes in oxen during dehydration, rehydration and overhydration. *Proc. 4th Int. Congr. Biometeorol.*

Castle, M. E. (1966) The European Grassland Federation. A report of the 1st general meeting. *Herbivoria* **4**, 2–6

Castle, M. E. (1966) The feeding and management of dairy cows in the 1964 grazing season; survey results. *Greensward* **8**, 8–12

Robertshaw, D. (1966) Methohexital anaesthesia in sheep and goats. *Vet. Rec.* **78**, 433

Robertshaw, D. (1966) Sweat gland functioning of the domestic ruminants. *Proc. 4th Int. Congr. Biometeorol.*

Castle, M. E. (1967) Grass farming down under – where the accent is on stocking rate. *Dairy Farm.* **14**(9), 26–27

Castle, M. E. (1967) Nitrogen path to plenty. *Farmer Stockbreed.* **81**(4064), 11

Castle, M. E. (1967) Some thoughts on New Zealand grassland. *Greensward* **9**, 35–38

Castle, M. E., Campbell, A. G. & Clayton, D. G. (1967) Rearing Jersey calves on concentrates. *Proc. Ruakura Farmers' Conf. Week*

Drysdale, A. D. (1967) A comparison of two sources of nitrogen for grassland, with special reference to the grass/clover ratio. *Proc. 10th Int. Grassl. Congr.*, pp. 255–258

Jenkinson, D. McE. (1967) On the classification of sweat glands and the question of the existence of an apocrine secretory process. *Br. Vet. J.* **123**, 311–316

Jenkinson, D. McE. (1967) Sweating and skin types in cattle. *Span* **10**, 163–165

Owen, E. C. (1967) The effects of other dietary constituents on the utilization of urea, in *Urea as a Protein Supplement* (Briggs, M. H., ed.), pp. 329–370, Pergamon, Oxford

Owen, E. C. (1967) Minerals and the goat. *Br. Goat Soc. Monthly J.* **60**, 48–50

Reid, D. (1967) Report on the 10th International Grassland Congress, 1966. *Greensward* **9**, 28–34

Waite, R. (1967) The influence of modern techniques of production on the composition of milk. *Proc. Int. Inst. Co-operat. Agric. Res. Symp.*, pp. 91–99

Bianca, W. (1968) Thermoregulation, in *Adaptation of Domestic Animals* (Hafez, E. S. E., ed.), pp. 97–118, Lea & Febiger, Philadelphia

Castle, M. E. (1968) Milk from grass in New Zealand. *SMMB Bull.* **15**(6), 10–12

Castle, M. E. (1968) Milk production and stocking rate. *Milk News* **9**(2), 11–13

Robertshaw, D. (1968) Some notes on the preparation of permanent fistulae in cattle and sheep, in *Animal Aspects of Pasture Evaluation in the Tropics* (Livingston, H. G. & McCulloch, J. S. G., eds), Longmans Green, London

Robertshaw, D. & Taylor, C. R. (1968) The control of equine sweat glands. *Proc. 24th Int. Congr. Physiol. Sci.*

Robertshaw, D. & Taylor, C. R. (1968) The control of sweating in the donkey. *J. Physiol. (Lond.)* **197**, 81P–82P

Robertshaw, D. & Taylor, C. R. (1968) The pattern and control of sweat discharge in some East African bovidae. *J. Physiol. (Lond.)* **198**, 89P–90P

Smith, J. A. B. (1968) Unique contribution of milk and milk products to world food supplies. *Proc. 2nd World Conf. Anim. Prod.*, session VI, pp. 192–201

Watson, J. N. (1968) Beef from the dairy herd. *SMMB Bull.* **16**(9), 9–11

Castle, M. E. (1969) Best grass for best cows. *Farm. Stockbreed.* **83**(4166), 43, 46

Castle, M. E. (1969) Hay, silage and dried grass for milk production. *Technol. Agric.* **2**, 192–193

Castle, M. E. & Reid, D. (1969) The British Grassland Society's summer meeting, 1969. The west of Scotland. *J. Br. Grassl. Soc.* **24**, 349–351

Czerkawski, J. W. (1969) Methane production in ruminants and its significance. *World Rev. Nutr. Diet.* **11**, 240–282

Holman, R. T., Egwin, P. O. & Christie, W. W. (1969) Substrate specificity of soybean lipoxidase. *J. Biol. Chem.* **244**, 1149–1151

Jenkinson, D. McE. (1969) Sweat gland function in domestic animals (with special reference to neural, hormonal and pharmacological control), in *The Exocrine Glands* (Botelho, S. Y., Brooks, F. P. & Shelley, W. B., eds), pp. 201–221, University of Pennsylvania Press

Okuyama, H., Lands, W. E. M., Christie, W. W. & Gunstone, F. D. (1969) Selective transfer of cyclopropane acids by acyl coenzyme A: phospholipid acyltransferases. *J. Biol. Chem.* **244**, 6514–6519

Smith, J. A. B. (1969) The Hannah Dairy Research Institute. *Scott. Agric.* **48**, 313–317

Smith, J. A. B. (1969) Milk and milk products, in *Proteins as Human Food* (Lawrie, R. A., ed.), pp. 213–224, Butterworth, London

Smith, J. A. B. (1969) Milk production. *J. Soc. Dairy Technol.* **22**, 63–72

Castle, M. E. & Watson, J. N. (1970) Formic acid as a silage additive. *SMMB Bull.* **18**(2), 8–10

Castle, M. E. & Watson, J. N. (1970) More gallons – less waste. *Milk Producer* **17**(4), 112–113, 115

Christie, W. W. (1970) Cyclopropane and cyclopropene fatty acids, in *Topics in Lipid Chemistry*, (Gunstone, F. D., ed.), vol. 1, pp. 1–49, Logos Press, Glasgow

Waite, R. & Abbot, J. (1970) The manufacture of instantly soluble dried full cream milk. *Int. Dairy Fed. Annu. Bull.* part VIII, 3–13

Alwash, A. H. & Thomas, P. C. (1971) The effect of the physical form of the diet and the level of feeding on the digestion of dried grass by sheep. *J. Sci. Food Agric.* **22**, 611–615

Castle, M. E. & Watson, J. N. (1971) Dairy cow grazing: a comparison between a paddock system and a 'Wye College' system. *J. Br. Grassl. Soc.* **26**, 191

Konar, A., Thomas, P. C. & Rook, J. A. F. (1971) The concentrations of some water-soluble constituents in the milk of cows, sows, ewes and goats. *J. Dairy Res.* **38**, 333–341

Manson, W. (1971) Preparative zonal electrophoresis, in *Methods in Microbiology*, vol. 5, pp. 549–571, Academic Press, London

Owen, E. C. (1971) Determination of xanthine oxidase. *Methods Enzymol.* **18**, 590–592

Owen, E. C. (1971) Identification of crystals of isoalloxazines by microscopy in polarized light. *Methods Enzymol.* **18**, 296–305

Owen, E. C. & West, D. W. (1971) Isolation and identification of 7,8-dimethyl-10 formylmethylisoalloxazine as a product of bacterial degradation of riboflavin. *Methods Enzymol.* **18**, 574–579

Owen, E. C. & West, D. W. (1971) Isolation and identification of 7,8-dimethyl-10-(2'hydroxyethyl)-isoalloxazine from natural sources. *Methods Enzymol.* **18**, 579–581

Rook, J. A. F. (1971) Discussion on nutrient utilisation by the lactating mammary gland, in *Lactation* (Falconer, I. R., ed.), pp. 333–338, Butterworth, London

Rook, J. A. F. (1971) Factors affecting the protein content of milk. *J. Soc. Dairy Technol.* **24**, 135–141

Watson, J. N. (1971) Silage additives. *Grass Today* (10), 13

Allsop, T. F. & Rook, J. A. F. (1972) Faecal magnesium output and plasma Mg concentration in sheep. *Proc. Nutr. Soc.* **31**, 65A–66A

Bannister, W. H., Dalgleish, D. G., Bannister, J. V. & Wood, E. J. (1972) Physiochemical and spectroscopic properties of human erythrocyte cupro-zinc protein (erythrocuprein). *Int. J. Biochem.* **3**, 560–568

Calvert, D. T., Findlay, J. D. & McLean, J. A. (1972) Quantitative thermoregulatory responses of the ox to anterior hypothalamic warming. *Biometeorology*, **5**, 40–41

Christie, W. W. (1972) The preparation of alkyl esters from fatty acids and lipids, in *Topics in Lipid Chemistry* (Gunstone, F. D., ed.), vol. 3, pp. 171–197, Elek, London

Czerkawski, J. W. (1972) Inhibition of methane in ruminants. *Vet. Rec. Inf. Suppl.* (103), 83–84

Dalgleish, D. G. (1972) The analysis of the far-ultraviolet circular dichroism spectra of proteins. *FEBS Lett.* **24**, 134–136

Dalgleish, D. G., Feil, M. C. & Peacocke, A. R. (1972) The circular dichroism of complexes of 2,7-di-tertiary-butyl proflavine with DNA. *Biopolymers* **11**, 2415–2422

Dalgleish, D. G., Peacocke, A. R., Acheson, R. M. & Harvey, C. W. C. (1972) The interaction of (+)- and (−)-9-sec-butylaminoacridines with DNA as measured by CD spectroscopy. *Biopolymers* **11**, 2389–2392

Ishaque, M., Thomas, P. C. & Rook, J. A. F. (1972) The effect of glandular tissue on digestion in the sheep. *J. Sci. Food Agric.* **23**, 595–603

Jenkinson, D. McE. (1972) Evaporative temperature regulation in domestic animals, in *Comparative Physiology of Desert Animals* (Maloiy, G. M. O., ed.), pp. 345–356, Butterworth, London

Jessen, C., McLean, J. A., Calvert, D. T. & Findlay, J. D. (1972) Anderungen der Kerntemperatur bei Kühlung und Warmung des Ruckenmarks am Ochsen. *Pfluegers Arch. Eur. J. Physiol.* **332**, suppl. r15

Konar, A., Thomas, P. C. & Towers, K. G. (1972) The composition of ovine mammary cells. *Comp. Biochem. Physiol.* **41A**, 195–203

Lawrence, P. & Thomas, P. C. (1972) Energy metabolism in sheep receiving diets of barley, hay and flaked maize. *Proc. Nutr. Soc.* **31**, 49A

Rook, J. A. F. (1972) Improving the efficiency of food utilisation in ruminants, in *Modern Animal Production and Veterinary Problems*, pp. 113–119, Royal (Dick) School of Veterinary Studies, Edinburgh

Rook, J. A. F. (1972) Symposium on 'Manipulation of rumen fermentation': chairman's introduction. *Proc. Nutr. Soc.* **31**, 125–126

Thomas, P. C. & Clapperton, J. L. (1972) Significance to the host of changes in fermentation activity. *Proc. Nutr. Soc.* **31**, 165–170

Waite, R. (1972) Artificial milk. *J. Soc. Dairy Technol.* **25**, 92–95

Waite, R. (1972) Milk powder manufacture. *World Rev. Anim. Prod.* **8**, 28–37

Weir, J. J. R., Cox, D., Moore, J. H. & Lawrie, T. D. V. (1972) Plasma lipoprotein lipase activity in primary type IV hyperlipoproteinaemia. *Scott. Med. J.* **17**, 214–216

Abbot, J. (1973) Present and future influence of EEC membership on the UK dairy industry. *Dairy Ind.* **38**, 555–564

Banks, W., Mazumder, N. K. & Spooner, R. L. (1973) Studies on the starch-degrading enzymes of bovine serum. II. The action pattern of amylase AmI-B. *Int. J. Biochem.* **4**, 125–131

Castle, M. E. (1973) Milk production from grassland at the Hannah Institute 1951–70. *SMMB Bull.* **21**(1) 8–11

Castle, M. E. (1973) Water for the grazing dairy cow. *Grass Today* (15), 20–21

Christie, W. W. (1973) *Lipid Analysis*, Pergamon Press, London

Clapperton, J. L. (1973) Utilization of food energy by the dairy cow. *SMMB Bull.* **21** (12), 8–11

Dalgleish, D. G., Dingsøyr, E. & Peacocke, A. R. (1973) A comparison of the interaction of proflavine with DNA and with deoxyribonucleohistone using circular dichroism spectroscopy. *Biopolymers* **12**, 445–457

Jenkinson, D. McE. (1973) Comparative physiology of sweating. *Br. J. Dermatol.* **88**, 397–406

Lawrence, P. R. & Thomas, P. C. (1973) A note on the utilization of the energy of the short-chain fatty acids in the fattening sheep. *Anim. Prod.* **17**, 209–212

Noble, R. C. (1973) Lipid metabolism in the young ruminant. *World Rev. Anim. Prod.* **9**, 18–27

Rook, J. A. F. (1973) Dietary factors in the aetiology of metabolic disorders, in *Production Disease in Farm Animals* (Payne, J. M., Hibbitt, K. G. & Sansom, B. F., eds), pp. 14–17, Baillière Tindall, London

Rook, J. A. F. (1973) Effect of diet on the yield and composition of milk. *SMMB Bull.* **21** (11), 8–9

Rook, J. A. F. (1973) Higher education in agriculture. *Agric. Prog.* **48**, 85–90

Rook, J. A. F. (1973) Milk composition – future requirements. *Ir. Grassl. Anim. Prod. Assoc. J.* **8**, 79–87

Taylor, M. M. (1973) Keeping quality of milk: solution offered to milk test anomalies. *Scott. Farm.* **81**(4230), 21, 26

Thomas, P. C. (1973) The protein nutrition of the dairy cow. *SMMB Bull.* **22**(1), 10–11

Thomas, T. P., Middleton, M. S. & Castle, M. E. (1973) Report on drinking-water studies of cows, winter 1972–73, in *Min. Agric. Fish. and Food, A.D.A.S. DH. HM5*, pp. 1–15

Thompson, G. E. (1973) Review of the progress of dairy science: climatic physiology of cattle. *J. Dairy Res.* **40**, 441–473

White, J. C. D. (1973) Standardization of the Gerber method. *Proc. Int. Circle Dairy Res. Leaders*

Wood, E. J. & Dalgleish, D. G. (1973) *Murex trunculus* haemocyanin. 2. The oxygenation reaction and circular dichroism. *Eur. J. Biochem.* **35**, 421–427

Abbot, J. & White, J. C. D. (1974) Dried milk – past, present and future. *Proc. 10th Anniv. Meeting, Inst. Food Sci. Technol.*

Banks, W. & Rook, J. A. F. (1974) Research and the dairy industry. *Dairy Ind.* **39**, 117–118

Banks, W., Clapperton, J. L. & Ferrie, M. E. (1974) The composition of milk from cows fed on various fat-supplemented diets. *Proc. Nutr. Soc.* **33**, 59A

Castle, M. E. (1974) Good silage from red clover. *Better Management* **17**(3)

Castle, M. E. (1974) Grazing systems for the dairy herd. *SMMB Bull.* **22**(4), 8

Chamberlain, D. G. & Thomas, P. C. (1974) The influence of the composition of the diet on the digestion of organic matter and nitrogen in sheep receiving diets containing hay, barley and flaked maize. *Proc. Nutr. Soc.* **33**, 59A

Davies, R. B., Abraham, E. P. & Dalgleish, D. G. (1974) Conformational changes in the extracellular β-lactamase I from *Bacillus cereus* 569/H/9. *Biochem. J.* **143**, 137–141

Dawson, R. R., Rook, J. A. F. & Wood, P. D. P. (1974) Some factors associated with the late-winter decline in the lactose content of herd bulk milks. *J. Dairy Res.* **41**, 317–329

Lloyd, D. H. & Dipeolu, O. O. (1974) Seasonal prevalence of flies feeding on cattle in Northern Nigeria. *Trop. Anim. Health Prod.* **6**, 231–236

Moore, J. H. (1974) Lipid biochemistry from forage to milk, in *Industrial Aspects of Biochemistry* (Spencer, B., ed.), pp. 835–863, North Holland, Amsterdam

Noble, R. C. (1974) Count correction of silicic acid suspension in scintillation gels by an internal vial. *Newsl. Applic. Nucl. Meth. Biol. Agric.* **2**, 8

Noble, R. C., Towers, G. E. & West, D. W. (1974) Scintillation counting of ^{32}P by a simple and direct method. *Newsl. Applic. Nucl. Meth. Biol. Agric.* **3**, 12

Proctor, S. D., Wheelock, J. V. & Davies, D. T. (1974) The heterogeneity of bovine α-lactalbumin. *Biochem. Soc. Trans.* **2**, 621–622

Smith, G. H., McCarthy, S. & Rook, J. A. F. (1974) Synthesis of milk fat from β-hydroxybutyrate and acetate in lactating goats. *J. Dairy Res.* **41**, 175–191

Strachan, N. H. & Rook, J. A. F. (1974) Site of magnesium absorption in the sheep. *Proc. Nutr. Soc.* **34**, A11

Thomas, T. P. & Castle, M. E. (1974) Watch winter water. *Dairy Farm.* **21**(2), 52–55

Thomas, P. C. & Rook, J. A. F. (1974) The partition of dietary energy in the rumen. *Proc. 6th Symp. Energy Metabolism of Farm Animals* (Menke, K. H., Lantzch, J. H. & Reichl, J. R., eds), pp. 429–434

Wright, T. A., Rook, J. A. F. & Wood, P. D. P. (1974) The responses in milk solids-not-fat and protein contents to improved feeding of cows receiving winter-stall diets and underfed for varying periods. *J. Dairy Res.* **41**, 155–161

Alexander, G. & Bell, A. W. (1975) Maximum thermogenic response to cold in relation to the proportion of brown adipose tissue and skeletal muscle in the body and to other parameters in young lambs. *Biol. Neonate* **26**, 182–194

Alexander, G. & Bell, A. W. (1975) Quantity and calculated oxygen consumption during summit metabolism of brown adipose tissue in new-born lambs. *Biol. Neonate* **26**, 214–220

Banks, W. & Greenwood, C. T. (1975) *Starch and its Components*, Edinburgh University Press

Boxer, D. H. & Clegg, R. A. (1975) A transmembrane location for the proton translocating reduced ubiquinone→nitrate reductase segment of the respiratory chain of *Escherichia coli*. *FEBS Lett.* **60**, 54–57

Castle, M. E. & Macdaid, E. (1975) The intake of drinking water by dairy cows at grass. *J. Br. Grassl. Soc.* **30**, 7–8

Castle, M. E. (1975) Additives can give you an improvement in end product. *Scott. Farm.* **83**(4312), II–III

Forrester, T., Harper, A. M., MacKenzie, E. T. & Thomson, E. M. (1975) Vascular and metabolic affects of systemic ATP on the cerebral circulation, in *Blood Flow and Metabolism in the Brain* (Harper, A. M., Jennett, B., Miller, D. & Rowan, J., eds), pp. 3.32–3.33, Churchill-Livingstone, Edinburgh

Garland, P. B., Clegg, R. A., Boxer, D., Downie, J. A. & Haddock, B. A. (1975) Proton-translocating nitrate reductase of *Escherichia coli*, in *Electron Transfer Chains and Oxidative Phosphorylation* (Quagliariello, E., Papa, S., Palmeiri, F., Slater, E. C. & Siliprandi, N., eds), pp. 351–357, North Holland, Amsterdam

Hankinson, D. J. (1975) Potential sources of copper contamination of farm milk supplies measured by atomic absorption spectrophotometry. *J. Dairy Sci.* **58**, 326–336

Holt, C. & Sellen, D. B. (1975) The statistics of chain sectioning experiments on cellulose fibres. *J. Polym. Sci.* **13**, 1–7

Jenkinson, D. McE. (1975) The skin surface: an environment of *D. congolensis*, in *Dermatophilus Infection in Animals and Man* (Lloyd, D. H. & Sellers, K. C., eds), Academic Press, London

Lloyd, D. H. (1975) The economic effects of Dermatophilus infection, in *Dermatophilus Infection in Animals and Man* (Lloyd, D. H. & Sellers, K. C., eds), Academic Press, London

Lloyd, D. H. (1975) Streptothricosis in the domestic donkey (*Equus asinus asinus*). III. Prevalence in Zaria, North-Central state of Nigeria. *Br. Vet. J.* **121**, 115–119

Lloyd, D. H. & Ola, O. M. (1975) Streptothricosis in the domestic donkey (*Equus asinus asinus*). II. Bacteriological and immunological relationships of the strains of *Dermatophilus congolensis* isolated. *Br. Vet. J.* **131**, 108–114

McLean, J. A. (1975) Loss of heat by evaporation, in *Heat Loss from Animals and Man* (Monteith, J. L. & Mount, L. E., eds), Butterworth, London

Moore, J. H. & Noble, R. C. (1975) Foetal and neonatal lipid metabolism, in *Digestion and Metabolism in the Ruminant* (McDonald, I. W. & Warner, A. C. I., eds), pp. 465–480, University of New England Publishing Unit, Armidale, Australia

Morrison, I. M. (1975) The degradation of hemicelluloses by rumen micro-organisms. *J. Sci. Food Agric.* **26**, 1435s

Morrison, I. M. (1975) The rate and products of degradation of xylans by rumen hemicellulases. *Biochem. Soc. Trans.* **3**, 992–994

Nettlefold, B. A. (1975) A course in communication and information retrieval for undergraduate biologists. *J. Biol. Educ.* **9**, 201–205

Noble, R. C. & Shand, J. H. (1975) The differential counting of ^{14}C and ^{32}P using a polyethylene mini vial. *Newsl. Applic. Nucl. Meth. Biol. Agric.* **4**, 3

Reid, D. (1975) Profitable use of nitrogen fertilisers on grass. *Scott. Farm.* **83**(4336), II–III

Steele, W. & Jenkin, H. M. (1975) The effect of octadecenoic acid on the growth of Japanese encephalitis virus in Novikoff hepatoma cells (39094). *Proc. Soc. Exp. Biol. Med.* **150**, 630–635

Thompson, G. E., Gardner, J. W. & Bell, A. W. (1975) The oxygen consumption of the liver in sheep during cold exposure. *Proc. Int. Union Physiol. Sci. 26th Int. Congr.*, vol. XI, p. 407

Bell, A. W. & Thompson, G. E. (1976) The energy metabolism of shivering muscle measured *in vivo* during cold exposure of young cattle. *Proc. 7th Symp. Energy Metabolism of Farm Animals*

Bray, R. C., Vincent, S. P., Lowe, D. J., Clegg, R. A. & Garland, P. B. (1976) Electron paramagnetic resonance studies on the molybdenum of nitrate reductase from *Escherichia coli* K12. *Biochem. J.* **155**, 201–203

Clapperton, J. L. & Basmaeil, S. (1976) The effect of changes in the rumen fermentation on the energy metabolism of sheep and goats. *Proc. 7th Symp. Energy Metabolism of Farm Animals*

Clarke, P. L., Thomson, E. M. & Thompson, G. E. (1976) Acute cold exposure and milk secretion in the goat. *J. Physiol.* (*Lond.*) **263**, 176P–177P

Clegg, R. A. (1976) Purification and some properties of nitrate reductase (E.C.1.7.99.4) from *Escherichia coli* K12. *Biochem. J.* **153**, 533–541

Gilbert, A. S., North, A. M., Parker, T. G. & Pethrick, R. A. (1976) Submillimetre and far i.r. studies of normal and branched chain hydrocarbons. *Spectrochemica Acta* **32A**, 931–936

Holt, C., Mackie, W. & Sellen, D. B. (1976) The configuration of cellulose trinitrate in solution. *Polymer* **17**, 1027–1034

Lloyd, D. H. (1976) Diagnostic features of Dermatophilus infection. *Trans. R. Soc. Trop. Med. Hyg.* **70**, 285

McLean, J. A. (1976) Effects of heat and exercise on energy metabolism of cattle, in *Progress in Biometeorology* (Johnson, H. D., ed.), pp. 184–188, Swets and Zeitlinger, Amsterdam

Rook, J. A. F. (1976) Feed from waste. *Chem. Ind.* (*Lond.*) **17**, 581–584

Rook, J. A. F. (1976) Nutrition of the cow and its effect on milk quantity and quality. *J. Soc. Dairy Technol.* **29**, 129–136

Rook, J. A. F. (1976) Nutritional influences on milk quality, in *Principles of Cattle Production* (Swan, H. & Broster, W. H., eds), pp. 221–236, Butterworth, London

Rook, J. A. F. (1976) Research and development in the 1980s. *Milk Ind.* **78**, 14–15

Shand, J. H., Gibson, P., Gregor, D. W., Cooper, R. J., Keir, H. M. & Postlethwaite, R. (1976) *Molluscum contagiosum* – a defective poxvirus? *J. Gen. Virol.* **33**, 281–295

Thomas, P. C., Kelly, N. C., Chamberlain, D. G. & MacDonald, L. (1976) Utilization of the energy and nitrogen of ensiled forages. *Proc. 7th Symp. Energy Metabolism of Farm Animals*

Thompson, G. E. (1976) The energy metabolism of cattle in the cold, in *Progress in Biometeorology* (Johnson, H. D., ed.), pp. 210–217, Swets and Zeitlinger, Amsterdam

Thompson, G. E. (1976) Principles of climate physiology, in *Beef Cattle Production in Developing Countries* (Smith, A. J., ed.), pp. 266–280, Centre for Tropical Veterinary Medicine, University of Edinburgh

Thompson, G. E. & Bell, A. W. (1976) The energy metabolism of the liver measured in vivo during cold exposure of sheep. *Proc. 7th Symp. Energy Metabolism of Farm Animals*

Vernon, R. G. (1976) Perirenal adipose tissue development in the neonatal lamb. *Biochem. Soc. Trans.* **4**, 228

Watts, P. R. & Goldstone, D. M. (1976) The assessment of smoke production by building materials and fires. Part 4. Large scale tests with wall lining materials. *Fire Res. Note 1013*

West, D. W. & Towers, G. E. (1976) A study of the enzymic dephosphorylation of β-casein and a derived phosphopeptide. *Biochim. Biophys. Acta* **453**, 383–390

Williams, A. G. & Wimpenny, J. W. T. (1976) Exopolysaccharide formation in non-proliferating suspensions of *Pseudomonas* PB1. *Biochem. Soc. Trans.* **3**, 893–895

Williams, A. G. & Wimpenny, J. W. T. (1976) Exopolysaccharide production by Pseudomonas PB1 in batch and continuous culture. The effect of growth conditions. *J. Appl. Chem. Biotech.* **26**, 326–327

Christie, W. W. (1977) Stereospecific analysis of trigylcerides and triglyceride biosynthesis. *Proc. Symp. Lipid Analysis*

Rook, J. A. F. (1977) Advances in the chemistry of milk and milk products. *J. Soc. Dairy Technol.* **30**, in press

Rook, J. A. F. (1977) Dietary manipulation of the composition and quality of milk and milk products. *J. Soc. Dairy Technol.* **30**, 169–177

Thomas, P. C. (1977) Ruminal fermentation and the flow of nitrogen compounds to the duodenum. *Proc. 2nd Int. Symp. Protein Metabolism and Nutrition*

Thomas, P. C. & Rook, J. A. F. (1977) Manipulation of rumen fermentation. *Proc. 11th Univ. Nottingham Conf. Feed Manufacturers*

Thompson, G. E. & Thomson, E. M. (1977) Effect of cold exposure on mammary circulation, oxygen consumption and milk secretion in the goat. *J. Physiol. (Lond.)* **272**, 187–196

Banks, W. & Kelly, M. E. (1978) Rheological properties of butters obtained by manipulating the diet of the cow. *Proc. 20th Int. Dairy Congr.*

Banks, W. & Wilson, A. G. (1978) The value of the constant to be added to the sum of fat, protein and lactose in order to obtain the total solids content of milk. *Proc. 20th Int. Dairy Congr.*

Chalmers, J. S. & Thomas, P. C. (1978) The effect of the level of feeding and composition of the diet on the yield and composition of protein in the milk of Ayrshire cows. *Proc. 20th Int. Dairy Congr.*

Chalmers, J. S. & Thomas, P. C. (1978) The effect of the ratio of forage to concentrates and of the composition of the concentrate mixture on rumen fermentation and milk secretion in cows given silage based diets. *Proc. 20th Int. Dairy Congr.*

Clapperton, J. L., Kelly, M. E. & Rook, J. A. F. (1978) Production of milk of high protein content and low fat content, the fat having a relatively high PUFA content. *Proc. 20th Int. Dairy Congr.*

Dalgleish, D. G. & Parker, T. G. (1978) Comparison of the kinetics of the precipitation of α_s- and β-caseins by calcium ions. *Proc. 20th Int. Dairy Congr.*

Davies, D. T. & Law, A. J. R. (1978) The composition of whole casein from large supplies of pooled milk from creameries in south-west Scotland. *Proc. 20th Int. Dairy Congr.*

Griffiths, M. W., Muir, D. D. & Phillips, J. D. (1978) The production of a heat-stable β-galactosidase. *Proc. 20th Int. Dairy Congr.*

Holt, C. (1978) The average radius, molecular weight and mineral composition of casein micelles under dissociation conditions. *Proc. 20th Int. Dairy Congr.*

Holt, C. & West, M. (1978) Ionic equilibria in the aqueous phase of milk. *Proc. 20th Int. Dairy Congr.*

Holt, C., Kimber, A. M., Brooker, B. E. & Prentice, J. H. (1978) The average radii of casein micelles. *Proc. 20th Int. Dairy Congr.*

Horne, D. S. (1978) Effect of modification of α_s-casein on the kinetics of its precipitation by calcium ions. *Proc. 20th Int. Dairy Congr.*

Horne, D. S. (1978) Fluorescence of mixtures of caseins labelled with Fluorescamine. *Proc. 20th Int. Dairy Congr.*

Horne, D. S., Dalgleish, D. G. & Parker, T. G. (1978) Aspects of the kinetics of the precipitation of α_s-casein by calcium ions. *Proc. 20th Int. Dairy Congr.*

MacDonald, L. & Thomas, P. C. (1978) The yield and composition of milk in cows given rationed amounts of silage with supplements containing barley and groundnut cake. *Proc. 20th Int. Dairy Congr.*

Muir, D. D. & Phillips, J. D. (1978) Bacteriological quality of milk going to manufacture in south-west Scotland. *Proc. 20th Int. Dairy Congr.*

Muir, D. D. & Sweetsur, A. W. M. (1978) Heat stability of concentrated milk. *Proc. 20th Int. Dairy Congr.*

Muir, D. D. & Sweetsur, A. W. M. (1978) Heat stability of caprine milk. *Proc. 20th Int. Dairy Congr.*

Parker, T. G. & Dalgleish, D. G. (1978) The binding of calcium ions to α_s-casein and their effect on the aggregation equilibrium. *Proc. 20th Int. Dairy Congr.*

Thomas, P. C. & Kelly, M. E. (1978) The effect of the level and frequency of feeding on milk secretion in the Ayrshire cow. *Proc. 20th Int. Dairy Congr.*